D0068972

FOUNDATIONS OF AGRICULTURAL EDUCATION

B. Allen Talbert
Agricultural Education
Purdue University

Rosco Vaughn
Agricultural Teacher Educator
California State University, Fresno

Barry Croom
Agricultural and Extension Education
North Carolina State University

Jasper S. Lee
Agricultural Educator
Georgia

PEARSON

Boston   Columbus   Indianapolis   New York   San Francisco   Upper Saddle River
Amsterdam   Cape Town   Dubai   London   Madrid   Milan   Munich   Paris   Montréal   Toronto
Delhi   Mexico City   São Paulo   Sydney   Hong Kong   Seoul   Singapore   Taipei   Tokyo

Editorial Director: Vernon Anthony
Senior Acquisitions Editor: William Lawrensen
Editorial Assistant: Lara Dimmick
Program Manager: Alexis Duffy
Director of Marketing: David Gesell
Marketing Manager: Stacey Martinez
Senior Marketing Coordinator: Alicia Wozniak
Senior Marketing Assistant: Les Roberts
Senior Managing Editor: JoEllen Gohr
Project Manager: Kris Roach
Senior Art Director: Jayne Conte
Cover Designer: Suzanne Duda
Cover Art: Shutterstock © violetkaipa
Full-Service Project Management and Composition: PreMediaGlobal
Printer/Binder: LSC Communications
Cover Printer: LSC Communications

Credits and acknowledgments borrowed from other sources and reproduced, with permission, in this textbook appear on the appropriate page within text.

Microsoft® and Windows® are registered trademarks of the Microsoft Corporation in the U.S.A. and other countries. Screen shots and icons reprinted with permission from the Microsoft Corporation. This book is not sponsored or endorsed by or affiliated with the Microsoft Corporation.

Many of the designations by manufacturers and sellers to distinguish their products are claimed as trademarks. Where those designations appear in this book, and the publisher was aware of a trademark claim, the designations have been printed in initial caps or all caps.

Library of Congress Cataloging-in-Publication Data

Foundations of agricultural education / B. Allen Talbert . . . [et al.].—3rd ed.
    p. cm.
  Includes bibliographical references and index.
  ISBN-13: 978-0-13-285960-8 (alk. paper)
  ISBN-10: 0-13-285960-2 (alk. paper)
  1. Agricultural education—Study and teaching. 2. Agricultural education—Textbooks. 3. Agriculture teachers—Training of. I. Talbert, B. Allen.
  S531.F68 2013
  630.71—dc23

                            2013007555

ISBN-10: 0-13-285960-2
ISBN-13: 978-0-13-285960-8

KV 01.14.2019 1529

FOUNDATIONS OF AGRICULTURAL EDUCATION

contents

7
Advisory and Citizen Groups   118

8
Curriculum Development   135

9
Student Enrollment and Advisement   149

10
Classroom and Laboratory Facilities   164

21
Using Laboratories 349

part four
Supervised Agricultural Experience, FFA, and Community Resources

22
Supervised Agricultural Experience 367

23
FFA 387

Preface

Foundations of Agricultural Education was previously published by Professional Educators Publications and continues to expand on the major goal of the book: to introduce future agricultural educators to their profession and support professional development of those now in the profession. Planning on the first edition of this book with the previous publisher began more than three years before its release. The second edition updated and expanded on the content of the first edition. Now, this edition enhances the useful features of the first two editions. The authors were determined to prepare a relevant book for the agricultural education profession and are pleased to be able to bring this book to you.

The audience for this book includes college students in agricultural teacher education programs, agriculture teachers, state supervisors, teacher educators, and others interested in agricultural education. The overall purpose is to provide a foundational resource, one that broadly covers each element necessary to be a teacher of agricultural education. The book is appropriate for introductory as well as advanced classes in agricultural teacher education. Incumbent teachers will also find information useful to them as they go about their roles as professionals in agricultural education.

Agriculture and education are both fast paced and ever changing. Agricultural education is a blend of each of these. This book focuses on current content, terminology, practices, and theory while giving historical and philosophical foundations to agricultural education. Examples and terms have been used that will help to keep the book current.

We strongly feel that the secondary agricultural education model of classroom/laboratory instruction, supervised experience, and FFA has withstood the test of time. This model, when properly followed, will result in enhanced student learning, a better-prepared agricultural workforce, more competent community leaders, and an agriculturally literate society. In recent years, emphasis on program and content standards has presented new challenges for agricultural educators. Overall, this emphasis has enhanced the quality of agricultural education throughout the United States.

Delve into *Foundations of Agricultural Education*. Review the Contents and thumb through the chapters. You will note a user-friendly organization, images that reflect successful local program practices, and a unified professional approach. You will also note wide regional representation across the United States. This reflects the backgrounds of the authors as well as nationwide practices. Your review of this book should tweak your interests and promote your professional enthusiasm for greater study. And, do the authors a favor: commit to an energetic and productive career as a teacher of agricultural education.

To access supplementary materials online, instructors need to request an instructor access code. Go to www.pearsonhighered.com/irc, where you can register for an instructor access code. Within 48 hours after registering, you will receive a confirming e-mail, including an instructor access code. Once you have received your code, go to the site and log on for full instructions on downloading the materials you wish to use.

Acknowledgments

We are grateful to many individuals who have directly or indirectly contributed to the production of this book. Some were high school teachers, and others were teacher educators who guided us in our own professional development. Others who should be acknowledged are our current professional associates, who daily enhance our knowledge and help mold our professional practice.

We appreciate the universities where we are teacher educators for allowing and supporting the development of this book. Purdue University; California State University, Fresno; and North Carolina State University are acknowledged in this regard. We are also grateful to other universities where we have studied or had close personal contact. These include Virginia Tech, Mississippi State University, New Mexico State University, University of Illinois, University of Florida, The Ohio State University, Texas Tech University, Clemson University, The Pennsylvania State University, Texas A&M University, Oregon State University, Cornell University, and University of Minnesota.

We wish to acknowledge the organizations affiliated with agricultural education, including the National Association of Agricultural Educators, the American Association for Agricultural Education, and the National Association of Supervisors of Agricultural Education. The staff of the National FFA Organization is also acknowledged for direct and indirect support of the authors.

We acknowledge the important roles of our families in this work. The extra time and effort required often took us away from family duties. We appreciate their understanding and hope that they view this product as worthy.

Several schools are acknowledged for extra efforts in supporting the book with images. These include Switzerland County, North Decatur, Eastern Hancock High School, Franklin Community High School, and Manual High School in Indiana; Eastern Randolph High School, Chase High School, and Chatham Central High School in North Carolina; Florin High School, Casa Roble High School, Ponderosa High School, Lemoore High School, West Central High School, Madera High School, Kingsburg High School, Elk Grove High School, and Clovis High School in California; Montrose High School in Colorado; Molalla High School, Canby Union High School, and North Clackamas High School in Oregon; Tallulah Falls School, Oconee County High School, Madison County High School, and Franklin County High School in Georgia; Sandra Day O'Connor High School in Texas; and Millsaps Vocational Education Center, Starkville Academy, and East Mississippi Community College in Mississippi. The assistance of students at Piedmont College, Georgia, as technical models is gratefully acknowledged. Education Images, P. O. Box 152, Sautee-Nacoochee, GA 30571, is acknowledged for its assistance as a source of many images used in this book.

We would also like to thank the reviewers for their thoughtful comments and suggestions. They are Thomas Dobbins, Clemson University; Gary Briers, Texas A&M University; Kristin Stair, New Mexico State University; and Gregory Miller, Iowa State University.

Formerly published by Professional Educators Publications (PEP), Inc., of Illinois, the book is now published by Pearson Education, Upper Saddle River, New Jersey. We express appreciation to PEP and its staff for supporting initial publication. The change in publishers brings exciting new opportunities. A thank you is extended to all individuals at Pearson Education who made this book possible.

FOUNDATIONS
OF AGRICULTURAL
EDUCATION

part one

Introduction to the Agricultural Education Professions

1

The Agricultural Education Professions

Agricultural education offers a number of important, challenging, and rewarding professional opportunities. You most likely have carefully investigated the possibilities for you as an agricultural educator. Some possibilities were likely obvious to you; others might not have been quite so obvious.

Teaching is the first opportunity many agricultural education graduates consider. It allows you to utilize the education you have received in teacher preparation and enjoy unique success as a teacher of agriculture. You may also want to consider other areas of agricultural education such as a state leader or university agricultural teacher educator. In addition, you may consider numerous opportunities in education (such as school administration or career counseling) and the agricultural industry (such as human resource directors or managers of staff development). Regardless, success in agricultural education will require setting goals and putting forth the needed effort to achieve your goals.

You are now likely enrolled in university-level classes that will prepare you to become an agricultural educator. You are probably pursuing a college degree that meets teacher credentialing requirements. You may be in the process of changing your thought perspective from that of a student to that of a teacher. You are likely thinking about professional matters related to being an excellent teacher. This chapter and others in this book are intended to help you develop into a highly competent, professional agricultural educator.

objectives

This chapter introduces the broad field of agricultural education with emphasis on becoming a professional agricultural educator. It has the following objectives:

1. Describe the meaning and importance of teaching
2. Explain teaching as a profession, including professionalism and the role of ethics
3. Discuss the meaning and scope of agricultural education
4. Identify relationships of agricultural education to education in the United States
5. Identify the roles and responsibilities of agriculture teachers
6. Relate teaching agriculture to the school community

terms

administration
agricultural education
agricultural education program
Carnegie unit
comparative agricultural education
continuing education
counseling
creed
demeanor

formal education
non-formal education
profession
professional
professional code of ethics
stakeholder
teach
teacher
teaching

1–1. An agriculture teacher is providing individual instruction during supervised study.
(Courtesy, Education Images)

TEACHING AGRICULTURE

People have different perspectives on the meaning of teaching. It does not make much difference about the subject area being taught. To be acceptable, a definition of teaching must acknowledge that the role of teaching is to promote learning. It is used in the education and training of individuals so that they have worthy knowledge, skills, and values that promote a gainful and successful life.

Its Meaning

Teaching is defined in terms of two other words: *teach* and *teacher*.

To **teach** is to guide or cause others to learn something. It also means to instruct others using various strategies, such as examples and experiences. The process is neither simple nor easy. To teach involves far more than contact with learners. It includes planning and evaluating the educational process as well as providing agricultural education program leadership.

Anyone who teaches is called a **teacher**. We may also think of a teacher as an individual whose professional occupation is to teach. Teachers are often further identified by the subject they teach, such as agriculture, horticulture, or veterinary science. An agriculture teacher is a teacher of agriculture, including closely related subjects.

Teaching is the art and science of directing the learning process (Lee, 2001b). This is an active role that involves many functions in helping others learn. Teaching can occur in a wide range of environments, such as in the pasture of a ranch, at a bench in a greenhouse, or in a school classroom. Teaching environments are discussed in detail later in this book.

Educators sometimes limit the definition of *teaching* to "instruction that occurs in an educational institution." That definition doesn't fit agricultural education very well. Think of the teaching that occurs in the agricultural industry outside the walls of a school building!

1–2. Agriculture teachers often use realia (real things or likenesses of real things) in teaching.
(Courtesy, Education Images)

Educators also sometimes limit the definition of *teaching* to "instruction provided by a credentialed, certified, or highly qualified individual." Every state has certification requirements. Some local school districts also have requirements beyond those of the state. In general, credentialing demands that a teacher be a graduate of a college teacher education program and meet other requirements a state may impose, such as minimum scores on tests.

Teaching is a profession of some 50 million people throughout the world. The field of education, however, includes more than teachers. It also includes administrators, curriculum specialists, counselors, and others at elementary, middle school, high school, and post–high school (college) levels. But, not all teachers are in education! Some teachers are employed by government agencies, agricultural businesses, and nonprofit organizations.

The roles in teaching agricultural education are many and diverse. Regardless, the overall aim is to direct the learning process of students—children and adults—efficiently and effectively toward educational objectives and goals.

Its Importance

The well-being of a nation requires an educated citizenry. The ability to read, write, use mathematics, and perform other basic skills is essential for people to lead productive and satisfying lives. Education promotes the acceptance of diversity and increases tolerance of individual differences in people. Educated citizens are less likely to commit crimes and engage in other illegal activities, and they enjoy a higher standard of living.

Some important reasons for teaching, with emphasis on agricultural education, are as follows:

- Promote overall development of youth and adults for productive and successful lives.
- Assure that people have sufficient agricultural literacy to be good consumers and citizens.

1–3. A teacher provides individual instruction to a student in preparing a landscape plan.
(Courtesy, Education Images)

- Promote agricultural practices to assure sufficient food and fiber supply to meet the needs of an increasing human population.
- Promote safe food handling and preparation practices.
- Develop knowledge and skills needed to enter and advance in an agricultural or related career.
- Provide preparation for students to pursue education in agricultural areas beyond high school.
- Promote the use of sustainable resource practices in agriculture, horticulture, forestry, and other areas.
- Provide related education to promote achievement in science, mathematics, reading, and other academic areas.
- Help individuals enjoy a higher standard of living.
- Promote an appreciation for the environment and its protection.
- Promote wise use of natural resources to achieve human goals.

Where Teaching Occurs

Teaching occurs in formal and non-formal settings. ***Formal education*** is training or education that is provided in an orderly, logical, planned, and systematic manner and often associated with school attendance. Formal settings are those in schools where structured teaching is used to assure that essential standards are met—formal education. ***Non-formal education*** is typically offered outside of schools and in settings other than those that involve the implementation of structured standards and fulfillment of diploma or degree requirements.

In agricultural education, formal education is provided through public and private schools at the primary, middle school, secondary, postsecondary, and college and university levels. Often referred to as agricultural education, the instruction and classes may help meet graduation requirements or other needs such as producer quality assurance.

1–4. An agriculture teacher is providing instruction in constructing a DNA model.
(Courtesy, Education Images)

Non-formal agricultural education may be provided by agencies, associations, and entrepreneurial businesses. A major provider of non-formal agricultural education is the Cooperative Extension Service. A collaborative effort of the federal, state, and local governments, the teaching is by local agents, specialists, or other qualified individuals. The National Institute of Food and Agriculture (NIFA) of the U.S. Department of Agriculture administers programs at the federal level. At the local level, extension agents are hired to organize and provide instruction in areas of agriculture, home and family, 4-H youth, as well as a few other areas related to community development and the environment. Non-formal education may meet certification needs (such as pesticide applicator requirements) and continuing education requirements but does not usually have credit toward a high school diploma or college degree.

Comparative Agricultural Education

Comparative agricultural education is the study of the theory and practice of agricultural education in different geographical or political entities or over time. Most often, comparative agricultural education is thought to address differences in how agricultural education is provided by foreign nations. How does Germany, for example, differ from the United States? Often, comparative agricultural education is applied to developing countries. International agricultural educators may attempt to integrate the model for agricultural education used in a developed country with a rudimentary model in a developing country so that an improved system for the delivery of agricultural education results.

Another approach with comparative agricultural education is the study of policy, curriculum, and practice among and between states and school districts in the United States. State and local school board policies and regulations result in some differences in agricultural education programs. State education agencies often seek curriculum guides and other materials developed in one state for use in another. A major factor in assuring some similarity among states and school districts is the National FFA organization. The programs and activities of the National FFA are typically applied with relative uniformity in all states. For example, judging teams that are preparing for national competition must be trained similarly in all states and local schools if they will experience success in national competition.

Agricultural education changes over time. The nature of agricultural education has changed markedly since the initial Smith-Hughes Act of 1917. Historical study should result in similarities and differences being highlighted. It should also allow individuals to determine those practices that were most efficient and effective in producing the desired educational outcomes.

TEACHING AS A PROFESSION

A *profession* is an occupation that requires a long, specific program of preparation in higher education and uses a code of ethics to guide the conduct of individuals in the profession. Testing may be used as an additional approach to assure that only individuals with high capability enter a profession. Most professions also have some degree of self-regulation and control over credentialing, though this area has sometimes drawn criticism as lacking in the education profession.

Professional Status

A *professional* is an individual who has acquired those traits ascribed to the profession practiced by the individual. This includes the appropriate education and competence in carrying out the duties of the profession. You might have had teachers who demonstrated that they were professionals in all regards. Unfortunately, you might have had a few teachers who did not demonstrate good professionalism.

Here are a few generally held expectations for agriculture teachers to become professionals and continue to enjoy high professional status:

- **Preparation**—The agriculture teacher has the credentials to be an outstanding professional. This includes having the education and experience to be an effective teacher and the desire to stay current in the profession. Most agriculture teachers have a college degree in agricultural teacher education plus additional formal and informal education. Some individuals become agriculture teachers following alternative routes of preparation such as having a degree in an area of agriculture followed by teacher education preparation. Certification as a teacher is typically provided by a state agency with the legal authority to do so. Agricultural teacher educators at a university or college with a teacher education program will be able to provide all details related to preparation.
- **Participation**—A professional agriculture teacher takes an active role in the agricultural education profession and in education. The teacher is a member of professional organizations and participates in professional activities to assure continued growth and development. A professional speaks positively about the profession and promotes its well-being. Agriculture teachers also interact professionally with other teachers in the school systems where they are employed.
- **Continuing education**—Staying up-to-date is an important responsibility of a professional. *Continuing education* is education after an individual has completed a general level of formal education. College courses, workshops, field days, reading, and other approaches are used to keep current. These may contribute to the requirements of an advanced degree or a higher salary level, or they may be for the general knowledge and skill development. Many states have inservice activities for agriculture teachers.
- **Demeanor**—*Demeanor* is the outward manner of an individual. It includes the way in which a person relates to others, as well as the manner in which he or she dresses and grooms, the skill with which the person uses language, and

the general state of organization and cleanliness in his or her environment. These attributes can be improved to increase professionalism.

- **Citizen Responsibilities**—Teachers should follow regulations, pay taxes, vote, meet financial obligations, and maintain honesty and integrity in all regards. Teachers who fail to do so discredit themselves as well as their fellow professional teachers.
- **Community Standards**—Though such standards are largely unwritten, agriculture teachers should meet acceptable moral and personal standards of the communities in which they live and teach agriculture. Teachers are typically held to higher standards than nonteaching citizens of the community. Most communities do not approve of teachers using recreational drugs or similar substances, engaging in deviant sexual activities, and violating laws established for the good of the overall population.

Codes of Ethics and Creeds

A *professional code of ethics* is a statement of ideals, principles, and standards related to the conduct of individuals within a profession. Codes of ethics establish rigorous ethical and moral obligations of individuals in a profession.

The education profession in the United States does not have a universally adopted code of ethics. Some state education agencies and local school districts have statements of principles and ideals for teachers. Associations related to the education profession may also have statements resembling codes of ethics. A code of ethics for the education profession was established by the National Education Association (NEA) and adopted by its representative assembly in 1975. This code of ethics has a preamble followed by two principles. Principle I focuses on commitment to the student. Principle II focuses on commitment to the profession of education. A limitation is that the NEA is not represented in all school districts and has limited visibility among some teachers. Lawrence Baines (2010) states that the NEA Code of Ethics is the closest facsimile of a national code of ethics for teachers. Some states and local school districts copy the NEA Code of Ethics verbatim and adopt it for their needs. (This code of ethics for the education professional may be viewed at the NEA Web site.)

Ag Teacher's Creed

I am an agricultural educator by choice and not by chance.

I believe in American agriculture; I dedicate my life to its development and the advancement of its people.

I will strive to set before my students by my deeds and actions the highest standards of citizenship for the community, state and nation.

I will endeavor to develop professionally through study, travel and exploration.

I will not knowingly wrong my fellow teachers. I will defend them as far as honesty will permit.

I will work for the advancement of agricultural education and I will defend it in my community, state and nation.

I realize that I am a part of the school system. I will work in harmony with school authorities and other teachers of the school.

My love for youth will spur me on to impart something from my life that will help make for each of my students a full and happy future.

1–5. Ag Teacher's Creed.

(Courtesy, National Association of Agricultural Educators)

Creeds may serve as guides for behavior of individuals in a profession. A *creed* is a statement of general beliefs of individuals in a profession or organization. The National Association of Agricultural Educators (NAAE) has developed an Ag Teacher's Creed.

AGRICULTURAL EDUCATION: MEANING AND SCOPE

Agricultural education is a program of instruction in and about agriculture and related subjects such as natural resources and biotechnology. It is most commonly offered in the secondary schools of the United States, though some elementary and middle schools and some postsecondary institutes/community colleges also offer such instruction. Subjects included are those of less than the four-year college level in plant and animal production, supporting biological and physical sciences, horticulture, natural resources and environmental technology, mechanics, and forestry. Some areas of food processing and distribution are included.

Agricultural education has been carried out, in one form or another, for centuries. In North America, its beginning is often equated with the arrival of first colonists. Much of the early emphasis on school based agricultural education at the federal level began in 1917 with passage of the Smith-Hughes Act, which promoted vocational education. Additional federal laws have been enacted since inception. These laws have included agricultural education as a part of vocational education. In recent years, vocational education has become known as career and technical education. Federal support continues to be provided in collaboration with states and local school districts. For many years, agricultural education in the secondary schools was referred to as "vocational agriculture." (More information is presented in Chapter 9.)

In recent years, the term "school based agricultural education" (SBAE) has gained relatively widespread use. This is to distinguish it from agricultural education that is not based in local schools. However, most individuals continue to use the term *agricultural education*.

Career Cluster

The National Association of State Directors of Career Technical Education Consortium has identified sixteen career clusters (see Chapter 9). One of these clusters is Agriculture, Food, and Natural Resources (AFNR). The AFNR cluster includes careers and instructional areas associated with the production, processing, marketing, distribution, financing, and development of agricultural commodities and resources, including food, fiber, wood products, natural resources, and other plant and animal products and resources.

The Agriculture, Food, and Natural Resources cluster structures the instructional content for agricultural education into eight pathways (seven were initially specified but an eighth pathway, Biotechnology Systems, was later added). The pathways and brief descriptions of their content are as follows:

- Plant Systems (the study of plants and cultural practices in plant production; areas include agronomy, horticulture, forestry, turf, viticulture, and soils)
- Animal Systems (the study of animals and their production and well-being; areas include large and small animals, wildlife animals, research animals, animal health, and others)
- Biotechnology Systems (the study and use of data and applied science techniques in the solution of problems associated with living organisms; areas include plant and animal biotechnology, microbiology and molecular biology, genetic engineering, and environmental applications)

- Power, Structural, and Technical Systems (the study and use of equipment, engines and motors, fuels, and precision technology; areas include power, structures, controls, hydraulics, electronics, pneumatics, and the like)
- Natural Resources Systems (the study of the management practices for soil, water, wildlife, forests, and air; areas include habitat conservation, mining, fisheries, soil conservation, and the like)
- Environmental Service Systems (the study and use of technology and instruments in environmental quality; areas include water and air quality, solid waste management, pollution prevention, hazardous waste management, sanitation, and the like)
- Agribusiness Systems (the study and use of economic and business principles related to economic systems, management, marketing, and finance; areas include agricultural sales and service, entrepreneurship, agricultural management, and the like)
- Food Products and Processing Systems (the study and application of science and technology to assure quality and wholesome food products; areas include processing, preserving, packaging, food safety, quality assurance, regulations, and distributing food)

The content standards for each of these pathways have been identified and published for use by agricultural educators and other individuals. These standards focus on the content or subject matter of agriculture instruction in grades 9 through 14 (high school and postsecondary school). In addition to the eight pathways, content standards have also been prepared for an area known as "life knowledge and cluster skills." The content standards are valuable in planning programs of study and developing overall curriculum guides by agricultural educators and others. Individuals who develop instructional materials make considerable use of these standards. (*The National Agriculture, Food and Natural Resources (AFNR) Career Cluster Content Standards* are available for downloading or printing at the Team Ag Ed website.)

1–6. A teacher is providing small group instruction in floral design using everlasting flowers.
(Courtesy, Education Images)

1–7. A state supervisor of agricultural education reviews local plan materials with a teacher of agricultural education.

(Courtesy, Education Images)

Teachers, Students, and Programs

A national study of agricultural education has been conducted every five years since 1994. This study has provided considerable insight into agricultural education programs, students, and teachers throughout the United States. In 2009, nearly 1.3 million secondary students were enrolled in agricultural education classes offered by 11,322 teachers in 8,038 schools. Enrollment per teacher has increased in the last decade as funding to support instruction has declined. Greater numbers of students take agriculture classes in schools with alternative scheduling. Further, the number of students per teacher is greater in schools with alternative scheduling. Selected findings of the most recent study, as presented in Table 1–1, help characterize agricultural education.

1–8. An agriculture teacher is providing small group instruction.

(Courtesy, Education Images)

Table 1–1
SELECTED FINDINGS OF THE NATIONAL STUDY OF AGRICULTURAL EDUCATION

- Agricultural education programs were offered in 8,038 secondary schools in the United States (7,262 schools in 2004).*
- Instruction was provided by 11,322 secondary teachers (10,711 in 2004).
- The percentage of one-teacher programs was 49.60 percent, which is a decline from 58.50 percent in 2004, 57.51 in 1999, and 74.67 in 1994).
- Enrollment was 1,294,212 students (1,098,233 in 2004, 971,328 in 1999, and 732,600 in 1994).
- Average enrollment per program was 161.00 students (150.19 in 2004, 139.60 in 1999, and 99.14 in 1994).
- Female students accounted for 40.0 percent of enrollment (35.4 percent in 2004, 31.2 percent in 1999, and 26.1 percent in 1994).
- The average years of teaching experience for teachers was 12.97 years (14.20 in 2004, 14.70 in 1999, and 14.90 in 1994).
- Male teachers accounted for 71.74 percent of the teachers. (No data available for prior years.)
- Enrollment per teacher in one-teacher programs was 111.2 students (108.4 in 2004, 91.0 in 1999, and 75.6 in 1994).
- Enrollment per teacher in programs with more than one agriculture teacher was 115.9 students per teacher (92.5 in 2004, 77.9 in 1999, and 59.6 in 1994).
- Enrollment per teacher by teacher gender was 118.38 students for female teachers and 108.81 students for male teachers in one-teacher programs. (Female teachers average teaching almost 10 more students per year than male teachers!) (No data available for prior years or multiple-teacher programs.)
- Female teachers have higher percentages of female students, with 49.3 percent of their students being female as compared to 36.7 percent female students for male teachers. (No data available for prior years.)
- Alternative (block) scheduling was used in 37.60 percent of schools (33.75 in 2004 and 40.69 in 1999).
- Average agriculture class enrollment in schools with alternative scheduling was 227.9 (157.9 students in schools not using alternative scheduling).
- Funding budgeted per teacher for instructional resources was $7,853 per year ($7,231 in 2004, $8,677 in 1999, and $4,561 in 1994).
- Funding available per student to buy instructional materials was $68.70 ($37 in 2004 and $52 in 1999).
- Agriculture classes were counted toward high school graduation academic requirements in 52.19 percent of schools (50.20 percent in 2004 and 71.70 percent in 1999).
- Only 9.4 percent of the teachers taught adult classes; of these, 81.8 percent were taught by male teachers; the average adult enrollment per program was 41.2 adult students. (No data available for prior years.)

*An **agricultural education program** is the total offering of agricultural instruction in a school. Programs range from one to several subject areas or classes and have one or more teachers on a full- or part-time basis. Supervised experience and student organizations, particularly the FFA, are included in an agricultural education program. Some programs include adult and young adult instruction.

Source: Lee (2009).

AGRICULTURAL EDUCATION IN AMERICAN EDUCATION

Agricultural education is much a part of American education. It is offered in the public and, sometimes, private secondary schools just as other required core curriculum and elective courses are offered. State-level curriculum guides, standards, and testing initiatives are typically provided for all accredited public secondary schools.

Education in the United States

Education in the United States is primarily offered in public schools and is universally available to all children. Child education is compulsory, with the ages for compulsory education ranging from ages 5–8 until ages 14–18, depending on state law. While the vast majority of children attend public schools, some attend private and parochial schools or may be home schooled (only about 10% attend private schools and 2.9% are home schooled). School governance has been viewed as a location function.

1–9. Local boards of education have major control over education, though state and federal regulations are encroaching on this control. (Courtesy, Shutterstock © Cynthia Farmer)

Most school policies are set locally by a board of education though control and support for education comes from three levels: federal, state, and local governments.

In the early years of the United States, education was viewed as a function of local and state governments. The federal government was not involved in education. This began to change in the early 1900s with passage of the Smith-Hughes Act of 1917. This act provided funds to the states and local schools to establish vocational education instruction. Agricultural education was part of vocational education, though agriculture classes had been offered in a number of schools for many prior years. Since 1917, several federal initiatives have supported local school education. The No Child Left Behind Act of 2001 (NCLB) was a major effort at federal influence over state and local education. Signed by President George W. Bush on January 8, 2002, in Hamilton, Ohio, NCLB has been controversial among state and local education officials. Overall, NCLB was intended to support standards-based education with measurable goals to improve education outcomes. Many educators felt that NCLB established unrealistic goals for the local schools and teachers. NCLB, with its far reaching provisions, replaced the Elementary Secondary Education Act (ESEA) of 1965.

The National Center for Education Statistics (NCES) of the U.S. Department of Education gathers and reports information about education in the United States. A report of NCES in 2011 indicated that there were 98,706 elementary and secondary schools in the United States. Of these, 24,348 were secondary schools (middle and high school). In addition, there were 1,690 two-year colleges (sometimes referred to as technical schools). The number of schools has greatly decreased in the past century due to school consolidation and increased size of school enrollments.

Schools throughout the United States are increasingly offering and requiring similar courses for graduation. This is much different from a century ago when local school districts had greater control over curriculum and other education matters. High school courses are typically measured as Carnegie units. A ***Carnegie unit*** is a standard for measuring school subjects by length of class participation, with 120 sixty-minute hours being the minimum requirement for schools to grant 1 unit of credit. One Carnegie unit is equivalent to one academic year of study. Course scheduling and period lengths vary widely. Many high schools have 180-day years.

The minimum courses in mandatory subjects in U.S. high schools are (one Carnegie unit is typically earned with each course) as follows:

- Science (three units or years minimum of science courses; normally biology, chemistry, and physics are required)
- Mathematics (four years minimum of mathematics courses; courses include algebra, geometry, precalculus, statistics, and, in some schools, calculus)
- English (four years minimum; including composition, literature, and oral language)
- Social studies (three years minimum, including history, government, and economics)
- Physical education (none to two years minimum)

Agricultural education courses (and all other career courses) are typically viewed as electives. This means that instruction and enrollment by students in agriculture is not required but is optional in most schools. In order to gain enrollment in agricultural education, the courses offered, the curriculum outlines followed, and the teaching approaches that are used must be attractive to students and provide highly useful learning opportunities to students. Students must feel that enrollment in agricultural education is beneficial to their educational and lifelong pursuits.

Agricultural Education Enrollment and Schools

Agricultural education (at the secondary school level, formerly known as vocational agriculture) is primarily offered at the high school level in the secondary schools of the United States. Some 15,500 public high schools (grades 9–12) enroll 14 million students. Of these schools, a few over eight thousand, or 52 percent, have agricultural education offerings. With 1.3 million student enrollments in agricultural education, only 9.3 percent of the total high school student population is enrolled in agricultural education. These data reflect a huge opportunity for agricultural education to increase enrollment and serve additional numbers of students.

In agricultural education, programs are offered in elementary, middle, and high schools as well as postsecondary institutions. The focus here is on instruction below the postsecondary level using data from the National Study of Agricultural Education (Lee, 2009). By far, the largest proportion of teachers (74%) are teaching in comprehensive high schools. Only slightly over 12 percent were teaching in middle schools and 6 percent were teaching in career and technical centers (formerly known as vocational-technical education centers). Nearly 2.5 percent were teaching in magnet or theme schools and academies. The remainder (about 5%) of teachers were teaching in correctional facilities, exceptional education schools, and related kinds of institutions.

The largest number of programs and student enrollments has been and continues to be at the high school level and in comprehensive high schools. A small increase in middle school programs has occurred in the past several years. In some cases,

1–10. A teacher and school administrator are reviewing the overall campus layout for a high school attendance center.

(Courtesy, Education Images)

teachers may teach high school and middle school classes. The number of agricultural education programs at career and technical education centers has declined slightly since 1994, and stood at 2.2 percent of such schools with agriculture classes in 2009.

SUCCESSFUL AGRICULTURE TEACHERS

Agriculture teachers have a number of roles and responsibilities that go far beyond meeting classes each day. They deliver programs of agricultural education, which include instruction, directing students in supervised experience, and advising the student organization—FFA. This is true in whatever area of agricultural education they may be teaching—horticulture, wildlife, food science, agricultural mechanics, animal care, veterinary assisting, or other area.

Roles of Agriculture Teachers

Some major roles and responsibilities of agricultural educators are as follows:

- **Being a school team member**—An agriculture teacher is a member of a team of teachers, counselors, administrators, and others who go about fulfilling the mission of a school. A teacher should participate and support the school. Attending school functions not related to the agricultural education program is always a positive step. Helping other teachers with activities and events is another step in developing as a team player in a school. Agriculture teachers should support the work of other teachers and programs in a local school and not isolate themselves.
- **Planning and developing a program**—Agricultural education programs have several major components. Nearly all have classroom and laboratory instruction, supervised experience, and student organization (FFA) components. Some have adult and young adult components and may include school farms, canneries, forestry labs, and similar kinds of facilities. Programs are planned to meet local needs as well as implement state-established standards and guidelines.

- **Preparing to teach classes**—Instructional planning includes making the necessary preparation for classes so that time and other valuable resources are used efficiently. It may involve developing lesson plans, organizing demonstrations, and preparing hands-on laboratory activities to achieve instructional objectives. Thorough planning is essential to provide students with the maximum time for learning.

- **Delivering instruction**—Instructional time is used to guide student learning in classroom and laboratory activities. Resources must be available and efficiently used. Students must have relevant materials if they are to achieve the maximum. These include textbooks appropriate to the grade level and with content of suitable depth. The materials should have high appeal to students. All activities should focus on learning and mastering the established objectives. Lack of such focus leads to off-task student behavior.

- **Evaluating student progress**—Assessing the progress students are making toward achieving the instructional objectives is a regular part of the teaching process. Information gained from such assessment can be used to re-teach as needed or to move to another instructional area.

- **Advising student organizations**—The major student organization in agricultural education is FFA. Agriculture teachers are also FFA advisors. In a school with one agriculture teacher, that teacher is the sole advisor. Teachers may divide the advisor role in a school with more than one agriculture teacher.

- **Supervising student experiences**—Agriculture teachers are responsible for promoting supervised experience and for helping students plan, carry out, and evaluate activities. This includes teaching record keeping as well as providing on-site supervisory visits.

- **Managing resources**—Agriculture teachers typically have a wide range of resources to manage. Resources are budgets, facilities, equipment, instructional materials, and other items, such as a pickup truck or van. These duties must be carried out efficiently and effectively.

- **Relating to publics**—Good relationships are needed with the publics important in agricultural education, including students, parents, prospective students, school personnel, local agriculture officials and businesses, and others.

1–11. An agriculture teacher is helping a student develop a supervised experience program.
(Courtesy, Education Images)

A good personality and the ability to meet and greet students and parents will be useful. A friendly, happy, and outgoing manner will serve a teacher well. Skills in planning good public relations, such as with newspaper and broadcast journalists, are also needed.

- **Lifestyle**—Agriculture teachers should follow lifestyles worthy of being modeled by students and others in the school community. Participation in civic clubs, agriculture organizations, and related groups is often beneficial in an agricultural education program. Holding to high moral standards is a typical expectation. Recreational drug use and involvement with illicit activities is never appropriate.

Agriculture Teacher Traits and Abilities

What are the qualities of a good teacher? Every individual has different ideas about what makes a good teacher. People don't always agree on the meaning of good teaching and the traits of individuals who are good teachers. In the 1980s, elaborate testing requirements were established in the states in order to distinguish potentially good teachers from individuals who would not be good teachers. Within a couple of decades it was determined that what a teacher scores on a test has little correlation to his or her effectiveness as a teacher (Baines, 2010). Many states still require passage of some form of the Praxis test to gain credentials as a teacher. Be sure to get full details from your university advisor on requirements for teacher certification.

In recent years, education agencies have attempted to define or relate good teaching (and teachers) to the scores their students gain on achievement tests. Controversial from the start, many factors are involved in student test scores. Some online testing opportunities have been widely used among the states and local school districts in assessing student achievement. Compromised integrity of the testing process has resulted in cheating scandals created by teachers, school principals, and other individuals. Achievement testing is available in areas of agriculture education, with

1–13. A student in a university-level agriculture teacher education program is reviewing her portfolio with her faculty advisor. (Courtesy, Education Images)

tests varying from those directly correlated to state instructional guides to testing in areas without regard to instructional guide content.

The Horace Mann: Reach Every Child lists eleven traits of good teachers, as follows:

- **Be Unsatisfied**—Good teachers are eager to learn new things and are lifelong learners.
- **High Expectations**—Good teachers hold high expectations for their students.
- **Create Independence**—Good teachers encourage students to seek answers on their own.
- **Knowledgeable**—Good teachers possess a deep knowledge of their subject matter.
- **Humor**—Good teachers have a sense of humor.
- **Insightful**—Good teachers provide quick and accurate assessment of student work.
- **Flexible**—Good teachers utilize the resources of the community.
- **Diverse**—Good teachers use a wide array of strategies to promote student learning.
- **Unaccepting**—Good teachers do not accept false excuses or less than the best work of students.
- **Unconforming**—Good teachers challenge and motivate students with a range of approaches.
- **A Communicator**—Good teachers are good communicators.

Research has been carried out in an attempt to identify the specific qualities (competencies, traits, abilities, and characteristics) of good and successful agriculture teachers. Several research reports have been published in the *Journal of Agricultural Education* and the *Journal of Career and Technical Education* identifying qualities of good agriculture teachers (see Chapter Bibliography). Different methods and sources of research data often result in findings that appear to support similar but varying traits.

1–14. Good communication skills are essential in teaching, such as the ability to name and explain relationships.
(Courtesy, Education Images)

Reading the reports has resulted in the following traits being identified as those of strong agriculture teachers:

- Plans and Manages the Agricultural Education Program (including budgetary areas)
- Follows Appropriate State and Local Education Policies and Guidelines
- Has Knowledge and Skill in Content Area(s)
- Plans Instruction, Both for Program, Courses, and on a Daily Basis
- Directs Learning Student Activities and Is On Task in the Classroom and Laboratory
- Evaluates Student Learning and Uses Findings to Re-Teach as Necessary
- Advises and Counsels Students Appropriately
- Manages the Learning Environment to Keep Students On Task
- Advises Student Organization
- Directs Supervised Experiences of Students (including student planning, supervision, record keeping, and reporting)
- Relates to Publics in the Community, Including Agriculture Groups and Other Stakeholders
- Promotes and Follows Professionalism
- Follows Appropriate Ethics and Moral Standards

These teacher qualities are discussed in more detail elsewhere in this book. Of course, local expectations may vary and teachers should know and follow these as they go about leading agricultural education programs.

THE SCHOOL COMMUNITY

Agricultural education is a part of a larger school community. It is not carried out in isolation from other classes, teachers, and students. A school community performs best when all members work together as a team to attain student achievement goals.

Community Elements

A school community at the secondary level has several elements. No one function or area is carried out in isolation from the others.

1–15. An agriculture teacher is discussing agricultural education program plans with her school principal.
(Courtesy, Education Images)

Administration   The board of education has the legal authority to take action and use resources in the operation of schools. Some of this legal responsibility may be delegated to hired personnel. Other duties, such as hiring teachers, cannot be delegated in most states. Overall, boards of education employ administrators to run the schools.

Administration is the management of the procedures used to implement policies in an educational organization, such as a high school or career center. Individuals who carry out administration are called administrators, with a superintendent being the top administrator of a school district and a principal the top administrator of an attendance center (school). Several assistants may help superintendents and principals in their work.

Overall, administrators are to be instructional leaders. Some individuals describe good administrators as cheerleaders for students and teachers. Administrators organize and facilitate teaching and learning processes. They also relate to the public; handle student discipline problems; employ, supervise, appraise, and discipline employees; and oversee all other areas, including purchasing, transportation, food service, and building maintenance.

Teaching   In a school community, the teachers are responsible for providing the instruction and implementing the curriculum. They plan and deliver learning activities to promote student mastery of the objectives established for the school and the classes offered. They appraise student achievement and provide reports on progress.

A larger school may have a cluster of teachers in a similar area of content, with a lead teacher, often called a department head. This approach is an extension of administration. Department heads often continue in the classroom but with a reduced teaching responsibility.

An agriculture teacher should support all teaching areas and roles in a school. Instructional content may be planned so that teachers in academic areas and those in career areas integrate instructional content. This means that a teacher in one area includes relevant content from other areas. An example would be a horticulture teacher articulating instruction with a biology teacher. Resources are sometimes shared, such as biology lab equipment being used in horticulture instruction and horticulture facilities being used in biology instruction.

Counseling   All schools have services available to counsel students and assist teachers. *Counseling* is the process of appropriately trained individuals helping persons with problems find satisfactory solutions.

Counseling may deal with personal problems, educational problems, and vocational or career problems and with the general advisement of students. Most teachers appreciate the role of counseling and refer students for help.

A school may have one or more counselors. Some schools have individuals in counseling-related positions, such as psychologists. Testing experts (psychometrists) may also be in larger schools or serve several schools within a district.

Extracurricular Staff   A school often has many activities for students outside the required curriculum. These are known as extracurricular activities. Staff members are hired and/or available to help with these activities. Supporting these activities helps promote the agricultural education program. (Note: Some activities or groups are considered cocurricular. This means that they are a part of the instructional program carried out in classes. Science clubs, FFA, and history clubs are cocurricular examples.)

Common extracurricular activities include the following:

- **Music**—Chorus, marching band, orchestra, and other music programs often involve directors and other individuals. Music programs may have flag performers and precision dance groups.
- **Drama and debate**—Drama clubs select, organize, prepare for, and give dramatic presentations, or plays. Some schools offer debate or speaking groups. These groups allow students to develop the ability to think on their feet and deliver presentations.
- **Dance and gymnastics**—Schools may offer dance and gymnastics activities. Teachers in these areas may collaborate with other school groups. For example, a dance teacher may assist the band with a precision dance group.
- **Athletics**—High schools often have several competitive athletic teams, such as soccer, tennis, swimming, and volleyball. Cheerleading may be included in the athletics area.

1–16. An agriculture teacher is advising an FFA committee on an activity it is planning.
(Courtesy, Education Images)

- **Student government**—A high school usually has a form of student government that allows students to develop leadership skills and promote an improved school environment. The organization is often called the student council. Students are typically elected through democratic voting.
- **Honorary and other curriculum-related and special-interest organizations**—High schools typically have many organizations for student participation. The honorary organizations, such as Beta Club and Golden Key, are important in promoting academic achievement. Organizations related to the curriculum are also common, such as science clubs and speech clubs.

Special Services   Instructional technologists, librarians, special education teachers, and language or speech specialists may be included in this element. In recent years, the number of instructional technologists and their importance has markedly increased. Often referred to as IT specialists, instructional technologists provide valuable support in the use of computer-based approaches in education and accountability. The agriculture teacher should maintain good relationships with all these. Good relationships lead to assistance when it is needed.

Support Personnel   Support personnel includes a range of positions that help a school operate efficiently.

Office Staff.   Secretaries, attendance clerks, registrars, purchasing agents, payroll clerks, and others are often employed in a school office or at the district level. They interface between students, parents, teachers, administrators, and other school and community individuals. Positive interactions of the agriculture teacher with these individuals promote a quality school community and help when problems arise in agriculture. These staff members are good resources for the agriculture teacher.

Custodial Staff.   The custodial staff is responsible for cleaning classrooms, hallways, restrooms, and other facilities and for assuring that these areas are in good order for students and teachers. The work often occurs after the end of the regular school day or early in the morning before a new school day begins. Custodial service should be available in the agriculture facility. However, the agriculture teacher and students are usually responsible for cleaning the agricultural laboratory.

Food Service Staff.   The food service staff is responsible for operating the school dining service (cafeteria). These workers plan meals, requisition food, prepare and serve food, and clean eating and cooking utensils. Meals are intended to provide adequate nutrition at a reasonable cost. These individuals may also prepare and serve meals for special functions, such as banquets, field trips, and receptions.

Transportation Staff.   The transportation staff includes bus drivers, schedulers, maintenance personnel, and others needed to assure that safe and efficient transportation is provided. This includes transportation on daily routes as well as for field trips and events. The agriculture teacher should follow all procedures in scheduling and using transportation.

Maintenance and Grounds Staff.   All school facilities require routine maintenance, such as replacing a light bulb or unstopping a drain. Scheduled maintenance, such as painting, reroofing, and the servicing of heating, ventilation, and air-conditioning systems, is also included. The grounds staff picks up trash, mows, establishes and maintains landscapes, and removes snow from sidewalks, among other duties.

Security and Safety Staff.   Security and safety officers promote a safe learning environment that is free of hazards. These individuals are sometimes associated with local law enforcement offices.

1–17. Good rapport with security personnel is needed in maintaining security of facilities, including farms and other laboratories. (Courtesy, Education Images)

Stakeholder Elements

A *stakeholder* is an individual or a group that has a stake or strong interest in an enterprise or program. Schools have a number of stakeholders. Parents are interested in the educational success of their children. Taxpayers are concerned about the efficient use of tax funds appropriated for public schools. Businesses want schools to prepare future employees who will be good workers. Some stakeholders have specific interests in agricultural education.

Family Members   Family members, particularly parents, have a great deal of interest in the quality of the education their children are receiving. Parents want their children to be well prepared for seeking additional education beyond high school and/or for entering occupations. They want their children to gain high scores on standardized tests for graduation, college admissions, and other purposes.

Family members are often very willing to assist with school activities. Sometimes they may help as chaperones for field trips. Other times they may serve as resource persons in class or provide supplies and materials for activities. They may also provide sites for supervised experience, help judge local FFA competitions, and help in other ways.

Relating well to family members helps build support for the school and agricultural education. Taking the initiative to speak to family members and showing sincere interest helps them to feel comfortable with school personnel.

Taxpayers   All citizens are taxpayers and want accountability in use of their tax money. Large property owners may pay more taxes than small property owners and be in positions to be more demanding. Good performance by a school demonstrates accountability. Excellence in the agricultural education program also shows strong evidence of accountability with funds it has received. Maintaining good public relations promotes a positive feeling among taxpayers.

Businesses   Businesses in the local community have a great deal of interest in the local school. The agricultural businesses can be particularly useful to the agricultural instructional program and to specific agriculture classes. They may readily offer their

1–18. Integration and application of mathematics skills are often a part of the agricultural education instructional program. (Courtesy, Education Images)

facilities for field trips, lend or donate equipment for educational purposes, and serve as supervised experience training sites.

Business owners and employees may have children enrolled in the local school. This provides increased incentive for supporting efforts of the school.

Other Educators   Other educators may be viewed as stakeholders. This includes administrators, counselors, and teachers. Good, positive relationships with "academic" teachers promote the value of agricultural education in helping students gain knowledge and skills in these. The integration of science, mathematics, and other areas into agricultural education has become increasingly important. Careful planning and implementation are needed.

REVIEWING SUMMARY

To teach is to guide others to learn. The process is not always easy and involves active, on-task approaches in teaching. Teaching is the process of directing the learning experiences of others. It is far more than meeting with students in a classroom. It involves planning and evaluating learning—managing the learning environment so that students are on task in mastering knowledge and skills. Consideration is given to the situations of the learners and the resources available for their use.

Agricultural education is education in and about agriculture and related subjects. It involves a cluster of careers and eight pathways. These pathways help organize instructional programs and the instruction offered.

Agricultural education is offered in 8,038 secondary schools by 11,322 teachers. Almost 1.3 million students are enrolled. Enrollments have been increasing more rapidly than the number of teachers and the funding available to support the instruction.

Teaching is important for many reasons. In agricultural education, teaching is viewed as a means of assuring adequate and safe supplies of food and fiber for the well-being of the people in our nation. Sustainable approaches are advocated to assure the ability to provide indefinitely for the needs of people.

An agricultural education teacher has many roles and responsibilities. Above all, he or she is in charge of a program of agricultural education. That program includes classroom and laboratory instruction, student organization (FFA), and supervised experience components. In some places, programs may have other components, such as adult and young adult education, school farms, canneries and food processing facilities, and land laboratories.

Teaching agricultural education is a profession. It involves meeting certain high standards of educational accomplishment, using various means of screening in only the best-qualified individuals, and following a code of ethics and a creed. Professional agricultural educators embrace continuing education and the importance of projecting an appropriate image in a community.

Agricultural education teachers are part of an overall school community. They do not go about their jobs in isolation. Being supportive of the entire school program helps build support for agricultural education. School personnel, as well as community stakeholders, appreciate an agriculture teacher who embraces all aspects of the school community, including extracurricular activities.

QUESTIONS FOR REVIEW AND DISCUSSION

1. What is teaching? How does the word *teaching* relate to *teach* and *teacher*?
2. Why is teaching important? List at least five reasons for teaching that emphasize agricultural education.
3. What are the two settings for education? Distinguish between the two.
4. What agency or service is the leading provider of non-formal agricultural education?
5. What is a profession?
6. What are the expectations for agriculture teachers to have and enjoy professional status?
7. How are codes of ethics used in professions?
8. What is a creed? What organization established the Ag Teacher's Creed? What impresses you the most about the Ag Teacher's Creed?
9. What are the expectations of an agricultural education teacher in becoming a professional and continuing to enjoy professional status?
10. What is agricultural education?
11. What are the pathways in the Agriculture, Food, and Natural Resources (AFNR) career cluster?
12. What are content standards? How are these standards useful? Why are they needed?
13. What have been the national trends in enrollment, teacher load, and funding for agricultural education?
14. How is agricultural education a part of American education?
15. How has school governance and control of education in the United States changed over the past hundred years?
16. What are two federal laws that have brought greater federal involvement into local education?
17. What is a Carnegie unit?
18. What are the minimum mandatory subject requirements in most U.S. high schools? Is agriculture one of these? Why?
19. What does "being a school team member" mean?
20. Name and briefly explain ten roles and responsibilities of an agricultural education teacher.
21. Why is continuing education important?
22. What is demeanor? How is it related to a teacher's image in a community?
23. What are the major elements in a school community? Briefly explain each.
24. Who are the major stakeholders in a school community? Briefly explain each.

ACTIVITIES

1. Investigate the professional code of ethics for educators. Use the Web site of the National Education Association for your investigation. Print out the code of ethics and incorporate it into your professional portfolio. Carefully read the code of ethics and, in one hundred words or less, summarize its meaning to a teacher.
2. Take a field trip to a modern high school agricultural education program. Observe the students and faculty, as well as the facilities, instructional resources, and other elements of the program. Prepare

a written report that describes what you observed and offer an evaluation of your observations. You may wish to use a digital camera and presentation software to prepare a report for your class.
3. Interview a secondary agriculture teacher and/or administrator of a school with secondary agricultural education. Prepare a list of questions ahead of time. Ask about the nature of the work, the trends that are occurring, the biggest challenges faced with agricultural education, the teacher's approaches that are most effective, curriculum/courses offered,

instructional resources available, and enrollment. Prepare a written report and give an oral report in class.

4. Explore comparative agricultural education. Investigate similarities and differences of agricultural education in another state as related to your home state. Prepare a report on your findings. An alternative is to investigate agricultural education in another nation and prepare a report on your findings about that nation.

CHAPTER BIBLIOGRAPHY

Baines, L. (2010). *The Teachers We Need vs. the Teachers We Have*. Lanham, MD: Rowman and Littlefield Publishing Group, Inc.

Harlin, J. F., T. G. Roberts, K. Dooley, and T. Murphrey. (2007). Knowledge, skills, and abilities for agricultural science teachers: A focus group approach. *Journal of Agricultural Education, 48*(1), 86–96.

Horace Mann: Reach Every Child. (2011). *Top 11 Traits of a Good Teacher*. Accessed November 18, 2011, at http://www.reacheverychild.com/feature/traits.html.

Lee, J. S. (2001a). Gaining high school achievement in agriscience. *The Agricultural Education Magazine, 74*(3), 12–13.

Lee, J. S. (2001b). *Teaching AgriScience*. Upper Saddle River, NJ: Pearson Prentice Hall Interstate.

Lee, J. S. (2001c). Using achievement testing in agriscience. *The Agricultural Education Magazine, 73*(5), 6–7.

Lee, J. S. (2005a). Maximizing accountability and student achievement. *The Agricultural Education Magazine, 78*(2), 11–13.

Lee, J. S. (2009). *Report of a National Study of Agricultural Education in the United States.* Sautee-Nacoochee, GA: Lee and Associates.

Newcomb, L. H., J. D. McCracken, J. R. Warmbrod, and M. S. Whittington. (2004). *Methods of Teaching Agriculture* (3rd ed.). Upper Saddle River, NJ: Prentice Hall.

Roberts, T. G., and J. Dyer. (2004). Characteristics of effective agriculture teachers. *Journal of Agricultural Education, 45*(4), 82–95.

Roberts, T. G., K. Dooley, J. Harlin, and T. Murphrey. (2006). Competencies and traits of successful agricultural science teachers. *Journal of Career and Technical Education, 22*(2), 1–11.

Sadker, M. P., and D. M. Sadker. (2005). *Teachers, Schools, and Society* (7th ed.). New York: McGraw-Hill, Inc.

Thompson, R., A. Whalen, J. Moore, et al. (2008). *Social and Cultural Founds of American Education.* Lexington, KY: Leserati Circle.

2

Entering the Profession and Advancing as a Professional

People can become agriculture teachers in different ways. Many think about earning a college degree in agricultural education as the way to enter teaching. Here are three examples:

- Ms. Morgan is 22 years old, a graduate of the state land-grant university in agricultural teacher education, and a standard-licensed teacher. From the time she was in the ninth grade, she knew she wanted to be a teacher.
- Mr. Bill has a B.S. degree in horticulture and has worked as a greenhouse and landscaping manager for the past ten years. He is a new agriculture teacher licensed under an alternative certification program.
- Mr. Berkeley has his B.S. degree in agriculture from a university in one state, took his education course work online from a university in another state, and has just completed student teaching in a third state.

What are the differences in the paths these individuals followed to enter the agricultural education teaching profession? What professional opportunities are available for agriculture teachers?

terms

American Federation of Teachers
Association for Career and Technical Education
career ladder
continuing renewal credits
credentialing
LEAP
National Association of Agricultural Educators
National Board for Professional Teaching Standards
National Council for Accreditation of Teacher Education
National Education Association
occupational specialist teaching license
Phi Delta Kappa International
reciprocity
salary schedule
standard teaching license
teaching license
transition-to-teaching program

2–1. A teacher education student is discussing course requirements for degree completion with her advisor.

(Courtesy, Education Images)

CREDENTIALING REQUIREMENTS

There are more than 11,000 agriculture teachers in the United States (Kantrovich, 2010). Each year around 700 new teachers are needed to replace retiring teachers and teachers leaving the profession, as well as to fill openings for additional teachers and for teachers in new programs. In a typical year, there is a 6 percent turnover of agriculture teachers in the country. How are these new teachers credentialed to teach agricultural education? This question will be answered in the following sections.

The responsibility of teaching agricultural education demands that the teacher be qualified to teach. One method of ensuring that teachers are qualified is credentialing. *Credentialing* is the process of determining and certifying that a person can perform to required standards and meets competencies necessary to be a teacher. The most typical route for credentialing is through a teacher education program at a university. Upon meeting the requirements of credentialing, a person receives a teaching license.

A *teaching license* is a certificate issued by a legally authorized state office documenting that a person is qualified to teach and the conditions under which that person may teach. Conditions typically include the subject area(s) the person can teach, the grade level(s) at which the person can teach, and the period covered by the license. Each state has its own unique credentialing requirements and licenses, although some components tend to be common among all states. For example, in agricultural education, a license may include career and technical education, agriscience, grades 5–12, and expires in five years as the conditions.

To become a licensed teacher, a person must demonstrate competence in content and pedagogical knowledge, dispositions, and performances. Potential teachers may demonstrate their competence by successful completion of college teacher education programs on either the baccalaureate or the master's degree level or through alternative means. Each of these is discussed next.

2–2. An agriculture teacher who initially gained certification as a science teacher is discussing plant growth in a greenhouse.
(Courtesy, Education Images)

Standard Licensure

A ***standard teaching license*** is a license issued upon successful completion of an accredited teacher education program and any other requirements of the state. An individual successfully completing a teacher education program may receive a baccalaureate or master's degree or a post-baccalaureate certificate, depending on the state and the university. For an agricultural education teaching license, other requirements may include passing national and/or state examinations, submitting a teaching portfolio, having no felonies or applicable misdemeanors on a criminal history check, verifying occupational work experience in agriculture, and documenting successful completion of required hours/weeks of student teaching internship. States may have other requirements such as a minor or degree in a content area, certification in cardiopulmonary resuscitation (CPR), or passing one or more standardized tests.

There may be more than one level of standard license. For example, a state may issue a beginning teacher a probationary license, sometimes also called an initial, induction, provisional, preliminary, or beginning license. A probationary license may be valid for one to five years, during which time the beginning teacher receives mentoring, is observed by experienced educators, and is evaluated by an administrator. These observations and evaluations, using approved forms, become part of the documentation to obtain the next-level license. Also, the probationary teacher may be required to submit a portfolio before obtaining the next-level license.

When held by an experienced teacher, this standard license is typically renewable every four or more years. Names for this license include professional, proficient practitioner, collegiate professional, and others denoting competency and professionalism. Some states may have higher-level licenses for individuals with advanced degrees, National Board certification, or other indicators of advanced accomplishment. The advantages of higher-level licenses may include a higher base salary or a longer renewal period.

Although some practicing teachers may hold "life licenses" that do not require renewal, most agriculture teachers must renew their teaching licenses every four to ten

2–3. Beginning agriculture teachers must demonstrate classroom competency and professional growth to renew their teaching licenses.
(Courtesy, Education Images)

years. Renewal requirements are in place to ensure that teachers stay current in their field for content, pedagogy, and technology. Most states allow teachers to apply graduate credits earned from college courses toward license renewal. Many salary schedules have steps that teachers can move up as they complete certain numbers of graduate credits, such as baccalaureate plus 15 credits, master's degree, master's plus 15, and master's plus 30. Other renewal options include *continuing renewal credits*, which are credits awarded for attendance and participation at workshops, conferences, and seminars and for participation in other inservice activities. Some states allow even greater flexibility in documenting professional growth and improvement that leads to license renewal.

Alternative Licensure

Alternative licensure allows individuals to be licensed without having completed traditional programs of teacher education preparation. Several approaches are presented. One or more of these may be used in your state.

Occupational Specialist   Many states, in addition to standard licenses, provide a separate route for teacher licensure for career and technical education subjects. An *occupational specialist teaching license* is a license for an individual with substantial occupational experience in the field he or she wishes to teach. Examples of such fields are automotive mechanics, cosmetology, electronics, and building trades. Many times the best people for teaching these subjects will not have baccalaureate degrees and almost always will not have degrees from teacher education programs in their subjects. For this reason, states provide an alternative route for these people to obtain teaching licenses.

Teachers licensed through the occupational specialist route typically have three or more years of work experience in their field and are recognized as technical and content experts. Once employed, they receive intensive instruction in teaching methodology, classroom management, and other topics necessary for effective teaching. They are assigned mentors and are regularly observed. Just as with standard licenses, occupational specialist teaching licenses may have different levels and require course work or continuing renewal credits in order to be renewed.

2–4. Most agriculture teachers are college graduates in agricultural teacher education.

(Courtesy, Education Images)

As a career and technical education subject, agricultural education in many states also allows for licensing of occupational specialist teachers. Common areas include horticulture, landscaping, animal and plant sciences, and natural resources. Depending on the state, these teachers may be restricted to teaching only the agricultural education courses that match their occupational specialization. In many cases, an agriculture teacher who holds an occupational specialist license has a baccalaureate degree or higher in agriculture.

Transition to Teaching   Another route to licensure is a system for people who have had careers outside of education and then want to be teachers. The programs go by different names but can generically be called transition-to-teaching programs. A ***transition-to-teaching program*** is a credentialing approach for people who hold content degrees and have pursued careers in the content field. Examples include laboratory researchers who want to become science teachers, retired military trainers who want to enter teaching, and agribusiness specialists who want to become agriculture teachers. A transition-to-teaching program typically consists of a limited number of education courses taught in an intensive manner, combined with an internship similar to student teaching. Standardized testing and criminal check requirements must still be met to obtain a teaching license. These programs may also have grade point average requirements and/or occupational experience requirements.

Emergency Certification   Some states allow teachers to teach outside of their field or even allow unlicensed people to teach on an emergency basis. For example, a licensed agriculture teacher may teach one biology class for a school year because a biology teacher left one week before classes started in the fall. Or, a licensed science teacher may teach a mathematics course. In other instances, people without teaching degrees or experience are hired because of a shortage of licensed teachers. Emergency certification is not desirable as a solution to the shortage of qualified teachers.

Troops to Teachers   Troops to Teachers (TTT) is a program operated by the U.S. Department of Education and Department of Defense to assist eligible military personnel in becoming school teachers upon leaving military service. TTT, established in 1994, provides financial support to eligible participants, placement assistance, and other services. The intent is to provide teachers for high-need schools that serve low-income families.

Distance-Based Certification Programs

LEAP, or Licensure in Education for Agricultural Professionals, is a distance-based certification program in agricultural education. The program is designed for people who already hold baccalaureate degrees in agriculture and meet the minimum qualifications.

The LEAP program appeals to people such as career changers who are unable to return to college full time. The required courses are delivered online so students have the flexibility to complete the courses at times and places convenient to them. LEAP is administered by the Agricultural and Extension Education Department at North Carolina State University. LEAP is accredited by the National Council for Accreditation of Teacher Education (NCATE).

The *National Council for Accreditation of Teacher Education* is an organization that accredits college and university teacher education programs. It uses a performance-based system and verifies that certain standards have been met. Completing a teacher education program at an NCATE-accredited institution has certain advantages, such as reciprocity among some states and institutions.

Successful completers receive teaching licenses from the state of North Carolina. They are then eligible to teach in almost all fifty states through a process called reciprocity. *Reciprocity* is the recognition by one state of the validity of a teaching license issued by another state. This allows teachers to take teaching jobs more freely in other states. Reciprocity does not relieve a teacher from state-specific requirements. For example, to continue teaching in the new state, the teacher may have to pass a state or national competency test or take additional courses.

AG*IDEA is an alliance of universities providing online courses and degrees. Students select a home university, which provides a base for advising, enrolling in courses, and obtaining a degree. An advantage of AG*IDEA is all courses cost the same tuition regardless of which institution originates the teaching. The Agricultural Education program within AG*IDEA can lead to a career as an agriculture teacher.

GAINING AN INITIAL POSITION

Once you have a license, you are ready to seek a position. Several important procedures are presented here that will help you to be successful in gaining an initial position.

Locating Available Positions

The first step in gaining an agricultural education teaching position is to locate schools that are seeking to hire agricultural education teachers. Finding out about openings is sometimes a challenge. You want to be sure that an opening exists before making a job application. Some individuals view it unprofessional to apply for a position before it has been declared open.

Several strategies in locating open positions are the following:

- **Consult with teacher educators**—Most teacher education programs maintain lists of vacancies. A program may post vacancy announcements in a bulletin or have a list on the teacher education Web site. Teacher education programs may also have Internet listserv capability and send regular position announcements to students' e-mail addresses. Always seek the help of your academic advisor with position openings and applications.
- **Consult with state supervisory staff**—State staff members in agricultural education are often the first to hear about position openings. Because they visit

local schools, they usually know quite a bit about local school situations in terms of support for agricultural education and the nature of the programs that have been in place.

- **School district Web sites**—Most school districts maintain Web sites. Among other information, position openings in a district are announced on its site. Regular checking of the Web site is needed, depending on how often it is updated.

- **Professional organizations**—The state agriculture teachers' organization may maintain a list of teaching vacancies. The information may be posted on a Web site, be available in a newsletter, or be obtained by contacting an officer.

- **Other organizations**—The state department of education may maintain a list of teaching vacancies. The information is typically posted on a Web site. Also, the National FFA Organization maintains an agricultural educator teaching positions Web site.

- **Newspaper advertisements**—Local school districts often advertise or list teaching vacancies in newspapers that serve their areas. Regularly check the papers for such announcements. It is somewhat difficult to keep up with advertised openings in a school district served by a newspaper many miles away.

- **Family and acquaintances**—People who work in a school or school district often hear about position openings. Teacher education students who took agriculture classes in high school often ask their former teachers about openings. School employees share the information with others, including family members. You will want to be sure that news about an opening is a fact, not a rumor, before you apply. Check out what you hear by calling the personnel director of the school system to see if an opening actually exists.

Searching for the Right Position

How does a newly licensed agriculture teacher decide what positions to interview for and which job offer to accept? Following are some considerations to help guide the selection process.

2–5. Program emphasis is an important consideration because you will need good preparation in the subjects you will be teaching, such as horticulture.

(Courtesy, Education Images)

Program Emphasis   It is important that beginning teachers find programs that are a good fit with their expertise and personalities. An agricultural education program that has a strong emphasis area, such as agriscience or horticulture, needs an agriculture teacher who has the expertise to teach in that area. A mismatch, without remediation, in this consideration can lead to frustration and ineffective teaching. Appropriate remediation can include mentoring, additional course work, and experiences such as internships and on-the-job training.

Program Characteristics and Resources   It is also important that beginning teachers decide whether they work better in single- or multi-teacher programs. Do you prefer to work by yourself or with someone else? Teachers who have had positive working experiences in multi-teacher programs tend to gain greater job satisfaction and potentially greater teaching effectiveness.

Another consideration is the resources and support that will be available for the program. These are important factors. A facility that is less than state of the art can often be improved. The same is true with inadequate instructional materials, equipment, and other resources.

Here are a few questions to answer in assessing program characteristics and resources:

- What is the history of the agricultural education program? How many changes in agriculture teachers have occurred in the past five years? How many years has the program been at the school?
- Does the program receive a budget each year? How are funds made available and expended?
- What do the facilities look like? Are they in good repair and appropriate for the program emphasis?
- Are the appropriate instructional materials available, or will the administration commit to obtaining needed materials?
- Are the proper equipment and tools available for the program emphasis?
- Is there evidence that the administration supports the local agricultural education program?
- What are the plans regarding the program? Are adequate fiscal resources provided for the program?

FFA and SAE Characteristics   FFA and supervised agricultural experience (SAE) are major components of the agricultural education program. You will want to assess these. Some teachers prefer to take a program that is "down" and build it "up." Other teachers prefer to go into a program that is in good condition.

Here are a few questions to answer:

- What are the expectations regarding FFA and SAE?
- Does the program have a history of success in career development events?
- What are school and community service-learning expectations?
- Does the program have more of a leadership, career, or personal development emphasis?
- Regarding the SAE program, what are the expectations for home/worksite visits, and what travel support is provided?
- How are travel, per diem, and lodging expenses funded for the agriculture teacher when he or she attends FFA events and activities?
- How are the FFA members' expenses funded?

School Characteristics   The physical and administrative layout of a school is somewhat important in selecting where to teach. Agricultural education programs located in comprehensive high schools, area career centers, and middle schools have different emphases and characteristics. The grade levels taught at a school and the proximity to schools of other grade levels influence the opportunities for the agricultural education program. The location of the agricultural education program within a school and the facilities associated with the program give an indication of the value placed upon the program.

Administrative support is an important consideration. Assess the support given teachers and the agricultural education program in the past. Gain an understanding of the current administrators' support by talking with the principal, career and technical education director, or other administrator.

Another consideration is the other teachers in the school. Included are those in agriculture as well as those in other subject-matter areas.

Here are a few questions to answer:

- Is the overall faculty experienced or inexperienced?
- Do teachers tend to stay at the school for several years, or is there frequent turnover of teachers?
- Do teachers socialize together?
- From what you can observe, such as from walking down the hallways, is there good teacher–student interaction in the classrooms?
- Is active learning occurring?
- Are classrooms well kept and inviting?
- Are students' work and accomplishments displayed within the classrooms and hallways?

Community Characteristics   Although a teacher does not usually begin and end his or her career at the same school, considering community characteristics when selecting a teaching position is still important. Some teachers like small towns, while others like urban or suburban areas.

Here are a few questions to answer:

- Is the rural, suburban, or urban location of the school what you want?
- If you have a geographic restriction, is the community in the part of the state or country where you want to live?
- Does the community have the shopping, cultural, social, and religious amenities you want? If not, are these within driving distance?
- Are job opportunities available for a spouse?
- What is the overall cost of living?
- How available is housing?

Position Benefits   Although important, salary should not drive your decision of where to teach. A high salary does not improve a bad situation or a mismatch. Salary and fringe benefits should be considered along with other factors when deciding whether you can be successful in a school situation.

Contract length is an indication of the value placed upon the total agricultural education program and the expectations for the agriculture teacher. Salary level compared with that of other school districts is usually an indication of the socioeconomic status of the community rather than the value placed on teaching. Some schools have contracts with extra days added beyond those in a regular teaching contract; others provide a stipend for FFA/SAE work. You will want to know if either or both are part of the master contract or of the agriculture teacher's contract.

Fringe benefits are an important part of the salary package and are often overlooked. Benefits must be included in assessing your overall compensation.

2–6. Note agricultural structures and businesses in a community.
(Courtesy, Education Images)

Here are some questions to answer about benefits:

- How is the retirement plan structured?
- What are the medical, dental, and vision care benefits?
- Are other types of insurance available at a group rate?
- What is the sick leave and personal days policy?
- Does the school have funds for reimbursing teachers for graduate course work, educational travel, or sabbaticals?

Most of these questions should be addressed to the school's personnel department, not asked during the job interview. In many cases, an interviewee receives an employee handbook and other materials that explain benefits as part of the initial interview.

Nonteaching Duties   Another consideration is the nonteaching duties required of teachers. Some schools expect teachers to perform duties beyond those in agricultural education.

Here are some questions to answer:

- Are teachers expected to perform hall, bus, or lunchroom duty?
- Are there expectations that teachers attend or work at a certain number of athletic or other events?
- Are new teachers expected to become athletic coaches or extracurricular club sponsors?
- Do teachers have homeroom, study hall, or other nonteaching duties during the school day?
- Is the agriculture teacher expected to be a certified school bus driver? (For an agriculture teacher, who many times holds or expects to get a bus driver's license, it is important to find out if he or she will be called upon as a substitute, field trip, or athletic team bus driver.)

Applications and Interviewing

Most school districts have standard application forms for teaching positions. Some districts accept only online applications. An application should be completed accurately, neatly, and promptly. It should be typed. Beyond the application form, an application packet usually includes a cover letter, college transcripts, letters of recommendation, a résumé, and verification of teaching license. Sometimes an application packet also includes a philosophy statement and other components. The application packet should be professionally organized and submitted in a timely fashion.

Portfolios, both in electronic form and on paper, are increasingly used for teacher education, initial licensure, and license renewal. You should check with your university teacher educators, agriculture teachers' association, and state agricultural education supervisors to determine whether a portfolio should be sent with your application or brought to an interview. The usefulness of portfolios in interviews varies among states, schools, and even individual principals and superintendents.

For many positions, the interview determines who receives the job offer. A job interview serves two purposes. For the school, it is a chance to verify that what appears on paper in the application packet is reality. Is the interviewee knowledgeable and articulate? Will this person be a good fit for the agricultural education program, the school, and the community? Is the interviewee energetic, creative, and passionate about educating young people? For the interviewee, the interview is a chance to determine whether he or she can be successful in this school and whether it is a good fit.

The interviewee should dress professionally for the interview, arrive early, and be prepared. He or she should have a firm handshake, look interviewers in the eye, and speak with energy and confidence. Nervous energy should be channeled into enthusiastic facial expressions and speech rather than exhibited as annoying mannerisms and fidgeting. This is helped by eating, sleeping, and exercising properly during the weeks before the interview. It is also helped by going through mock interviews and practicing answers to probable interview questions.

Interview questions may cover a wide variety of topics. In general, interviewers want to discover how the interviewee will work with students, parents, and colleagues. They want to find out what instructional and classroom management techniques the interviewee will employ. They also want to determine why the interviewee is interested in the particular position and what makes this person better than the other candidates for the position. The interview is also a good time for assessing the interviewee's oral communication skills. Avoid poor grammar, incomplete sentences, and bias in your speech.

Some interviewers like to create situations to get a feel for how the interviewee will respond to students with special needs or in certain circumstances. Others try to determine the interviewee's philosophical orientation—that is, how the person thinks about and approaches the teaching situation. Still others have a practical orientation and want to determine what the interviewee knows and is able to do.

At the conclusion of the interview, the interviewee should thank the interviewers for their time. Typically, the interviewers should explain the timeline for making a decision and give instructions for whom the interviewee should contact if further questions arise. The interviewee should avoid being pushy or demanding. If a teacher is selected for a position, approval by the board of the school administrative unit is required before a contract can be offered.

2–7. Application forms and other materials are usually posted on school district Web sites.
(Courtesy, Education Images)

COMPENSATION AND OTHER BENEFITS OF TEACHING

Teacher compensation is typically based on years of experience and educational level. These are combined into a salary schedule. A *salary schedule* is a list or table that shows the salary at each experience level and educational level. Increments are added for each year of teaching experience, though some schedules may not go beyond fifteen or twenty years. Compensation may be based on teacher performance using a system commonly known as merit pay. Some merit pay systems do not use a salary schedule.

Compensation

Teacher compensation is composed of base salary, extended days, extra duties, and benefits. Merit pay can be given as a one-year bonus or included in the base salary. Most teacher contracts are for a 180- to 190-day school year. The year includes 180 or more instructional days, one or more inservice days, one or more preparation days, and other times, such as evening parent–teacher conferences. Every teacher receives this same base contract. For the 2006–07 school year, the U.S. average beginning teacher base salary was $35,284, and the average teacher base salary was $51,009 (AFT, 2012).

Agriculture teachers often receive higher salaries because of job duties and extended contracts. Experience and advanced degrees also increase the salaries. Salaries for experienced agriculture teachers in the $70,000- to $100,000-a-year range are not uncommon around the United States. Teachers who have completed National Board certification may receive several thousand dollars a year more.

Some school districts may use collective bargaining. Representatives of the teachers' association negotiate salary and benefits with school administrators. After negotiations are concluded, a master contract is in force for a set time period. You may ask for a copy of the master contract to help you in making a position decision.

Extended-Day Contracts

Not all states treat extended days for agriculture teachers the same. Some states require that every agriculture teacher be employed on a twelve-month contract. This type of contract provides for vacation days and holidays and boosts teacher pay greatly over that of the base salary.

A different type of yearlong contract is the 240-day contract. The agriculture teacher works 240 days and is off the remaining days of the year. This is the equivalent of receiving 21 vacation days and holidays (365 days in a year minus 104 Saturdays and Sundays leaves 261 days eligible for working). The agriculture teacher must ask how evening and weekend work as well as overnight trips count toward days worked.

Other extended-day contracts are written for any number of days between 180 and 240. Some states allow agriculture teachers to direct supervised experience or teach courses (including adult and young adult) in the summer. These also increase pay over that of the base salary.

Most master teacher contracts also include a schedule for paying teachers for extra duties. These duties include serving as athletic coaches, department heads, class sponsors, or directors or advisors of band, chorus, and certain other teams and clubs. This extra pay is in recognition of teachers working with the respective students and activities outside of the normal school day. In some cases, the agriculture teacher receives extra-duty pay for FFA, SAE, or other agricultural education duties. Extra-duty pay is calculated in several ways. Most of the systems place all extra duties into categories and pay for every duty within a category at the same rate. In some systems the category rates are set, in others they are adjusted for years of experience, and in still others the pay is included in calculating a teacher's salary and salary increases.

Benefits can add 25 percent or more of base salary to the total compensation package. Most teachers are a part of the Social Security/Medicare system. For these teachers, the employer pays 7.65 percent of their salaries into the system. This is one component of the retirement benefit for teachers. Another component is the school- or state-sponsored retirement program. In most cases, a set percentage of the teacher's salary is paid into the retirement program each pay period. A final component of the retirement benefit is optional retirement and savings programs, such as 401(k), 403(b), and other such plans. Contributions to these may be entirely from the employee's pay or matched by the employer. Teachers typically receive medical insurance and, possibly, dental and vision care insurance. The number of insurance plans and combinations are too numerous to discuss in this section.

Other benefits are available depending on the school system and the state. Schools are typically large employers within a community. Because of this they are able to negotiate for reduced group rates for insurance, such as disability, term life, and long-term care. These are optional for teachers to purchase, depending on their own needs and circumstances. Some school systems reward their teachers for pursuing graduate degrees and other certifications. These rewards may be in the form of tuition reimbursements, higher steps on the salary schedule, and educational travel funds. Other school systems have paid or unpaid sabbatical programs that allow teachers to take a year to pursue graduate degrees or other educational opportunities and return to their jobs after the sabbaticals.

ASSISTANCE PROVIDED TO BEGINNING TEACHERS

A beginning teacher is expected to assume the duties of a teacher from the first day of school. To assist with the transition from college student to teacher, school systems, universities, and others have developed various beginning-teacher assistance programs. These programs may be required or voluntary, formal or informal, and subject-specific or general in nature.

A school-level assistance program typically involves mentoring, inservice, and observation of teaching performance, with feedback provided to the teacher. Ideally, the beginning agriculture teacher is paired with an experienced agriculture teacher as a mentor. This is not always practical, as the beginning teacher may be the only agriculture teacher in the school system. In this case, another career and technical education teacher, a science teacher, or another knowledgeable teacher can serve as the assigned mentor.

A beginning agriculture teacher should feel comfortable in seeking out experienced teachers as informal mentors. Inservice workshops may be held specifically for beginning teachers, but typically they will be school- or system-wide inservices. The beginning teacher should participate in these inservices with the attitude of gleaning information that will help make teaching and learning better for the students. Observations may be stressful, but their purpose is to help the teacher improve in pedagogical practice in order to enhance student learning. Ideally, the beginning teacher is observed several times throughout the first year of teaching.

State-level assistance programs typically involve documenting progress toward teaching standards and possibly attendance at subject-specific workshops and conferences. In agricultural education, opportunities include summer, fall, and/or spring state or area workshops; the National FFA Convention; the NAAE Convention; and various NAAE workshops, activities, and meetings. Several states have agricultural education consultants available to visit and observe beginning teachers. These visits should be stress-free for beginning teachers and include helpful advice and suggestions.

University assistance programs are typically associated with graduate-level courses and/or teamed with state-level programs. The graduate courses tend to be practical and designed to be flexible to meet the needs of the group of beginning agriculture teachers. In many cases, these courses either are taught by distance education technologies or are conducted at central locations within a state. Whether sponsored by the university or a state education agency, a valuable component of these assistance programs is that camaraderie and socialization are encouraged. Beginning agriculture teachers can be isolated both within their schools and geographically, so they need a chance to meet with other beginning agriculture teachers who are going through the same experiences they are.

In many states, the agricultural education leadership, called Team AgEd, will provide a mentoring program for beginning agriculture teachers. This may be formal or informal, but its purpose is to assist the beginning teacher in those areas that are unique to agricultural education. To find out about mentoring provided in your state, contact university teacher educators, state staff, or the teachers' professional association. A Local Program Success (LPS) Specialist can be a resource for beginning teachers and their mentors. Contact information for LPS Specialists is on the National FFA Organization Web site.

PROFESSIONAL ORGANIZATIONS

An individual who enters a profession sets goals to advance and move to higher levels with increased compensation. Agriculture teachers have several organizations to help them grow as professionals.

National Association of Agricultural Educators (NAAE)

The *National Association of Agricultural Educators* is the national professional organization for agriculture teachers (NAAE, 2012). It has more than 7,650 members in fifty state associations grouped into six regions. The NAAE was founded in 1948 in Milwaukee, Wisconsin, as the National Vocational Agricultural Teachers' Association. It changed its name to the National Association of Agricultural Educators in 1997 to better reflect changing terminology and to include agricultural educators

2–8. The NAAE Web site has useful information for agricultural education teachers.
(Courtesy, Education Images)

on all levels and in all arenas. Its national office is in Lexington on the campus of the University of Kentucky.

The NAAE provides leadership, advocacy, and service for agricultural education and for all agricultural educators (NAAE, 2012). The NAAE maintains governmental relations with the U.S. Congress, U.S. Department of Education, and U.S. Department of Agriculture. It also has representation on agricultural education and career and technical education boards and committees. In addition, the NAAE provides numerous opportunities for member involvement and service to its members.

NAAE members receive professional liability insurance as a part of their dues package. They also receive newsletter, e-mail, and Web-based communications, updates, and legislative alerts. Members and their agricultural education programs are eligible for various awards and recognitions. These are offered on state, regional, and national levels. The NAAE sponsors upper division scholarships for junior, senior, or post-baccalaureate students in agricultural education who intend to become agriculture teachers. It also works to obtain scholarship funding for entering undergraduates who want to become agriculture teachers.

The NAAE holds a national convention each year. National business is conducted at the convention, and regional meetings are held. Recipients of awards are recognized, and an exhibitors' expo is featured. Workshops, seminars, and other meetings round out the schedule.

The NAAE hosts a Web-based collaboration tool called "Communities of Practice" (CoP). There are numerous communities including technical agriculture topics such as animal sciences and floral design as well as pedagogical ones such as classroom dynamics and middle school programs. CoP provides for discussions, sharing of documents, and collaborating on projects.

Association for Career and Technical Education (ACTE)

The *Association for Career and Technical Education* is the national professional organization for advancing the education of youth and adults for careers (ACTE, 2012). It has more than 27,000 members representing teachers, administrators, researchers, guidance counselors, and others within the field of career and technical education. ACTE was founded in 1926 as the American Vocational Association. Its name was changed in 1998 to better reflect the nature of the educational segments it serves. The association has thirteen divisions, of which agricultural education is one. The fifty states, the District of Columbia, and U.S. territories are organized into five regions.

ACTE offers its members several benefits. It has award and recognition programs. ACTE serves as the major legislative advocate for career and technical education by attempting to influence public policy on the national level. Its headquarters is in Alexandria, Virginia. ACTE has an executive director and other staff members. It has several print, e-mail, and Web-based publications. ACTE also provides its members with partnership, professional development, and career services benefits.

National Education Association (NEA) and American Federation of Teachers (AFT)

The *National Education Association* was founded in 1857 and boasts more than 3 million members (NEA, 2012). It is the largest national organization for educators. The NEA has a broad membership that includes all categories of public school workers, college educators, college students studying to become teachers, and other educational personnel.

The NEA operates on the local, state, and national levels. In many cases, the local NEA affiliate is the bargaining unit for teacher contracts and salary schedules. State-level NEA affiliates provide numerous benefits to members, including legal assistance, discount programs, insurance, and state government monitoring and lobbying. On the national level, the NEA lobbies Congress and federal agencies to influence educational policies and legislation. Many university teacher education programs will have an NEA-affiliated student organization.

The *American Federation of Teachers* is also a national education organization. It has more than 1.5 million members (AFT, 2012). The AFT was formed in 1916, starting in the Midwest. It differs from the NEA in that the AFT is a recognized union and an affiliate of the AFL-CIO. The AFT, like the NEA, has different member groups. These include teachers, other school personnel, higher education faculty and staff, health care professionals, and state and municipal employees. It provides benefits to its members similar to those of other unions.

Phi Delta Kappa (PDK) International

Phi Delta Kappa International is a professional organization for educators on all levels, including undergraduates preparing to become teachers (PDK, 2012). It is an international association with hundreds of local chapters. Many of the local chapters are based on college campuses that have graduate programs in education. PDK chapters sponsor seminars, educational service efforts, and other professional development opportunities. On the national level, PDK sponsors several print publications, educational research, and member services.

CAREER PATHS FOR PROFESSIONAL ADVANCEMENT

Agricultural educators have a number of options in their careers. Some understanding of the career ladder is appropriate.

A *career ladder* is the steps in an individual's general progression in a career from entry until retirement. Typically, an individual enters on a lower level and moves upward based on gaining additional education, expanded career responsibilities, experience, or other conditions. Individuals can remain as classroom teachers and advance, or they can move into other areas in or related to agricultural education.

Advancement in Teaching

Agriculture teachers who wish to remain as teachers for their entire careers have several paths for career advancement. One of the most recognizable is obtaining advanced degrees. As discussed in the "Standard Licensure" section earlier in this chapter, most school systems provide salary steps for obtaining master's and doctorate degrees. Graduate degrees may also make teachers eligible to host student teachers and to serve as master teachers. In some states, master teachers may also teach college-level courses. In addition to the financial and professional benefits, the process of obtaining advanced degrees allows agriculture teachers to deepen their knowledge of pedagogy and the subject matter, reflect on classroom practices, and enhance the learning environments of their students.

The state and national agricultural teachers' associations provide additional avenues for professional advancement. Most state associations have area, district, regional, or section officers. Becoming one of these officers allows an agriculture teacher to get involved after only a few years of teaching. The teacher can then advance to become a state officer, the highest of which is state president. The NAAE provides opportunity for its members to serve as delegates, regional officers, and national officers. The experiences and activities associated with these leadership positions can provide much accomplishment, energy, and growth for an agriculture teacher.

Agriculture teachers may also be able to advance administratively. Agriculture teachers with the required education and experience are eligible to become department heads. Depending on the school and program size, an agriculture teacher may be head of just agricultural education or of all the career and technical education programs in a school. This position may give an individual greater responsibility and can lead to increased involvement in school and system committees.

Advancement Outside Teaching

Some agriculture teachers chose to move their careers into other areas in or related to agricultural education. Within agricultural education, some possible career moves include becoming state agricultural education staff personnel, teacher educators at colleges and universities, or agriculture curriculum specialists. This means that these individuals are no longer directly teaching secondary students. Their responsibilities

2–9. Inservice education prepares teachers to deliver instruction in new areas.
Courtesy, Shutterstock © Goodluz

usually have a major impact on the instruction, however. A few secondary teachers move into postsecondary schools to teach agriculture. They must usually gain additional credentials to do so. Sometimes they must take additional course work at the graduate level in a technical subject area.

Some agriculture teachers obtain licenses to become school administrators. Doing so can qualify them for positions as assistant school principals, principals, curriculum coordinators, assistant superintendents, and, possibly, superintendents. Agriculture teachers may also seek credentialing to become school counselors, instructional technologists, or other officials.

Each year a few agriculture teachers assume positions with agricultural agencies, businesses, and associations and with other agriculturally related enterprises. Their preparation for a career in agricultural education usually serves them well.

NATIONAL BOARD CERTIFICATION

The *National Board for Professional Teaching Standards* (NBPTS) is an independent, nonprofit, nonpartisan, nongovernmental organization that provides advanced certification to teachers who meet its standards. NBPTS began in 1987 and by 2012 had certified more than 91,000 teachers.

National Board certification is based on five core propositions (NBPTS, 2012), as listed below:

1. Teachers are committed to students and their learning.
2. Teachers know the subjects they teach and how to teach those subjects to students.
3. Teachers are responsible for managing and monitoring student learning.
4. Teachers think systematically about their practice and learn from experience.
5. Teachers are members of learning communities.

There are twenty-five certificate areas. These are typically in varying subject matters and grade levels. Early Adolescence through Young Adulthood Career and Technical Education is the area in which agriculture teachers would become certified. A teacher who holds at least a baccalaureate degree and has three years of teaching experience is eligible to apply for National Board certification. Once issued, the certificate is valid for ten years. The assessment process consists of a portfolio of classroom practice and an assessment of content knowledge. The process is rigorous and expensive. Many school systems subsidize the application fees for their teachers going through the process. Depending on the state and school system, teachers who receive National Board certification are given incentive pay, such as bonuses, and/or are placed on the salary schedule at the highest level for their years of experience. Most states recognize National Board certification teachers as meeting the definition of highly qualified teacher under the No Child Left Behind Act.

ATTRIBUTES NEEDED FOR SUCCESS

Agriculture teachers who are successful tend to exhibit certain attributes. These can be categorized into demeanor, knowledge of the subject matter, and professionalism.

Demeanor

Successful agriculture teachers have a certain manner and appearance that sets them apart. First, they have an attitude that they are "teachers of individual students." Agriculture teachers who display this attitude have care and concern for each of their students. These teachers are committed to student learning but are also concerned about students as persons.

2–10. A teacher should dress and groom to demonstrate professionalism.
(Courtesy, Education Images)

Successful agriculture teachers radiate an air of confidence and positive self-concept. Even in topics for which they are not experts, their confidence shows as they approach these topics with an attitude of experimentation and "we can learn this together." They have an energy and enthusiasm that is easy to catch and draws students to them. Successful agriculture teachers are creative and are able to see potential in students and usability in things.

An agriculture teacher's character and ethics must be without question. Students look up to their teacher as a role model, so the agriculture teacher must display ethical behavior in all circumstances. Agriculture teachers should watch their language and habits. Of special importance is having a pleasing, positive personality. An agriculture teacher with these characteristics has the potential to influence numerous students in a positive way.

Finally, an agriculture teacher's appearance should be neat and professional. Clothing should be attractive and appropriate for the situation. It should be neat and in good repair. Clothing can be protected in the laboratory with a shop coat or coveralls. Hair, skin, and overall appearance are also important, as they can influence initial impressions of students, parents, and others. The agriculture teacher should do everything possible to maintain good health, physical fitness, and stamina. Remember, agriculture teaching is a demanding, potentially stressful profession, so proper nutrition, exercise, and rest are critical to optimal performance in the classroom.

Knowledge of the Subject Matter

Agriculture teachers must have sufficient knowledge of the subject matter in their area of responsibility to be effective teachers. This knowledge is to include concepts, theory, and practice, including specific "how-to" skills. Neither experience nor a college degree alone is sufficient. To be successful, an agriculture teacher must understand the theory and concepts of agriculture and have practical experience in the field. Does this mean that the only people who can be agriculture teachers are those who grew up in an agricultural environment and were themselves in secondary agricultural education? No!

People who want to be agriculture teachers can gain occupational experience while in college through internships, through summer jobs, and in other ways. They

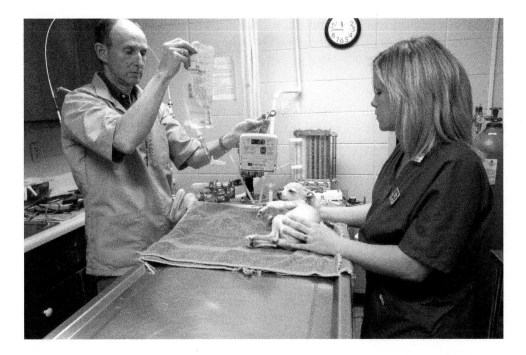

2–11. Practical experience, such as in a veterinary clinic, can be gained through summer and part-time work and internships.
(Courtesy, Education Images)

will also need to take advantage of every opportunity to become involved with secondary agricultural education activities.

Choosing the type of experience to get can be difficult. Persons with no agricultural experience at all may need experiences in production agriculture to gain the fundamental knowledge to which all agricultural occupations have some connection. Others who know they want to specialize in certain areas, such as horticulture and landscaping, will want to concentrate their experiences in those fields. Still others may want to have a wide variety of agricultural experiences to make themselves marketable for the greatest number of teacher openings.

Potential agriculture teachers should also remember that no one enters any profession fully knowing all the subject matter. You will learn much through your teaching experience and should have the attitude of being a lifelong learner. Even a person who knows much upon entering a profession will have their knowledge become obsolete quickly if they do not continually update.

Professionalism

Successful agriculture teachers view teaching as a profession and act accordingly. They accept the responsibility of staying current in pedagogy and subject matter. They act according to the highest ethics, always keeping the interests of their students in first place. They strive for excellence and attempt to give their best effort in all that they do. Finally, they support the work of other teachers by refraining from gossip, spitefulness, and negative comments regarding their work, especially in front of students.

REVIEWING SUMMARY

Credentialing provides assurance that a teacher meets the minimum qualifications deemed necessary by a state to be an effective educator. A standard teaching license is typically issued based upon a college degree, passing scores on standardized tests, and completion of a student teaching internship. States may also add further requirements. Agriculture teachers typically must also verify that they have occupational work experience in agriculture. Standard teaching licenses must be renewed every four to ten years.

Increasingly, alternative methods are being used to license teachers. One method common to career and technical education subjects is occupational specialist licensure. A more recent method is transition to teaching. This method is designed to bring career changers quickly into the teaching profession. Another method, although not desirable as a long-term solution to teaching shortages, is emergency certification. Two distance-based certification programs in agricultural education are LEAP (Licensure in Education for Agricultural Professionals) and AG*IDEA.

Applying for a teaching position involves several steps. People wishing to teach agricultural education should first ask themselves a series of questions designed to determine which positions are good fits for them and their areas of expertise. Once these are answered, applying for positions comes next. Applications should be completed accurately, neatly, and timely. Interviewing is the next step. An interview allows the school to crosscheck how the person is in real life against what is on paper and allows the interviewee to determine if the school is a good fit.

Salary and benefits are important considerations when deciding on a teaching position. Are the resources sufficient, such as extended contract days, for the agriculture teacher to be successful? Obtaining the highest salary should not be the driving force behind obtaining a particular teaching position.

Once individuals are in the profession, assistance is available for beginning teachers, and avenues are available to advance as professionals. The local school system, the state, and the university will all offer opportunities and activities to assist beginning teachers. These may be inservices, workshops, mentoring relationships, and courses. Professional organizations can help beginning teachers become successful and can also assist experienced teachers in advancing as professionals.

To be successful, agriculture teachers need proper demeanor, agricultural knowledge and experiences, and professional attitudes and actions. Agriculture teachers serve as role models to many young people and as such should hold themselves to the highest morals, ethics, and conduct. Agriculture teachers who are healthy have more energy and stamina, so proper nutrition, rest, and exercise are important. The better that agriculture teachers can relate content to the world of agriculture, the greater student learning. This requires teachers to have real-world, practical experience in agriculture. Finally, teachers who view themselves as professionals and act accordingly are respected and treated as professional educators.

QUESTIONS FOR REVIEW AND DISCUSSION

1. Why is teacher credentialing important?
2. What are the requirements for obtaining and renewing a standard teaching license?
3. Why are alternative pathways to teacher licensure available?
4. What are the factors to consider when selecting an agricultural education teaching position?
5. Why is it important that a job application be completed accurately, neatly, and timely?
6. What are the advantages and disadvantages of including a teaching portfolio in the application packet?
7. What are some hints for interviewing for a teaching position?
8. How do teaching salaries compare with the salaries of other professions?
9. What are the different methods for calculating extended-day contracts?
10. What are the professional organizations that are most applicable for an agriculture teacher?

ACTIVITIES

1. Investigate the pathways to teacher certification for agricultural education in your state. Also, explore what subjects and grade/developmental levels an agricultural education license allows the holder to teach.
2. Explore the structure, dues, activities, and benefits of your state's professional agricultural education organization. Write a report about the impact the professional organization has had on agricultural education in your state.
3. Using the *Journal of Agricultural Education, The Agricultural Education Magazine,* and the proceedings of the National Agricultural Education

Research Conference, write a report on research about beginning agriculture teachers. The terms "induction," "first year," and "new" may be used in addition to "beginning." Current and past issues of these journals can be found in the university library. Selected past issues can be found online at the American Association for Agricultural Education and the National Association of Agricultural Educators websites.

4. Investigate employment application procedures by referring to the Web sites of at least two school districts. Determine the information needed and the procedures to follow. Prepare a short report that summarizes your observations.

5. Join the NAAE Communities of Practice. Explore resources for helping beginning agriculture teachers through the induction process. Join the Pre-Service Teachers community or contact one of your teacher educators about setting up a community for your university.

CHAPTER BIBLIOGRAPHY

AG*IDEA. (2012). *AG*IDEA Web Site*. Accessed on October 3, 2012, at http://www.agidea.org.

American Federation of Teachers (AFT). (2012). *AFT Web Site*. Accessed on November 5.

Association for Career and Technical Education (ACTE). (2012). *ACTE Web Site*. Accessed on November 5, 2012.

Binkley, H. R., and R. W. Tulloch. (1981). *Teaching Vocational Agriculture/Agribusiness*. Danville, IL: The Interstate Printers & Publishers, Inc.

Frick, M., and S. Stump. (1991). *Handbook for Program Planning in Indiana Agricultural Science and Business Programs*. Unpublished manuscript, Purdue University.

Indiana Department of Education. (1999). *Teacher/Local Team Self-study of Standards and Quality Indicators for Agriscience and Business Program Improvement*. Indianapolis: Author.

Kahler, A. A., B. Morgan, G. E. Holmes, and C. E. Bundy. (1985). *Methods in Adult Education* (4th ed.). Danville, IL: The Interstate Printers & Publishers, Inc.

Kantrovich, A. J. (2010, October). *The 36th Volume of a National Study of the Supply and Demand for Teachers of Agricultural Education 2006–2009* (online). Accessed on November 30.

Lee, J. S. (2000). *Program Planning Guide for AgriScience and Technology Education* (2nd ed.). Upper Saddle River, NJ: Pearson Prentice Hall Interstate.

Moore, G. E. (n.d.). *Licensure in Education for Agricultural Professionals*. Accessed on November 30, 2011.

Moore, G. E. (Ed.). (1998). Theme: A primer for agricultural education (Entire Issue). *The Agricultural Education Magazine, 71*(3).

Moore, G. E. (Ed.). (2000). Theme: Reinventing agricultural education for the year 2020 (Entire Issue). *The Agricultural Education Magazine, 72*(4).

National Association of Agricultural Educators (NAAE). (2012). *NAAE Web Site*. Accessed on November 5, 2012.

National Board for Professional Teaching Standards (NBPTS). (2012). *NBPTS Web Site*. Accessed on November 5, 2012.

National Education Association (NEA). (2012). *NEA Web Site*. Accessed on November 5, 2012.

National FFA Organization. (2003). *Local Program Resource Guide: 2003–2004* [CD-ROM]. Indianapolis: Author.

National FFA Organization. (2011). *Local Program Success* (online). Accessed on November 30, 2011.

Newcomb, L. H., J. D. McCracken, J. R. Warmbrod, and M. S. Whittington. (2004). *Methods of Teaching Agriculture* (3rd ed.). Upper Saddle River, NJ: Prentice Hall.

Perry, S. E. (2003, July). Professional LEAP toward education. *NAAE News & Views*. Accessed on December 3, 2003.

Phi Delta Kappa (PDK). (2012). *PDK International Web Site*. Accessed on November 5, 2012.

Phipps, L. J., and E. W. Osborne. (1988). *Handbook on Agricultural Education in Public Schools* (5th ed.). Danville, IL: The Interstate Printers & Publishers, Inc.

True, A. C. (1929). *A History of Agricultural Education in the United States, 1785–1925*. U.S. Department of Agriculture, Miscellaneous Publication No. 36. Washington, DC: GPO.

3

Philosophical Foundations of Agricultural Education

Vocational administrator Sharon Davis was producing a report on the school year for the district's superintendent and was ready to write about the agricultural education program. To prepare herself, she looked at enrollment data and course offerings. She decided to take a quick tour of the agricultural education facilities and assess what the teacher, Tommy Walters, was doing.

Was she surprised! Mr. Walters was in the classroom with eleven students in advanced horticulture telling them a story about a hunting trip. The students appeared bored, and one had his head down, obviously asleep. The books were on shelves and out of date. The greenhouse lab facility had a few wilted plants that were apparently left from last year. When Mr. Walters saw Ms. Davis, he acted surprised and told her he was tutoring students for the state achievement test.

Ms. Davis became quite concerned. As she walked back to her office, she contemplated the future of the program. She thought about how resources were being wasted on teacher salary and other areas. She further thought about how student time was being wasted because of a lack of instruction. Mr. Walters' approach to teaching and learning was not the approach these students needed. Ms. Davis decided it was time to make a change in agriculture teachers. She reached for the phone to call the principal.

terms

Aristotle	Plato
Charles Sanders Peirce	pragmatism
idealism	progressivism
Johann Amos Comenius	realism
Johann Heinrich Pestalozzi	Socrates
John Dewey	Thales
philosophical foundations	theorem
philosophy	William James

3–1. A teacher's approach to teaching and learning is guided by basic beliefs.
(Courtesy, Education Images)

MEANING AND IMPORTANCE OF PHILOSOPHICAL FOUNDATIONS

The search for meaning is as old as humankind. Humans have always had an unquenchable desire for meaning and truth, and the story of this search has been written across the ages. The ancient Greek philosopher **Thales**, around 580 BCE, found that all matter could be reduced to the quintessential element water. Watson and Crick's discovery of the DNA molecule as the smallest package of life and the discovery of subatomic particles certainly weakened Thales's argument, but humankind's insistence on knowing the truth has not been diminished.

Philosophy is a discipline that attempts to provide general understanding of reality and interpret the meaning of what is observed. It seeks to draw meaning of truth and knowledge. People turn to philosophy because of their natural curiosity about the meaning of the world around us. Overall, *philosophy* is a complex term that describes the efforts of people to understand some of why we do what we do. Further, people don't agree on its meaning or on the philosophies studied. But, if someone asks about your philosophy of agricultural education, you can share the basic beliefs you hold and cite those of well-known philosophers, along with the commonly accepted principles of the profession.

Philosophical foundations are the basic beliefs that guide people's actions. In agricultural education, philosophical foundations guide how programs are planned and delivered. Much of education is built on a broad base of philosophical foundations. These foundations emerged from the efforts of individuals (referred to as philosophers) to develop a rationale for education as related to the meaning of life and the surrounding world.

For the casual observer, the study of philosophy appears to be at odds with scientific principles. Nothing could be further from the truth. In their landmark text *Principia Mathematica*, Bertrand Russell and Alfred North Whitehead establish a relationship between mathematics and logic, thus advancing the philosophy of logic far beyond Aristotle's wildest imagination. Scientists and philosophers search for both truth and meaning, and the conceptual vehicles they use to arrive at the truth often cross paths.

3–2. An agricultural educator's philosophy is demonstrated in the dual role of teacher and chapter FFA advisor.

(Courtesy, National FFA Organization)

It is important for the educator to understand that before teaching occurs, an established need for the knowledge must exist. The knowledge must be arrived at in some fashion acceptable to those who will use it. Thus, the study of philosophy should precede any discussion of teaching and learning to help show relevance in how we go about education.

PHILOSOPHICAL FOUNDATIONS

Until the late 1900s, agricultural education focused primarily on training young men to be better farmers or farm workers. Teachers were originally required to have farm experience before they could teach agricultural subjects, and the curriculum focused on the skills and abilities needed by young men for effective service on their home farms. In the last few decades, the number of people training for farm service has diminished, and more young men and women are going on to higher education to fulfill the basic requirements for positions in the agricultural industry. This shift has brought the philosophy of education closer to a realist perspective.

Early Philosophers Who Influenced Modern Education

Mention the word *philosophy* to some individuals, and they will immediately conjure up a mental picture of ancient men in robes lounging about the columns and steps of the temple in Athens. A discussion of Socrates, Plato, and Aristotle today would probably be based upon some desire to be amusing in the presence of colleagues instead of an attempt to cite a reference in a scholarly argument. However, one should not have any significant discussion about philosophy without considering the influence of the great thinkers of Athens, who arrived on the scene around 470 BCE.

Socrates is perhaps the best known of the early Greek philosophers, although we have no manuscripts by him to study. Most of what we know about Socrates is found in the writings of his students, particularly Plato. Perhaps he is best known as the creator of the "Socratic method," or dialectic method, of teaching, in which the teacher posed a series of questions designed to help the learner arrive at the deeper truth and understanding of an idea. The general result of such questioning was to prove consistently that the first answer to the first question posed was wrong. Socrates believed that it was impossible for people to know all the answers to life's questions, and this often infuriated those around him. Not only did Socrates's questions reveal a deeper understanding of ideas, but they also had the often unpleasant result of revealing one's ignorance. The questions asked by Socrates eventually embarrassed the Athenian leaders, who arranged for his trial on a charge of corrupting the youth of Athens and blasphemy against the gods. His trial and execution are vividly portrayed in Plato's *Crito*, *Apology*, and *Phaedo*.

As a pupil of Socrates, **Plato** carried on the tradition of the dialectic through his writings and teaching. Plato's *Republic*, which attempted to define the nature of the ideal state, is his most famous work. Plato established the first known college of higher education in his home, the Academy, where he taught his concept of *idealism*. For Plato, the essence of existence was the idea. The idea for something is the purest and ideal form. The material or physical forms of an idea are merely decaying copies of the original idea. For instance, the idea of a chair is perfect and indestructible because it exists outside of the physical world. Anyone who has ever viewed a saloon fight scene in a Hollywood western movie knows that the physical form of a chair can be destroyed easily by striking a cowboy across the back with it. Regardless of how we design or build a chair, its physical form will eventually end up in the local landfill. Plato suggests that it is impossible to kill an idea regardless of the many forms it takes in the physical world. As Plato puts it, "Everything in the world is always becoming something else, but nothing is ever permanently the same" (Magee, 2001).

Around 335 BCE, the world was introduced to Plato's pupil, **Aristotle**. If Plato was the father of a system of education, then Aristotle was the father of scientists. As the founder of the Lyceum, a school in ancient Athens, he taught such noted luminaries as Alexander the Great and influenced generations of scientists for more than 2,000 years. Aristotle differed from Plato in that he believed in *realism*—that is, that the truth of an idea rested in its form. For instance, a pile of rubber, plastic, metal, and circuitry might be the essential parts of an automobile, but until these parts are

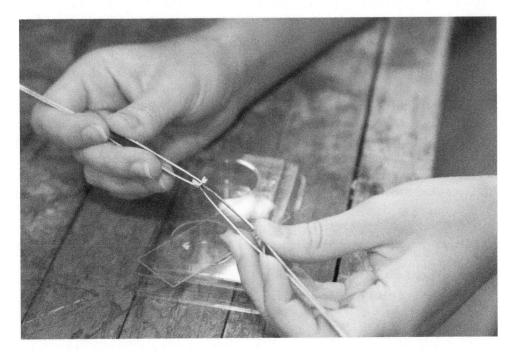

3–3. The addition of biotechnology to the agricultural education curriculum illustrates the influence of realism on agricultural education.
(Courtesy, Education Images)

assembled in the correct order and fashion, the car does not exist. Aristotle's realist philosophy urged individuals to seek the cause of a concept, for it is in the cause that we find the true meaning of the concept. The work of Aristotle yielded what scientists know today as the scientific method and what scores of agriculture teachers know as the problem-solving method of instruction.

These three early philosophers, Socrates, Plato, and Aristotle, developed the foundation for education. However, many other philosophers contributed to the philosophy of education along the way. ***Johann Amos Comenius*** (1592–1670) advocated for modern education, lifelong learning, and extracurricular activities for children. It was Comenius who brought forth the idea that teaching should promote a student's natural tendency to learn and that learning should proceed from easy subjects to more difficult ones. Comenius was one of the first to promote education for career preparation. As a survivor of the Thirty Years' War, he was much attuned to human rights issues and fought for the education of all people, including women.

Johann Heinrich Pestalozzi (1746–1827) gave us the principle of informal education and the need to make children feel welcome and comfortable in the learning environment. He also gave us the concept of hands-on learning in the sciences. By the mid-1800s, the search for truth had run headlong into the rapidly evolving American social, moral, and industrial structure of the pre– and post–Civil War periods. The new economy demanded the search for new answers to new problems.

Emergence of Pragmatic Thought

The philosophy of ***pragmatism*** centers on the concept of "knowing by doing." Knowledge is a practical activity, and questions about meaning and purpose are understood within this context (Magee, 2001). It is a relatively young philosophy when compared with the ancient thoughts about idealism and realism.

Charles Sanders Peirce   Although a number of other philosophers and scientists had worked on concepts relating to pragmatic thought, the first well-known proponent of this new philosophy was ***Charles Sanders Peirce*** (1839–1914), who argued for pragmatism in his 1878 paper "How to Make Our Ideas Clear" (Peirce, 1878). Peirce served as a scientist with the U.S. Coastal Survey upon graduation from Harvard University and used a scientific approach to his study of philosophy.

Peirce believed that true meaning evolves through a three-stage process. The first stage of true understanding begins at a rudimentary level he refers to as the "first grade of clarity." This stage is defined as having an unconscious or unreflective understanding of a concept. For instance, most people do not think about the concept of air as they perform their daily activities. However, their lack of reflective thought about air does not mean that they are unaware of its existence, nor does it indicate an aversion to the concept. Most people simply breathe normally and give air no conscious thought at all. They take for granted the fact that air will exist for their personal survival in their offices, in their cars on the way home from work, and in their homes at night.

The second stage of clarity is being aware of the characteristics of the concept such that one can explain it to others. For instance, the air that humans breathe is approximately 21 percent oxygen and 79 percent nitrogen, give or take a few stray gases here and there. Humans inhale a mixture that holds 21 percent oxygen and exhale a mixture that includes approximately 16 percent oxygen. Understanding this second level of clarity puts a person in position to understand the final phase.

The third stage of clarity is reflected in one's ability to apply what one knows about a concept to other situations involving that concept. If one knows the concept of air, the third stage of clarity might be to distinguish between poor air quality and good air quality. Anyone who has ever crossed the street behind a city bus knows that air laced with diesel fumes is neither healthy nor desirable, and has achieved the third stage of clarity for that concept.

Unfortunately for Charles Sanders Peirce, questions about his moral and ethical behavior suppressed the widespread dissemination of his philosophy. It would take a medical doctor by the name of William James to introduce pragmatism to the world at large.

William James   *William James* (1842–1910) entered Harvard Medical School with some degree of reluctance about his study. The field of medicine could not abate his intellectual hunger, so he engaged himself in studies of philosophy, physiology, and psychology. Because of his friendship with Charles Sanders Peirce, he lectured on pragmatic theory in 1898, but he departed from Peirce's definition of the concept. James asserted that pragmatism is about searching for truth in contradiction to Peirce's assertion that pragmatism is about finding the meaning of a concept. James believed that the test of an idea is in its truth or workability. Does the idea withstand criticism? As a result, James's position on pragmatism has often been described as "Whatever works is true." This, however, is not exactly how James intended pragmatism to be viewed. The workability of an idea is often independent of its practicality. Still, the work of Peirce and James set the stage for John Dewey's work.

John Dewey   Plato's philosophy on education conceptualized the purpose of education as being the discovery of the inherent strengths of the individual and training him or her for mastery in those strengths. In mastering one's intellectual strengths, the individual is most useful to society. It was upon this foundation that *John Dewey* (1859–1952) based his work. As a pragmatist, Dewey believed that the importance and value of vocational education stemmed from the ability of the individual to "learn by doing." Dewey believed that inquiry was an orderly and scientific process by which the learner experienced a felt need, created a hypothesis or problem statement that accurately defined the problem, developed a series of potential solutions to the problem, and attempted to solve the problem using these solutions. The end result of the process identified those solutions that worked and those that did not. Once the problem was sufficiently solved, then the learner could move forward to solve other problems.

Dewey's problem-solving model serves a dual role in education—it encourages students to be creative in solving problems while using relevant theory to help them master competencies in their field of endeavor. Dewey believed that a student's interest in a subject was a powerful motivating force to help that student learn the subject matter more thoroughly. He founded the Laboratory School in Chicago in 1896 to test his theories about teaching and learning, and he served as its director until 1904. In time, Dewey's ideas became known as *progressivism*, a philosophy upon which the Montessori method of instruction is based.

The philosophical ideas of Peirce, James, and Dewey are the foundation upon which career and technical education is constructed. Understanding and appreciating the intellectual meanderings of learned scholars is one thing, but putting educational theory into practice is quite another. In the early twentieth century, the industrial revolution was indeed revolutionizing the way America did business. Henry Ford's assembly-line production method for automobiles in 1913 soon spread to other industries. The entrance of America into World War I hastened the expansion of industry as America geared up for the war effort. The need for workers in this new industrial age created the need for vocational education.

The Need for Vocational Education

In the early 1900s, vocational education generally meant the acquisition of skills or competencies through some type of instructional program or apprenticeship. The purpose of learning these skills was to develop a ready supply of experienced workers for farming or industry. It was the result of efforts to adjust to the changing demands of the economy (Prosser & Quigley, 1949; Snedden, 1920).

These early writers were very much concerned about the nature of vocational education in the early twentieth century. Vocational education in that era consisted primarily of apprenticeship programs that trained young men and women on the job. Unfortunately, this training consisted mostly of observation and imitation without any explanation of the underlying theory and practice necessary for full understanding of the work being done. Apprentices would follow their tradespersons around and help them as they completed their tasks. Often, the tradespersons would not fully explain the techniques or skills the apprentices were observing. The quality of the apprentices' work was highly related to the quality of the tradespersons' work. This left young workers with a "hit or miss" education in the industrial arts.

The inappropriateness of this method of training young men and women for vocations was beginning to be felt in the economy. The advancement of technology in industry was delayed by a lack of well-trained workers prepared for emerging occupations. A new system of vocational education was needed.

The new system of vocational education was infinitely better than the old apprenticeship system (Prosser & Quigley, 1949). Vocational education now provided an organized and systematic method of instruction. Students were no longer limited to apprenticeship experience. In agricultural education, Rufus Stimson pioneered the development of supervised farming practice, which is the precursor to supervised experience in agriculture today. Young men developed farm projects under the supervision of an agriculture teacher, who closely monitored the students' operations.

To define the tenets of vocational education further, Prosser and Allen (1925) created their sixteen theorems of vocational education (see Figure 3-4). A **theorem** is a statement or proposition that can be demonstrated as a truth. Theorems are often part of larger theories.

THE SIXTEEN THEOREMS APPLIED TO AGRICULTURAL EDUCATION

Prosser and Allen's (1925) sixteen theorems define the meaning and purpose of vocational education and, in a similar way, agricultural education. However, these ideas were introduced more than eighty years ago during a unique period in American history. How do these sixteen theorems relate to agricultural education today? Are they still relevant? For agricultural education, Prosser and Allen's work can be condensed into seven major categories.

Authentic Instruction

Agricultural education mirrors the current state of technology in the agricultural industry. Students use the same types of equipment, perform the same tasks, and are exposed to the same risks as someone who is currently employed in an agricultural profession. It is wasteful and impractical to teach students skills that are outdated or not essential to

3–4. *Sixteen Theorems of Vocational Education*
(*Source:* Prosser & Allen, 1925)

Sixteen Theorems of Vocational Education

Theorem 1

Vocational education must provide a replica of the environment in which students will be hired to work.

Theorem 2

Vocational education must teach the skills needed in the industry under the same conditions as experienced by workers in the industry.

Theorem 3

Vocational education must capitalize on the students' vocational interests and needs.

Theorem 4

Vocational education must be provided only to those students who want and need it.

Theorem 5

Vocational education must train students to use effectively the cognitive and psychomotor skills required in their chosen occupational fields.

Theorem 6

Vocational education must provide learning experiences that allow students to blend cognitive and psychomotor skills so that the students are adequately prepared for gainful employment.

Theorem 7

Vocational education should provide a student with at least the minimum skills needed in an occupation.

Theorem 8

Vocational education should provide students with training in the skill processes needed by specific occupations so that they understand the need for each step in a skill process.

Theorem 9

Vocational education must provide learning conditions that train learners to meet the demands of the market in their respective industry.

Theorem 10

Vocational education should provide instruction that is content-specific to occupations.

Theorem 11

Vocational education will be effective only if there is a highly skilled instructor who has mastery in the vocational subjects he or she teaches.

Theorem 12

Vocational education should be well grounded in the content within specific occupations, and students should learn from those instructors who have high degrees of experience in those occupations.

Theorem 13

Vocational education should help all enrollees develop vocational skills.

Theorem 14

Vocational education should provide for the social and physical development of students in addition to their intellectual development.

Theorem 15

A vocational education program should have a viable plan for change and be flexible enough to accomplish change when it is needed.

Theorem 16

Vocational education should be provided in a school only when the appropriate fiscal resources are available to assure the level of instruction needed to prepare students for careers in their chosen occupational fields.

their chosen fields of endeavor. Instruction goes beyond general training in agricultural occupations and provides training specific to jobs in the local community.

Authentic Self-Evaluation

The agricultural education curriculum in a school provides courses that capitalize on student interests and abilities. Students who are interested in the subject matter and see it as an essential tool for helping them become established in the careers of their choice will be motivated to learn. Agricultural education is not a good choice for every student, and only those students who are interested in pursuing careers in agriculture or related areas should enroll.

Higher-Order Thinking Skills

The agriculture profession requires highly skilled workers. Agricultural education provides learning experiences that encourage higher-order thinking. Teachers should prepare learning experiences that cause students to analyze and critically appraise information and to be creative. Students should be able to make connections between knowledge acquired through cognitive means and psychomotor skills related to that knowledge. For instance, a student must understand the principles of electricity before he or she can safely use a shielded metal arc welding machine.

Psychomotor skills learned in the agriculture classroom are directly related to current occupational skills and must be authentic in nature. To engage students effectively in the process of learning, the learning experience should be exactly the same as one might expect to find in the industry. Furthermore, these psychomotor skills must be practiced to the point of mastery so that good work habits are established and skills are ingrained in long-term memory. Students who are not prepared to the minimum level expected of gainful employment in the industry are not well served by the agricultural education program.

Agricultural Education Curriculum

The agricultural education curriculum is driven by the market demand in agricultural occupations. The instructional program should provide learning experiences that prepare students for the entry point into agricultural jobs in the community, even if the skills needed by the agricultural industry in that local economy are less efficient and less technical than in other communities. For instance, if the local economy needs workers to operate farm machinery, then instruction should prepare students for jobs of that type. If the local community requires workers highly skilled in biotechnology subjects, then the curriculum should prepare students accordingly.

3–5. Agricultural education laboratories allow students to practice psychomotor skills.

(Courtesy, Education Images)

The Agriculture Teacher

Agricultural education in the local school community will be only as successful as the skills and abilities of the agriculture teacher will allow. The teacher is essential to the success or failure of the program and must be highly qualified, well trained, and enthusiastic about the profession of teaching. Teachers must not only master the art and practice of teaching, but they must also stay current in the technical content of the profession. Teachers must have professional development plans that allow them to stay abreast of recent developments in the field of agriculture. Even the best teachers become ineffective when the technical content of their lessons becomes outdated.

FFA and Agricultural Education

Agricultural education must also consider the whole student in the development of the curriculum. Students must have experiences that allow them to grow intellectually, physically, and socially. For a well-rounded instructional program, FFA is needed to provide experiences in teamwork, leadership, cooperation, conflict resolution, management, and interpersonal communications. The agricultural industry requires workers who can cooperate with other individuals in carrying out the goals of an organization. Those individuals who are deficient in social skills are at a disadvantage.

3–6. FFA Career Development Events provide opportunities for teamwork, cognitive development, and psychomotor skill development.

(Courtesy, National FFA Organization)

Planning for Change in the Agricultural Education Program

The agricultural education program should have a plan for continuous improvement. To accomplish change when needed, the program should be flexible. Too often, an agricultural education program fails to change the curriculum to match the changing market needs in the agriculture profession because it has invested heavily in certain technologies and cannot afford to make changes in the curriculum when needed. At some point, administrators will have to decide whether to continue to offer an agricultural education program in a school, and the decision is usually determined by the ability of the program to provide relevant and high-quality learning experiences in an economical fashion.

THE PHILOSOPHY OF AGRICULTURAL EDUCATION TODAY

By the mid-1980s, there was a tremendous groundswell of support for education reform in America. Government agencies with responsibility for public education were spurred into action to create a world-class education system. In *Understanding Agriculture: New Directions for Education*, the National Research Council (1988) recognized the need for agricultural subject matter at all levels of education. The National Research Council's report significantly changed the traditional focus of agricultural education.

For the most part, agricultural education was originally meant for those students who planned to enter the farming profession. With the passage of the Vocational Education Act of 1963, the scope of agricultural education was broadened to include training in nonfarm agricultural occupations. Students were now receiving instruction in agricultural sales and service, horticulture and natural resources, and other nonfarm occupations. This was a major legislative enactment shaping new curriculum areas in agricultural education.

Understanding Agriculture recommended broadening the scope of agricultural education even further by concluding that agricultural literacy was an essential ingredient of instruction. Agriculture became a universal subject, and agricultural education was charged with the responsibility for agricultural literacy among all students. Agricultural education is a subject worthy of inclusion in every school at all grade levels—from kindergarten through the twelfth grade. The positive effect that agricultural education programs have had on thousands of youth is something to be shared with all young people.

To become a voice for agricultural literacy at all grade levels and in every classroom requires significant change on the part of the agricultural education profession. The program must change to meet the needs of a diverse student body, and the current technology of agriculture must be visible in the instructional program. A local agricultural education program that is hindered by outdated equipment, outmoded instructional techniques, and a narrowly focused curriculum will drive students away. Many high school students expect to attend some form of postsecondary institution, and the curriculum has shifted somewhat to incorporate academic subject matter within the context of the agricultural sciences. Accordingly, supervised experience programs should be focused on knowledge acquisition with an appreciation for financial earnings. This has paved the way for research and inquiry-based supervised experiences.

The principles that guide agricultural education today are still founded in the work of Prosser, Allen, and others, but there are a few changes that bear mentioning. In the 1970s, vocational education was deemed a public necessity, with programs,

3–7. Agricultural education instructors provide individual instruction, as is shown here during supervised experience.
(Courtesy, Education Images)

including agricultural education, open to everyone. In those years following the turbulent 1960s and the Vietnam War, vocational education was seen as a method of eliminating class distinction. For agricultural education to exist in this environment, it was necessary to complement the general academic subjects through curriculum integration. To create the skilled workers of the 1970s, much theoretical and practical knowledge was essential in their career paths. This reinforced the demand for well-prepared and highly skilled teachers in the agrisciences and refocused the need for individualized instruction in agricultural education (Barlow, 1970).

The 1980s saw a continued emphasis on the individual student of vocational education. The purpose of vocational education, and agricultural education specifically, was to improve the efficiency of the worker by providing a total educational package. This package included individualized instruction, career guidance and counseling, vocational assessment, and evaluation. Agricultural education was charged with the responsibility of helping each student find a good job and keep that job through relevant and specific technical training. Taking this one step further, agricultural education was to help students develop higher-order thinking skills that would allow them to transfer skills and knowledge to new demands in the workplace. It was not enough to teach students how to perform certain job tasks. To create new products and processes, students needed to be able to think independently, analyze data, and synthesize new methods, products, and processes. The technology explosion in the 1980s needed workers who could adapt to their changing work environment (Paulter & Bregman, n.d.).

As the twentieth century came to a close, agricultural education kept building upon its philosophical foundation by continuing to concentrate efforts on career guidance. The focus of agricultural education has shifted from training in specific jobs to training in career clusters or pathways. The great technological advances of the 1990s rendered training for specific jobs nearly obsolete. The emerging focus of agricultural education is to provide learning experiences clustered around general agricultural areas, such as horticulture, engineering, and natural resources (Findlay, 1993). The advances in biotechnology have changed the face of agriculture in the United States and around the world. To address proactively the rapid advance of technology, agricultural educators have had to look beyond the jobs and careers of today and focus on the careers of the future. Students entering the agricultural education classroom will choose among emerging career opportunities in the agricultural industry.

In order to prepare students for careers and jobs that may not yet exist in the economy, agricultural education will become more dynamic, and focused on encouraging exploration and innovative thinking. Education will focus on the whole student as opposed to his or her specific occupational goals in agriculture. That is, the curriculum in agricultural education will continue to move toward using a blend of academic and applied science elements to craft a curriculum that broadens and deepens a student's knowledge (Biggs, Hinton, & Duncan, 1996; Kincheloe, 1999).

The focus in this contemporary approach to agricultural education will shift from preparing workers for entry-level positions toward a longer-term career focus. The curriculum will be sufficiently broad, yet demanding in skill development. Skills will emphasize the development of creativity and innovation. Partnerships between secondary and postsecondary institutions will provide strong and seamless connections in learning experiences (Rojewski, 2002). This new curriculum will likely develop with elements from arts, sciences, mathematics, and language. Job-specific preparation will give way to career-specific preparation.

REINVENTING AGRICULTURAL EDUCATION FOR THE YEAR 2020

In 1996, the National Council for Agricultural Education initiated a project entitled Reinventing Agricultural Education for the Year 2020. The purpose of this initiative was to create a framework by which the profession could develop a vision for the future direction of food, fiber, and natural resources education. This visioning process examined all facets of the American condition, including the changing nature of agriculture, the national and international global marketplace and its influence on the national economy, and the needs of traditional and nontraditional students enrolled in agricultural education. From this vision, strategic plans were created for moving the profession closer toward its preferred future. A national dialogue about the future of agricultural education grew out of the Reinventing Agricultural Education project, and the profession continues to work toward a unified vision of agricultural education for the year 2020.

The world that high school graduates will enter is one where markets for agricultural products reach around the globe and where fierce competition for resources and markets are fueled by information transfer at the speed of light. The labor needs of the next hundred years require workers with creative and innovative approaches to problem solving. The agriculture industry is in a constant state of flux, with businesses restructuring, upsizing, and downsizing in search of the optimum state of efficiency and profit. The world of work now requires skill in the management of knowledge just as much as the management of people (Rojewski, 2002). The worker in this new economy will not be hired to perform discrete tasks, but be engaged in work teams with ever-changing leadership and roles. How, then, do we prepare graduates for their professional debut in this new world?

REVIEWING SUMMARY

Philosophy is an attempt to draw meaning from the world around us. Out of this effort, general guiding principles serve to mold educational practices, including those in agricultural education. These form philosophical foundations, or the basic beliefs that guide people's actions.

Early philosophers whose influence has had significant impact on modern education include Socrates, Plato, and Aristotle. Later, Comenius and Pestalozzi were important in shaping education and how it is carried out. The development of pragmatism had a large influence on the nature of vocational education as it emerged in the late 1800s and early 1900s in the United States. Peirce was a leading proponent of pragmatism.

Prominent educational philosophers included William James and John Dewey. Many agricultural educators cite Dewey as the father of the problem-solving method. James, Dewey, and Peirce were proponents of ideas that served as the foundation of career and technical education. Snedden, Prosser, and Quigley are well known for their shaping of vocational education in the 1900s. Among efforts by Prosser and Allen are the sixteen theorems of vocational education. These have often been interpreted in agricultural education in areas such as authentic instruction and authentic assessment.

Today's philosophical foundations of agricultural education are shaped by efforts such as those of the National Research Council in the 1980s. Later, Reinventing Agricultural Education for the Year 2020 helped create a framework for a future vision of food, fiber, and natural resources education.

QUESTIONS FOR REVIEW AND DISCUSSION

1. Identify the impact of each of the following individuals on the emergence and practice of vocational education in the United States:
   a. Aristotle
   b. Charles Sanders Peirce
   c. Johann Amos Comenius
   d. Johann Heinrich Pestalozzi
   e. John Dewey
   f. Plato
   g. Socrates
   h. William James
2. What are the basic tenets of each philosophy listed below?
   a. Idealism
   b. Pragmatism
   c. Progressivism
   d. Realism
3. What are the three stages of clarity in pragmatism? Give an example of each stage.
4. What was the major impact of the National Research Council in its report entitled *Understanding Agriculture: New Directions for Education*?
5. What was the major purpose of the initiative entitled Reinventing Agricultural Education for the Year 2020?
6. Prosser and Allen stated sixteen theorems of vocational education. What is a theorem? Which theorem is or has been most important in agricultural education? Why?

ACTIVITIES

1. Complete a brief research project on the differences and similarities between progressivism and pragmatism. How are these two philosophies interrelated?
2. Obtain a copy of the report prepared by your state as part of the Reinventing Agricultural Education for the Year 2020 initiative. Assess recommendations in the report, and relate whether you feel progress is being made in implementing the recommendations.
3. What is your philosophy of agricultural education? Write a half-page statement that presents your understanding of agricultural education and methods of program delivery.

CHAPTER BIBLIOGRAPHY

Barlow, M. L. (1970). *Principles of Vocational Education: A Review of the Historical Background with a Focus on the Present.* Washington, DC: American Vocational Association.

Biggs, B. T., B. E. Hinton, and S. L. Duncan. (1996). Contemporary approaches to teaching and learning. In N. K. Hartley and T. L. Wentling (Eds.), *Beyond Tradition: Preparing the Teachers of Tomorrow's Workforce.* Columbia, MO: University Council for Vocational Education.

Findlay, H. J. (1993). Philosophy and principles of today's vocational education. In C. Anderson and L. C. Ramp (Eds.), *Vocational Education in the 1990s II: A Sourcebook for Strategies, Methods and Materials.* Ann Arbor, MI: Prakken Publication.

Kincheloe, J. L. (1999). *How Do We Tell the Workers?* Boulder, CO: Westview Press.

Magee, B. (2001). *The Story of Philosophy* (1st American ed.). New York: DK Pub.

National Research Council (U.S.), Board on Agriculture, Committee on Agricultural Education in Secondary Schools, and Upchurch Collection (North Carolina State University). (1988). *Understanding Agriculture: New Directions for Education.* Washington, DC: National Academy Press.

Paulter, A. J., and R. Bregman. (n.d.). *Occasional Papers of the Western New York Educational Service Council.* Buffalo: University of New York.

Peirce, C. S. (1878). How to make our ideas clear. *Popular Science Monthly, 12*(January 1878), 286–302.

Prosser, C. A., and C. R. Allen. (1925). *Vocational Education in a Democracy.* New York: Century.

Prosser, C. A., and T. H. Quigley. (1949). *Vocational Education in a Democracy.* Chicago: American Technical Society.

Rojewski, J. W. (2002). *Preparing the Workforce of Tomorrow: A Conceptual Framework for Career and Technical Education.* Columbus, OH: National Dissemination Center for Career and Technical Education, Ohio State University.

Snedden, D. (1920). *Vocational Education.* New York: Macmillan.

4

History and Development of Agricultural Education

The history of agricultural education in the secondary schools of the United States is closely aligned with that of career and technical education. Common legislation and similar missions have held the two together over the years. Agricultural education was created under the overall umbrella of vocational education, much as it is a part of career and technical education today.

Several individuals emerged as early leaders and top officials in forming and shaping vocational education. Simeon De Witt, Justin Morrill, Jonathan Baldwin Turner, Hoke Smith, Dudley Hughes, and Charles Prosser are a few of those leaders recognized for establishing agricultural education in the United States. Prosser had a great deal of insight and proposed overall direction for the programs. He is well known for his statements on the purpose and mission of vocational education. His writings reflect a strong sense of understanding about promoting improved levels of living through gainful employment.

objectives

This chapter introduces the history and development of agricultural education in the United States. It has the following objectives:

1. Describe situations in the United States that contributed to the development of agricultural education

2. Describe the creation of land-grant colleges

3. Discuss legislation that created and expanded secondary school agricultural education and the roles of key individuals in gaining the legislation and guiding program implementation

terms

A. C. True
Carl Perkins
Carroll Page
Charles Prosser
Dudley Hughes
Hatch Act
Hoke Smith
Jonathan Baldwin Turner

Justin Morrill
Morrill Act of 1862
Morrill Act of 1890
Simeon De Witt
Smith-Hughes Act
Smith-Lever Act
Vocational Education Act of 1963

4–1. Early American agriculture used animals as power for many farm activities.
(Courtesy, Shutterstock © Jim Parkin)

PRE-COLUMBIAN PERIOD

American Indians were farming in the Americas as early as 5000 BCE, and as such were the first farmers in what became the continental United States. American Indians domesticated several species of vegetables and by 1000 CE had developed a sophisticated system of agriculture based on vegetables and grains. The methods by which American Indians and Alaskan Natives transferred knowledge about agricultural practices from one generation to the next was a form of agricultural education. Tribal elders taught youth how to cultivate plants, and this knowledge was handed down as oral tradition over hundreds of years. American Indians taught early European colonists how to plant and cultivate maize, pumpkins, squash, beans, and tobacco (Cochrane, 1979). Indians were adept at finding suitable arid land for crop production, but lacked advanced methods of fertilizing crops. These grains and vegetables were supplemented by wild game to round out the diet.

COLONIAL AND EARLY NATIONAL ERA

Farming was the chief economic engine of the new English colonies in America. Nearly every colonial family raised their own food and sold or bartered agricultural products. As with American Indians, colonial farmers used a method of trial and error with crop and livestock production. Knowledge about agriculture was transmitted orally through the generations. In the colonial period, farmers began to rely on farmers' almanacs for knowledge about agricultural practices. These almanacs provided advice based on astrological phenomena, and as such American agricultural education was guided in its infancy more by superstition than science (Hurt, 2002).

Farming by the astrological signs as proposed by the farmers' almanacs may have been sufficient for subsistence farming. However, even the earliest European colonists knew the value of farm products, and sought to barter or trade them with each other and the Native Americans. The early European farmers in America may have started production to first meet the needs of the family, but that was one of

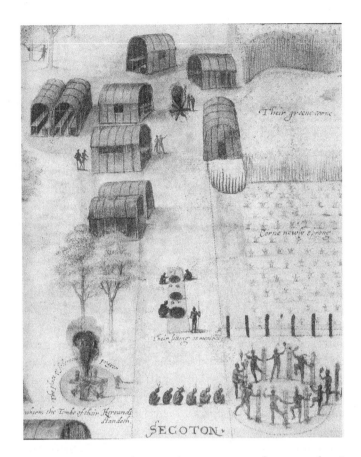

many intermediate goals along the way to using agriculture production as a means to improve the quality of life for the settler's family. Subsistence agriculture placed beef, grains, and vegetables on the dinner table, but the barter and sale of agricultural products bought salt, gunpowder, nails, fabrics and linen, farm implements, and oxen for improved agricultural production.

Most efforts to colonize the North American continent were motivated by economic forces. The settlers who established the colony in Jamestown, Virginia, were entrepreneurs seeking to replicate the discovery of wealth experienced by Spain in its colonization of South America; in many ways, they were successful. Colonist John Rolfe learned how to cultivate a plant the Virginia Native Americans called *uppowoc* (Kell, 1966). The dried leaves of this plant had both medicinal and recreational uses as demonstrated by the natives. Rolfe's production of *uppowoc* (tobacco, *Nicotiana tabacum*) made him a very rich man (Kell, 1966). Before long Rolfe's fellow yeomen were growing tobacco, which significantly added to growing trade between the New World and Europe. Trading companies in Europe financially supported the growth and development of American agriculture in these early years.

As the sophistication of the colonial American farmer grew with each successful growing season, there arose a need for more substantial training in agricultural methods. Agricultural societies were established to promote and develop interest in productive farming methods. This was largely a European idea. The colonies in America took their cue from agricultural societies in Europe. The Edinburgh Society of Improvers in the Knowledge of Agriculture was organized in Scotland in 1723. The Society for the Encouragement of the Arts, Manufactures and Commerce was established in England in 1754, and the first veterinary school was established in France in 1766 (True, 1939).

In the United States, the Philadelphia Society for Promoting Agriculture was formed in 1785. Its purpose was to draw attention to agriculture with the goal of increasing productivity and improving rural life. The motto of the society was "Venerate the Plough." The society published articles in newspapers about agriculture and encouraged experiments that led to greater productivity in all aspects of agriculture. A secondary purpose of the Philadelphia Society was to encourage development of other similar agricultural societies throughout the United States. Furthermore, the society also distributed awards to members for essays produced on subjects of importance of the society. In 1788, for instance, the society awarded essayists for papers on the best experiments in crop production, animal husbandry, innovative cultivation methods, pest control, food preservation, and soil conservation (True, 1939).

In 1792, the Massachusetts Society for Promoting Agriculture was established to explore new methods of improving agriculture. Throughout the 1800s, agricultural societies disseminated research in agriculture through publications, newspaper articles, and lectures (True, 1969). These societies urged legislative action to create formal schools of agriculture. In 1855, the Agricultural College of the State of Michigan was founded. Yale University added three professors of the agricultural arts in 1845. A Farmers High School was created in Pennsylvania by the state legislature in 1854.

These societies and schools encouraged the development of innovative agricultural practices, and fostered the inventiveness of the American people. New agricultural technologies were being developed in the late 1700s and early 1800s. A major processing technology in the late 1700s involved the cotton gin, which was first developed in a very simple form by Eli Whitney in 1793. This invention greatly reduced the hand labor needed to separate seeds from cotton lint.

Other developments followed, such as the grain reaper by Cyrus McCormick in 1842 and the steel moldboard plow by John Deere in 1837. McCormick's invention was the first major advance in developing what we refer to today as harvesting equipment. Deere's contribution was significant because it allowed heavy soils to be tilled, opening vast land areas for agricultural production.

CIVIL WAR ERA

The early 1860s saw the nation embroiled in a great civil war. With every able-bodied man on the front line or engaged in running the war industries, agriculture in both the North and the South had to adopt new technologies for production. The early 1860s saw a change from hand-powered agriculture to horse-powered agriculture as the first American agricultural revolution got underway (U.S. Department of Agriculture, 2012). As markets for agricultural exports increased in the 1860s, so did the need for good methods of transporting those products to market. The era saw a great expansion of railroads across the nation.

Economic fluctuations caused by wartime inflation, recession, and recovery fueled the fires of ambition for many Americans. In 1862, the Homestead Act was passed, which granted 160 acres of land to those Americans willing to carve out a niche for themselves in the West. The western territories developed rapidly under the agrarian interests of farmers and ranchers. The late 1860s were the days of rapidly expanding cattle production in the western United States.

Industrial Growth

The 1860s was a period of rapid industrial growth. A major new direction was making steel and iron for the railroads and other development. Railroads were built to the western United States, opening up a new era in transportation.

The American population was also on the move, shifting westward across North America. John Deere's steel plow made it possible for farmers to cut through the hard

Midwestern soil for planting. To protect the interests of agrarian opportunists and improve the emerging commercial agriculture, a number of farm organizations and societies were formed. These rapidly expanded to promote the causes of agriculture. There were almost a thousand agricultural organizations in the United States by the mid-1860s (U.S. Department of Agriculture, 2012).

During this period of westward expansion and wartime industriousness, some leaders with foresight saw the need for a nationwide system of agricultural, mechanical, and military education.

Creation of Land-Grant Colleges

Today, the land-grant colleges are major forces shaping agricultural education and research in the United States. These were established and have prospered as the result of legislation known as the Morrill Acts.

The concept of agricultural colleges was being discussed in the United States as early as 1819, when **Simeon De Witt** (1756–1834) of New York proposed colleges for teaching agricultural subjects and conducting experimental research (Cross, 1999).

De Witt may have broached the idea of agricultural colleges for the populace, but **Jonathan Baldwin Turner** (1805–1899), a professor of classical literature at Illinois College, brought the idea to American awareness. Turner transformed the idea of agricultural and industrial colleges into an issue worthy of attention by influential people.

Jonathan Baldwin Turner was born and raised on a farm in Massachusetts and educated at Yale University. His classical education at Yale led him to a teaching position at Illinois College. Turner was a fiery orator who espoused his populist views with much vigor and zeal. His passion for the common people stirred the public at a time when the national economy teetered between recession and depression. In the late 1840s, Turner's ideas about how to serve the educational needs of the people crystallized into the concept of making higher education in the practical arts and sciences available at low cost. Turner believed that the federal government should pay for the establishment of universities to provide such education. His argument

4–3. The University of Minnesota is a land-grant university with branch campuses throughout the state. (The Crookston Campus serves the northwestern area of Minnesota with a number of programs.)
(Courtesy, Education Images)

eventually convinced the Illinois legislature, which sent a resolution to Congress asking for the development of land-grant colleges (Cross, 1999).

Justin Morrill (1810–1898) entered the U.S. House of Representatives in December 1855 as the new Congressman from Vermont. In the first few months of his first term, Morrill distinguished himself as a proponent of legislation designed to help the American farmer and the working-class American. However, as the representative from Vermont was to discover, his colleagues from the South did not share his zeal for the methods he was proposing. His resolution to establish a board of agriculture and national agricultural schools failed because of southern opposition to any legislation deemed threatening to the rights of states (Cross, 1999).

Not to be deterred, Morrill continued to work on land-grant legislation through his first term in office. On December 17, 1857, Morrill introduced a bill granting lands for agricultural colleges. After a year of parliamentary wrangling, the bill passed the House by a vote of 105 to 100. The Senate passed its version in February 1859, and the bill eventually found its way to the desk of President James Buchanan.

Unfortunately, President Buchanan vetoed the bill. Under pressure from Southern Democrats, Buchanan chose to avoid another fight on state's rights with land-grant legislation. Morrill was forced to wait until the 1860 elections for perhaps a more favorable audience with the next president of the United States. Abraham Lincoln would support the legislation, but Morrill had once again to navigate the political minefield to get his bill on the president's desk. Southerners opposed to the bill would not take kindly to its return to the congressional agenda in 1861, and the battle for land-grant colleges looked to be a tough one.

In January 1861, Morrill received some rather unintended but not altogether unexpected support from President Lincoln. With Lincoln's inauguration that year, the southern states began to secede from the Union, starting with South Carolina and ending with North Carolina in May 1861. With the southern states gone from the House and Senate chambers, conditions were favorable for the bill's return. On December 16, 1861, Morrill introduced a revised land-grant bill for consideration by the House of Representatives. After some additional editing by both houses, the bill finally passed the Congress and went to the White House for President Lincoln's signature in July 1862.

The Morrill Act of 1862

On July 2, 1862, President Abraham Lincoln signed the *Morrill Act of 1862* into law. With the stroke of his pen, he not only created a system of higher education that would free millions of people from the poverty of ignorance but also set into action a course of events that would establish the United States as the leader among all nations in agricultural production. The creation of colleges for the teaching of the agricultural, mechanical, and military arts opened new opportunities in the United States. (It should be noted that 1862 was also highlighted by President Lincoln's issuing the Emancipation Proclamation. This action freed black Americans from slavery and allowed them to enjoy the privileges of higher education and other aspects of culture.)

The Morrill Act of 1862 kindled the flames of expansion for agriculture and industry in the United States. The industrial revolution of the early 1900s can trace its roots back to Congressman Justin Morrill's ability to shepherd legislation through the U.S. Congress. While Justin Morrill may forever be associated with the creation of land-grant colleges, the idea for land-grant colleges can be traced back to a number of sources.

Major Provisions   The Morrill Act of 1862 set aside 30,000 acres of land for each member of Congress from those states then incorporated into the United States of America.

4–4. The Morrow Plots at the University of Illinois in Urbana were set up in 1867. (They are the oldest continuous experimental plots in the United States.)
(Courtesy, Education Images)

Upon the sale of public lands, the proceeds were to be used for the creation in each state of at least one college where the leading subjects would be agriculture and the mechanical and military arts. Because the Civil War was in full swing in 1862, the Congress of the United States sought to exclude those southern states in the Confederacy by inserting this phrase into the act: ". . . [N]o state while in a condition of rebellion or insurrection against the government of the United States shall be entitled to the benefit of this act."

For the states to receive the proceeds from the sale of public lands, the legislatures had to accept the provisions of the act within five years. Furthermore, the act provided for free public land to be made available for each new state as it entered the Union.

RECONSTRUCTION AND INDUSTRIAL ERA

The U.S. government and agricultural colleges and universities supported the agricultural education movement through agricultural short courses made available to farmers (Croom, 2008). In 1868, Kansas Agricultural College provided training to farmers on the application of modern agricultural practices. In the summer of 1884, the Alabama State Agricultural College conducted meetings with farmers regarding agricultural problems. Colleges in Massachusetts, Illinois, Iowa, and New Hampshire also adopted similar institutes, and state boards of agriculture also conducted farmers' institutes (True, 1969).

In 1888, the Office of Experiment Stations in the U.S. Department of Agriculture recognized the value of farmers' institutes, and began collecting data and researching the work of the institutes. By the end of the nineteenth century, agricultural education had expanded outward from farmers' institutes and university short courses into public schools across the nation. In 1906, public school officials in Michigan, Arizona, and Georgia would invite institute speakers to speak to the students on agricultural subjects (True, 1969).

The Morrill Act of 1890

Justin Morrill was eventually elected to the U.S. Senate and continued to work for legislation that supported the land-grant college movement until his death in 1898.

The *Morrill Act of 1890* set aside funds for teacher education in agricultural and mechanical arts. However, these funds were restricted to those colleges where no distinction was made on the basis of race or color in student admissions. It was acceptable at the time for states to establish separate colleges for Caucasian and African American students under the principle of a separate but equal education.

The second Morrill Act provided funds that were generally used to establish African American agricultural colleges. Many students assume that the second Morrill Act created black land-grant colleges. It did not. The first Morrill Act provided funding that could be used for African American institutions, and in some cases it was indeed used for that purpose. Alcorn State University was established in 1871, and Hampton University was established in 1872, almost two decades before specific legislation would be passed to assist in the development of agricultural colleges for black students (Cross, 1999).

The Morrill Act of 1890 was a little more complex than the first one. For the purpose of providing instruction in feed and agricultural sciences, this act appropriated funds from the sale of public lands, beginning with a single appropriation of $15,000 in 1890, then increasing in annual increments of up to $50,000.

The Beginning of Agricultural Education in the Secondary Schools

In 1821, Robert Hallowell Gardiner and his neighbors in the town of Gardiner, Maine, petitioned the Maine legislature for a school for agricultural and mechanical subjects. Gardiner had supplied the land and buildings for the prospective school, so the State of Maine obliged Mr. Gardiner's request for a school on the banks of the Kennebec River (Stevens, 1921). Gardiner's Lyceum was likely the first school of agriculture in the United States, and it combined liberal arts subjects with studies in agriculture and mechanical arts. The school eventually became a high school academy and operated as such until the building was destroyed by fire in 1870. Other small schools like the Gardiner Lyceum operated in that time period almost exclusively as community schools. In 1845, Theodore S. Gold established his Cream Hill agricultural school in Cornwall, Connecticut, for the purpose of educating young men on the virtues of farming and rural life (Cornwall Historical Society, 2012). In Massachusetts, two similar schools were established in 1858 for agricultural education (Hamlin, 1962).

For almost three centuries after European settlers arrived, no nationwide curriculum or program in agricultural education existed. Local schools that offered agricultural subjects were left to their own means. The passage of the Hatch Act— though primarily a research law—was the first significant unified nationwide effort for agricultural education at less than the college level. The *Hatch Act* of 1887 established agricultural experiment stations and educated the public about the implications of the research conducted at these experiment stations. In 1889, a successful school was established on the grounds of the agricultural experiment station near the University of Minnesota to teach agricultural subjects to students below the college level. Other states soon adopted similar arrangements for the dissemination of information arising out of agricultural research in the experiment stations (Moore, 1988).

New York's legislature appointed the Cornell University Agricultural College in 1897 to supervise and establish agricultural education in public schools. The Farm Life Act of 1911 in North Carolina created boarding schools for agriculture and home economics. The state school superintendent and the farm-life school advisory board established the curriculum, which had to include practical farm work (Stimson & Lathrop, 1942).

4–5. Land-grant universities maintain a number of experiment stations that focus on particular problems. The experiment station at Montana State University operates an agronomy farm.
(Courtesy, Education Images)

Eventually, the federal government would recognize the need and importance of agricultural education and create legislation that specifically encouraged states to develop agriculture teacher training programs and fund local agricultural education programs.

An early proponent of agricultural education in public schools was **A. C. True**. As director of the Office of Experiment Stations in the federal government, he wrote extensively on the subject. Director True believed that schools in farming communities could provide instruction for rural farm youth, and he devoted a significant portion of his time to promoting the cause of agricultural education to citizens, colleges, and the federal government (True, 1969).

True's work led to increased funding for agricultural education and to its continued growth in schools below the college level. In the years before the passage of the Smith-Hughes Act, it was obvious that agricultural education was well established in schools. By 1917, thirty states had agricultural education in schools (Hamlin, 1962). The task still remained of organizing agricultural education so that it would be available to every student who wished to study it.

The Smith-Lever and Smith-Hughes Acts

For a number of years, proponents of national legislation for vocational education had worked to create a consensus on what this legislation should do. Senator **Carroll Page** of Vermont had worked for a number of years prior to 1917 to establish federal support for vocational education, but to no avail. In spite of the efforts of his associate, Charles Prosser, no legislation seemed to be forthcoming.

Efforts to build support for vocational education were unsuccessful in the Republican-controlled Congress in the early 1900s. During the presidency of Theodore Roosevelt, the United States became a strong world power. The era of progressivism yielded significant improvements in food safety, farming, and conservation. Roosevelt modernized the federal government and began efforts to clean up illegal business trusts and the economy.

Theodore Roosevelt held to his pledge not to run again for the presidency in 1908, and the nomination and subsequent election went to William Howard Taft. Taft was able to carry out many of Roosevelt's initiatives, but he was unable to rally the people behind the Republican platform during the 1910 congressional elections. In 1910, the Democrats gained control of Congress for the first time in almost two decades (Morison, 1965). Momentum began to build for federal legislation on a number of fronts. In 1912, the election of President Woodrow Wilson resulted in increased federal legislation that would have far-reaching effects across the nation into the farms and towns of America.

For southern farmers, federal assistance could not come too soon. Life had been hard since the Civil War. By 1910, more than 80 percent of all southerners were living in rural communities, and many of these were in poverty. Unfortunately, half of all farms in the South were unimproved. Southern farmers lacked the skills and resources necessary to improve farmland and lacked knowledge of the best practices in animal science. In 1910, the average southern farm was valued at half the national average for farms.

The U.S. Congress wasted little time in passing legislation designed to help the American farm family. The **Smith-Lever Act** of 1914 created the Cooperative Extension Service. This law established a partnership between the federal government and the land-grant colleges for the purpose of extending knowledge about the best practices in agriculture to rural communities. However, this federal law was not passed without its share of debate.

The Hatch Act of 1887, the federal legislation that established agricultural experiment stations, paved the way for legislation that established federal support for extension and vocational education. The Page-Wilson bill was the first solid attempt by Congress to address the need for vocational training and extension. However, the agriculture lobby would not support it until the agricultural extension legislation was adopted (Hillison, 1996). The efforts to establish federal legislation for vocational education proved to be an uphill battle for its proponents.

Senator Carroll Page had a passion for vocational education. He believed that it was the duty of the federal government to provide for job-specific training for young people. From an economic viewpoint, it made good sense to produce skilled workers who could step into the American workforce without having to go through an apprenticeship period. Vocational education had the potential to improve job training for millions of youth, and Senator Page was motivated to sponsor legislation that provided the necessary federal support.

The chief motivator of Senator Page was **Charles Prosser**. Dr. Prosser knew from firsthand experience the condition of education in the years preceding the Smith-Hughes Act. As an elementary school teacher, principal, and school superintendent in New Albany, Indiana, he knew the kind of education being provided and the kind of education needed in the workforce. Because of his strong stance on education, he was elected to serve as president of the Indiana Teachers Association and later became the deputy superintendent for industrial education (Moore & Gaspard, 1987).

A few years later, Charles Prosser became the secretary of the National Society for the Promotion of Industrial Education. His association with this organization placed him in contact with legislators who could be persuaded to sponsor legislation friendly to vocational education. This, of course, included Senator Carroll Page, along with Senator Hoke Smith of Georgia and others.

The same year that Senator Page came to Washington, an attempt was made to provide federal legislation for vocational education by Senator Jonathan Dolliver of Iowa and Representative Charles Davis of Minnesota. The Dolliver-Davis bill languished in Congress as attempts were made to amend it into an acceptable form. The death of Senator Dolliver, a champion for vocational education, in 1910 effectively removed hope of the passage of the bill.

Many in Congress did not like the Page-Wilson bill. The main argument against the bill seemed to be that it was a competing measure to the more popular Smith-Lever bill, already making its way through Congress. Since the Page-Wilson and Smith-Lever bills were similar in nature, it was possible that one could swallow the other during the debate (Barlow, 1976). Senator Page and Charles Prosser negotiated several amendments to the Page-Wilson bill to make it more palatable to its opponents, but the Smith-Lever bill would eventually win the fight. Senator **Hoke Smith**, whose name is indelibly linked to the Smith-Hughes Act, was a sponsor of the Smith-Lever Act creating the Cooperative Extension Service.

As a skilled politician and parliamentarian, Smith successfully blocked the passage of the Page-Wilson bill until his legislation to create the Extension Service had passed both houses. Smith's chief argument against the Page-Wilson bill was that it was poorly written and underbudgeted to meet the needs of vocational education. Smith believed in vocational education, but he also believed that the Page-Wilson bill was not the correct measure to establish successfully a national system of vocational education (Barlow, 1976). In 1914, the Smith-Lever Act establishing agricultural extension passed both houses of Congress and was signed into law by President Wilson.

After Smith-Lever was signed into law, President Wilson established the Commission on National Aid to Vocational Education to study the need for federal aid for vocational education. The membership of the commission was stacked in favor of vocational education, so the committee's report was characteristically pro-vocational education. Specifically, the report recommended that the Congress of the United States create legislation for vocational education. This commission's report established guidelines that later became the framework for the Smith-Hughes Act.

With Smith-Lever safely passed, Smith turned his attention to the matter of vocational education. Acting on the recommendations of the Commission on National Aid to Vocational Education, Senator Smith introduced his vocational education bill on December 7, 1915. Representative **Dudley Hughes** introduced companion legislation in the House of Representatives two days later. The Smith-Hughes bill was born.

Over the next year, the Smith-Hughes bill would be amended through committee work and floor debate. In time, the Congress of the United States finally reached general agreement on the merits of the bill, and it passed both houses of Congress on February 17, 1917. Six days later, President Wilson signed the bill into law. After six years of effort by Charles Prosser, Senator Carroll Page, Senator Hoke Smith, and countless others, the Vocational Education Act of 1917, forever known as the **Smith-Hughes Act**, was law.

Provisions of the Smith-Hughes Act

Funding for Vocational Education   The Smith-Hughes Act provided funding to the states for the purpose of training teachers in agricultural education, industrial arts education, and home economics education. Furthermore, the act paid the salaries of teachers in these subjects. The act provided funding for the establishment of teacher education programs in colleges and universities and funded the hiring of supervisors to manage the expenditure of funds at the school level. These supervisors gave direct assistance to teachers in the teaching of their respective subjects.

State Boards for Vocational Education   Smith-Hughes also created a state board for vocational education in each state receiving funding under the act. There was some flexibility in how the board was set up in each state. In some cases, the responsibilities of the state board for vocational education were absorbed into the work of the state board of education.

For the states to receive and continue to receive funding under Smith-Hughes, a provision of the law required that the minimum appropriations for the training of teachers be expended before any appropriations for salaries could be released.

Federal Board for Vocational Education   The Smith-Hughes Act also created the Federal Board for Vocational Education. The purpose of this board was to see that the provisions of the act were carried out according to the law. The members of the federal board were appointed by President Wilson on July 17, 1917, and met for the first time on July 21, 1917, with Secretary of Agriculture David Houston as chair. On August 15, 1917, Charles Prosser was appointed director of the board. The first members of the board were

- David Houston, Secretary of Agriculture
- William Redfield, Secretary of Commerce
- William B. Wilson, Secretary of Labor
- P. P. Claxton, U.S. Commissioner of Education
- Charles Greathouse, Agriculture Representative
- Arthur E. Holder, Labor Representative
- James Munroe, Commerce Representative

The federal board was authorized to hire a staff to carry out the administrative tasks of the board. The state boards for vocational education were required to submit annually to the federal board detailed plans of how vocational education funds would be expended.

The Beginning of Agricultural Education for African Americans

Prior to the twentieth century, public schools were not of great importance in the Southern United States. The prevailing mood among Southerners was one of skepticism toward strong central governments, and government programs included public education. In the antebellum South, schools were the responsibility of the family and, to some extent, the church.

Consequently, a weak system of schooling prevailed, starved to near death by a lack of public tax support (Cash, 1954). After the American Civil War, schools did not fare much better. Public taxes went into rebuilding those parts of the South destroyed by the war. There was inadequate funding for school construction, the hiring of teachers, and the equipping of schools. Southern states, because of segregationist legislation, had established a dual system of schooling. Whites attended schools supported by public taxes, and blacks attended schools that were supported by donations and taxes derived largely from the African American community. African Americans occupied the poor lower working class where what little money could be earned was spent on the immediate subsistence needs of the family, and more often than not their schools were significantly underfunded.

In 1865, 90 percent of the Southern black population was functionally illiterate, yet efforts were underway by supporters of education to mitigate this problem. During the American Civil War, teachers from the North arrived in the South behind the advancing Union armies. These teachers set up shop in old stores, churches, and homes and began the process of educating those who arrived each day in school. The funding for these schools was largely through the Bureau of Refuges, Freedmen, and Abandoned Lands (Freedmen's Bureau) and through donations of the members of the American Missionary Society and other Christian missionary groups. These societies often paid for teacher salaries and helped secure teaching materials for teachers.

Table 4-1
MAJOR PHILANTHROPIC ORGANIZATIONS SUPPORTING AGRICULTURAL AND INDUSTRIAL EDUCATION FOR AFRICAN AMERICANS

Name	Year Established	Amount of Initial Funding	Donor
General Education Board	1903	$1,000,0000	John D. Rockefeller
Phelps-Stokes Fund	1911	$1,600,0000	Olivia and Caroline Phelps-Stokes
Anna T. Jeanes Foundation	1907	$1,000,0000	Anna T. Jeanes
Julius Rosenwald Fund	1917	$3,030,0000	Julius Rosenwald
John F. Slater Fund	1882	$1,000,0000	John F. Slater

Source: Butler (1931).

By the end of the work of the Freedmen's Bureau in 1870, there were approximately 10,000 teachers for black schools in the South; most these were white Northerners (Wright, 1949).

As it became evident that public funding of African American schools for agricultural and industrial education would not materialize from state legislatures or the federal government, educational leaders began to lobby vigorously for the support of these schools through other means, and for other reasons. In 1901, Charles Dabney, educator and scientist, wrote,

> Everything in the South waits upon the general education of the people. Industrial development waits for more captains of industry, superintendents of factories, and skilled workmen. . . . We must educate all of our people, blacks as well as whites, or the South will become a dependent province instead of a coordinate portion of the nation (Dabney, 1901, p. 208).

Dabney voiced the concerns of a number of prominent clergymen, political figures, and wealthy industrialists, and these individuals began to mobilize financial resources to build schools, hire and train teachers, and provide agricultural and industrial education for African American students (see Table 4–1).

THE GROWTH AND DEVELOPMENT OF SUPERVISED AGRICULTURAL EXPERIENCE

Supervised agricultural experience in America likely began as youth apprenticeship in the colonial period or earlier. The concept of learning agricultural subjects under the direction of a mentor or skilled practitioner is an old one indeed. Evidence of apprenticeship can be found among Native Americans in the North American archeological record (Struck & Wright, 1945). As the population of the American colonies grew, the economy supported the growth of the apprenticeship. In early American schools, students learned basic skills in reading, mathematics, history, Latin, and Greek (Urban & Wagoner, 2000), and then went home to the farmstead to learn the concepts of animal husbandry, crop science, and engineering.

In the early 1900s, Rufus W. Stimson, principal of the Smith Agricultural School, developed the project method by which students learned the basics of agricultural production methods and applied these methods on their home farms (Moore, 1988). Emphasizing the benefits of a practical hands-on instructional program, Stimson insisted that these projects be completed away from the school campus and on the home farm (Stimson, 1919). Stimson thought that school projects were impractical, unrealistic, and involved too many students in one project. Students had no personal ownership in school projects. Stimson's model for farm projects specified that projects be realistic and useful on the farm, and be measured by specific learning goals. The two-pronged curriculum at Stimson's school included the study of production agriculture and individual project work.

The Beginning of the Integrated Model of Agricultural Education

Agricultural education is typically organized into interrelationships between three major concepts: classroom and laboratory instruction, supervised agricultural experience, and the FFA youth organization. How did these three concepts become linked together as the model for agricultural education?

Dr. Glen C. Cook wrote the *Handbook on Teaching Vocational Agriculture*. First published in 1938, the text was subsequently revised in 1947, 1952, 1980, and 1988. This textbook represented more than five decades of agriculture teacher education, and provides some insight into the priorities of agricultural education in its formative years.

In the 1938 handbook, Cook identified not three but four components of agricultural education. These were classroom instruction, supervised farm practice, farm mechanics, and extracurricular activities. Cook (1938) described supervised farm practice as an integral part to agricultural education, but stopped short of making the same judgment about the FFA. While FFA activities were acceptable for agricultural education students, Cook did not limit agricultural education to a sole partnership with the FFA. Extracurricular activities also included 4-H and agricultural clubs in addition to the FFA.

Cook's 1947 *Handbook on Teaching Vocational Agriculture* cleared up the role of the FFA in agricultural education. Cook identified the five major phases of agricultural education to include classroom instruction, supervised farming programs, farm mechanics, community food preservation activities, and FFA activities. Cook describes farm mechanics as an essential subcomponent of supervised farming programs. He further identifies community food preservation activities as essential during the World War II era of rationing and food shortages (Cook, 1947).

Later editions of Cook's textbook (Phipps, 1965, 1966, 1972, 1980; Phipps & Cook, 1956; Phipps & Osborne, 1988) continued to proscribe the three-component model of agricultural education. In earlier versions of the text, the component devoted to agricultural youth organizations included the New Farmers of America and Young Farmers. References to the New Farmers of America in the agricultural education model disappeared in editions of the text written after 1965. By the 1988 edition of the handbook (Phipps & Osborne, 1988), references to young farmers in the model had disappeared, and the four instructional components became classroom instruction, supervised experience, laboratory instruction, and vocational student organization.

Cook's textbook explained the agricultural education model, but did not reduce the model exactly to the present-day three-component version. In the 1970s, the FFA began a series of teacher development programs designed to improve the quality of agricultural education programs. The 1975 FFA Advisors Handbook was prepared by the National FFA Organization with emphasis on teacher development, and it included the three-component model of classroom and laboratory instruction, supervised experience, and FFA (National FFA Organization, 1975). Page 7 in the handbook shows the Venn configuration of three overlapping circles portraying these three components. The handbook justifies the integral nature of FFA with the instructional program, and explains that FFA activities require both supervised agricultural experience and instruction to be useful.

The three-component model continues to be the prevalent method for explaining the concept of agricultural education. The question remains, is the model truly of an integral nature? SAE and FFA activities have become optional elements in some agricultural education programs, and FFA activities may not always reflect the goals of the instructional program.

"High school vocational agriculture has been compounded of three parts: classroom teaching, supervised practice, and the Future Farmers of America organization. These parts have developed separately and unfortunately, they still remain to a considerable extent discrete." (Hamlin, 1949)

Federal Legislation from 1918 to 1963

As the nation grew, so did the need for vocational education. More importantly, the complexity of the American education system grew during the twentieth century. As of 2012, there were more than 14.9 million students enrolled in more than 30,000 public high schools in the United States (National Center for Education Statistics, 2012). The school curriculum evolved into a comprehensive curriculum with career and technical education subjects being provided alongside the traditional academic subjects. The U.S. Office of Education became the U.S. Department of Education in 1980. The Vocational Education Act of 1963 superseded the Smith-Hughes Act, and the Federal Board of Vocational Education no longer exists.

Much of the federal education legislation enacted in recent years has a broader and more comprehensive purpose. For instance, the Elementary and Secondary Education Act (ESEA) was enacted in 1965 as part of President Lyndon Johnson's Great Society program. Its purpose was to provide resources so that every child has an opportunity for a quality education. All other acts of the same title that followed this 1965 legislation typically amend it or reauthorize funding for it. The Improving America's Schools Act of 1994 and the No Child Left Behind Act of 2001 are reauthorizations of the original ESEA of 1965. Both of these acts provided funding to schools with agricultural education. The 1994 reauthorization of the ESEA increased federal support for technology in education. The 2001 reauthorization of the ESEA increased accountability for teachers and sought to find new methods for improving student performance in all subjects.

The U.S. government usually provides direction and guidance to states on the improvement of educational programs. The federal government provides only 10 percent of the funding for public education in the United States, but it is a very important partner in funding education. The federal government's funding share is typically used to handle shortfalls in local and state funding. For instance, the U.S. Department of Education (ED) works with the U.S. Department of Agriculture to provide the school lunch program, and with the Department of Health and Human Services to provide the Head Start early childhood program. The ED provides funding where the most good can be derived. This has been the traditional response of the federal government to the improvement of education, and the most significant period of federal funding for education began in the early 1900s.

Between 1918 and 1963, a number of federal laws were passed to modify or increase funding for vocational education. Senator Walter George of Georgia was the principal sponsor for many of these acts. The "George Acts," as they are sometimes called, amended the Smith-Hughes Act by increasing funding in existing vocational education programs and expanding the reach of federal legislation into new areas of vocational education.

Here is a brief summary of four "George Acts":

George-Reed Act of 1929—Increased federal support for vocational education and gave home economics independent status as a division.

George-Ellzey Act of 1934—Repealed the George-Reed Act. This legislation provided additional funding for vocational education and implemented new funding for distributive education.

George-Deen Act of 1936—Increased funding for vocational education in four areas: agriculture, home economics, trade and industrial education, and distributive education.

George-Barden Act of 1946—Significantly increased annual appropriations for vocational education and altered the formula for distributing funds in favor of agricultural education. This act also established area vocational schools for training students in vocational subjects. Because the federal funding provided by this act allowed for vocational guidance and the purchase of equipment, vocational programs grew substantially in the years following its passage.

National Defense Acts   The National Defense Acts are reauthorized periodically, as they provide funding and leadership policy for the defense of the United States. For instance, the 1940 National Defense Act and the 1958 National Defense Education Act provided funding for the creating of a highly skilled workforce in times of war. These acts also provided training for all workers in war production, including women. The 1958 National Defense Education Act provided financial assistance for students enrolling at colleges and universities to prepare for the highly skilled and technical jobs of the future.

The Vocational Education Act of 1963

By the 1960s, vocational education under the Smith-Hughes Act was in need of revision to meet the changing needs of the American economy. More than any other law since Smith-Hughes, the *Vocational Education Act of 1963* broadened agricultural education and made great strides in moving the program forward. It particularly provided that instruction could be in nonfarming areas of agriculture.

The 1960s was a turbulent period in America's history. Matters related to civil rights and America's involvement in the Vietnam War were lightning rod issues for most Americans. Underneath all this, the face of American agriculture was changing. The Fair Labor Standards Act extended federal minimum wage requirements to employers of most farm workers, and the United Farm Workers began unionizing California farm workers. Great strides were made in farm mechanization, and the need for skilled mechanics and technicians increased exponentially. Agricultural commodity groups significantly increased their lobbying efforts in Congress. Agricultural exports made up almost one-fifth of total U.S. exports.

To meet the new challenges of the agricultural industry in the last half of the twentieth century and to meet the growing demand for a highly skilled workforce in all vocational disciplines, the federal government's presence in the classroom needed to change. The Vocational Education Act of 1963 replaced the categorical funding of specific vocational programs, such as agricultural education and trade and industrial education, with a population-driven funding formula to the states. Federal funding under this act was distributed proportionally to the states by the number of students within a certain age group. Each state now played an important role in how federal funds would be distributed.

4–6. The Vocational Education Act of 1963 expanded agricultural education in the secondary schools to include a wide range of nonproduction agriculture, such as horticulture.
(Courtesy, Education Images)

The increased federal funding provided by the Vocational Education Act of 1963 made vocational education more flexible and open to emerging trends. This act, sponsored by Senator **Carl Perkins** (1912–1984) of Kentucky and Representative Wayne Morris of Oregon, opened the door for vocational training in other areas, such as business education. In agricultural education, it expanded the options for youth to include careers beyond the farm in such areas as marketing, horticulture, agribusiness, and natural resources. The act further provided funding for training youth with physical disabilities, creating work-study programs, and establishing vocational training centers.

The Vocational Education Act of 1963 also changed supervised experience in agriculture to include more than production agriculture. Unfortunately, this expanded variation of supervised experience led some states to abolish or reduce the value of it as part of the agricultural education model. The act also abolished the supervisory role of the state agricultural education staff and relegated the staff to a consulting role.

The Vocational Education Act of 1963 was amended in 1968 and again in 1976. The 1968 amendments increased federal funding for vocational education and allowed for vocational guidance, career counseling, and construction of additional vocational schools. Fifty percent of the funding in vocational education was to be used on students with disadvantages or disabilities and for employment services to students. The 1976 amendments increased federal funding and established guidelines for eliminating gender discrimination.

The Perkins Acts

The federal Vocational Education Acts of 1984, 1990, 1996, 1998, and 2006 bear the name of Carl Perkins. Carl Perkins was a member of the U.S. Congress from Kentucky, who served from 1949 to 1984. He was a strong advocate in Congress for vocational education legislation. These laws attempted to modernize vocational education further and to expand its emphasis as career and technology education. The 1984 act created federal legislation to support the efforts of vocational education in making the United States more competitive economically. The primary focus of the 1984 act was to improve the quality of vocational education for all students and to improve the accessibility of vocational education for all students who wanted it.

More than half the funding provided by the first Perkins Act was earmarked for special student populations—students with disadvantages or disabilities, adults, and single-parent families.

The Carl D. Perkins Vocational Education Act of 1990 increased funding for vocational programs to meet advances in technology. This act also increased funding for curriculum integration efforts between vocational and academic subjects. It sought to move vocational education away from training for specific jobs and toward education in the more general aspects of careers. The act further provided for the establishment of articulation agreements between secondary schools and colleges and universities.

In 1996, the third Carl D. Perkins Vocational Education Act reauthorized funding for vocational education, but the emphasis was now even stronger on curriculum integration and articulation agreements between secondary and postsecondary institutions. The 1998 reauthorization of Perkins retained the same emphasis on integration and required that students enrolled in career and technical education courses be provided with the same rigorous curriculum as other students.

1862	U.S. Department of Agriculture was founded by President Abraham Lincoln. (Note: Agriculture was given a federal level office some two decades before it became a Department.)
1862	Morrill Act was signed by President Lincoln establishing a system of land-grant colleges.
1887	Hatch Act was passed, creating a system of agricultural experiment stations in conjunction with the land-g-rant colleges.
1890	Second Morrill Act was passed, setting a side funds to establish land-grant colleges in 17 states where admission to higher education was open to all people.
1914	President Woodrow Wilson signed the Smith-Lever Act, creating the Cooperative Extension Service as part of land-grant colleges.
1917	President Woodrow Wilson signed the Smith-Hughes Act, creating a system of public school vocational education.
1963	Vocational Education Act of 1963 was enacted to provide for broadened areas of instruction in vocational education and to expand the mission of secondary school agricultural education.
1984	Carl D. Perkins Vocational Education Act of 1984 was passed to improve the quality and accessibility of vocational education for all students, with half the funding ear marked for special student populations.
1990	Carl D. Perkins Vocational Education Act of 1990 was enacted to provide more vocational education funding to meet technological advances in business and industry, promote curriculum integration, and establish articulation agreements between secondary schools and institutions of higher learning.
1994	President William J. Clinton signed the Improving America's Schools Act, which provided expanded higher education opportunities for Native Americans.
1996	Carl D. Perkins Vocational Education Acts of 1996 and 1998 reauthorized funding for vocational education. With increased emphasis on curriculum integration and on the holding of vocational education students to the same rigorous standards as all other students.
2002	President George W. Bush signed the reauthorization of the Elementary and Secondary Education Act of 2001 (also known as the No Child Left behind Act), which emphasizes high achievement test scores and, in the opinion of some individuals, posed potential threats to agricultural education.
2006	The 2006 version of the Carl D. Perkins Vocational and Technical Education Act provided approximately $1.3 billion for career and technical education. This law also used for the first time in federal legislation the term "career and technical education."

4–7. Prominent legislative events in agricultural education.

The latest revision of the Carl D. Perkins Vocational and Technical Education Act was enacted on August 12, 2006. The new act increased the emphasis on academic achievement of career and technical education students, and sought to enhance state and local measures of continuous improvement in career and technical education, including agricultural education. The new law uses for the first time in federal legislation the term "career and technical education." This revision of existing federal law provided approximately $1.3 billion for career and technical education (U.S. Congress, 2006).

Morrill Revisited

One of the first federal laws to establish some form of agricultural education specifically suited to supervised experience was the Civilization Fund Act of 1819, which provided funding to teach Native Americans "the mode of agriculture suited to their situation" (Fraser, 2001, p. 47). For a number of years afterward, the U.S. government made additional attempts to assimilate American Indians into Western culture (Stahl, 1979), but the most significant legislation to provide agricultural education for American Indians was the Equity in Educational Land-Grant Status Act of 1994. This act conferred land-grant status to twenty-nine tribal colleges, for the purpose of teaching agricultural and mechanical subjects in accordance with the original Morrill Act of 1862. This third "Morrill Act" was the result of a successful lobbying effort by the American Indian Higher Education Consortium (AIHEC). In 1992, this group formed in order to secure the necessary legislation to provide land-grant funding to the tribal colleges. This "1994 Morrill Act" created an endowment fund for Native American education at these twenty-nine colleges, and provided matching grants to improve buildings and laboratories. Most of these colleges are two-year technical schools, but at least three offer baccalaureate degrees.

Land-Grant Institutions Today

Today, a land-grant institution has a three-pronged mission: teaching, research, and service. Teaching is carried out through the ongoing instructional program. Research is carried out through an experiment station as well as other research thrusts that may be established. Service is carried out through the Extension Service (known in the USDA as the Cooperative Extension System) and other service initiatives of the land-grant institution. Local offices of the Extension Service are found in nearly every county or parish in the United States.

Each state has at least one land-grant institution. Seventeen states also have institutions created by the Morrill Act of 1890, plus Alabama has a third institution, Tuskegee University.

Elementary and Secondary Education Act of 2001

The Elementary and Secondary Education Act of 2001, also known as the No Child Left Behind Act, authorizes most of the federal funding for elementary and secondary education. The act was to supplement local and state educational funding for the purpose of improving student achievement. The standards imposed by the act are challenging. To receive continued funding, schools must meet selected goals. Schools that fail to meet rigorous standards have sanctions levied against them. There is a concern that the act will harm career and technical education because of the focus on high-stakes testing and accountability. Will schools find it necessary to cut vocational programs in order to focus resources on helping students meet annual academic goals? This act is a politically charged issue among educators, and only time will tell whether it meets its goals to improve student achievement in American schools.

REVIEWING SUMMARY

The founders of vocational education are gone, but their legacy lives on. Today, millions of young people have the opportunity to have the careers of their choice because of career and technical education. Millions of Americans have learned useful skills, acquired technical jobs, raised families, and, as responsible citizens, contributed to the overall well-being of the United States. There can be no better epitaph for these great visionaries.

Agricultural education in the United States came about primarily through the efforts of individuals responding to needs within their local communities. Before the Smith-Hughes Act of 1917, much of the leadership for agricultural education was found within the individual states and often came from visionaries who worked under the authority provided by federal legislation for experiment stations and land-grant universities. The Vocational Education Act of 1917 (Smith-Hughes Act) provided the organization for federal aid for vocational education until 1963, when a new vocational education act replaced it.

The new focus for agricultural education is in curriculum integration and articulation agreements between secondary and postsecondary institutions. Today, career and technical education is available for any student who wishes to participate in it, can profit by it, and needs it to meet career goals.

QUESTIONS FOR REVIEW AND DISCUSSION

1. What were some of the chief provisions of the Morrill Acts of 1862 and 1890?
2. What did the Smith-Hughes Act accomplish?
3. What was the primary purpose of the "George Acts"?
4. What trends do you notice in federal legislation for vocational education in the 1900s?
5. What were major provisions of the Vocational Education Act of 1963 as related to agricultural education?
6. Briefly identify the following individuals: Justin Morrill, Dudley Hughes, Hoke Smith, Charles Prosser, Carl Perkins, and Jonathan Baldwin Turner.

ACTIVITIES

1. Complete a biographical research project on one of the people mentioned in this chapter. Specifically, how did that person's background, interests, and vocation lead the individual to support legislation for career and technical education?
2. Investigate the history of secondary agricultural education in your state. Who were the early leaders? What were the first schools to initiate agriculture classes? What was the nature of the instruction in these early classes?
3. Investigate the structure of the land-grant system in your state. Prepare a report that provides the names and locations of the colleges or universities and discusses how the teaching, research, and extension mission is accomplished.

CHAPTER BIBLIOGRAPHY

Barlow, M. L. (1976). *The Unconquerable Senator Page: The Struggle to Establish Federal Legislation for Vocational Education*. Washington, DC: American Vocational Association.

Butler, J. H. (1931). *An Historical Account of the John F. Slater Fund and the Anna T. Jeans Foundation*. San Francisco: University of California, San Francisco.

Cash, W. J. (1954). *The Mind of the South*. Garden City, NY: Doubleday.

Cochrane, W. W. (1979). *The Development of American Agriculture: A Historical Analysis*. Minneapolis: University of Minnesota Press.

Cook, G. C. (1938). *Handbook on Teaching Vocational Agriculture*. Danville, IL: The Interstate Printers and Publishers, Inc.

Cook, G. C. (1947). *A Handbook on Teaching Vocational Agriculture*. Danville, IL: Interstate Printing Company, Inc.

Cornwall Historical Society. (2012). *The Eight Cornwalls*. Retrieved October 1, 2012.

Croom, D. B. (2008). The development of the integrated three-component model of agricultural education. *Journal of Agricultural Education, 49*(1), 110–120.

Cross, C. F. (1999). *Justin Smith Morrill: Father of the Land-Grant Colleges*. East Lansing: Michigan State University Press.

Dabney, C. W. (1901). *The Public School Problem in the South*. Knoxville: University of Tennessee.

Fraser, J. W. (2001). *The School in the United States: A Documentary History* (1st ed.). Boston: McGraw-Hill.

Hamlin, H. M. (1949). *Agricultural Education in Community Schools*. Danville, IL: Interstate.

Hamlin, H. M. (1962). *Public School Education in Agriculture: A Guide to Policy and Policy-making*. Danville, IL: Interstate Printers & Publishers.

Hillison, J. (1996). Agricultural education and cooperative extension: The early agreements. *Journal of Agricultural Education, 37*(1), 9–14.

Hurt, R. D. (2002). *American Agriculture: A Brief History* (Rev. ed.). West Lafayette, IN: Purdue University Press.

Kell, K. T. (1966). Folk names for tobacco. *The Journal of American Folklore, 79*(314), 590–599. doi: 10.2307/538224

Moore, G. E. (1988). The forgotten leader in agricultural education: Rufus W. Stimson. *Journal of the American Association of Teacher Educators in Agriculture, 29*(3), 50–58.

Moore, G. E., & C. Gaspard. (1987). The quadrumvirate of vocational education. *Journal of Career and Technical Education, 4*(1), 3–17.

Morison, S. E. (1965). *The Oxford History of the American People*. New York: Oxford University Press.

National Center for Education Statistics. (2012). *Digest of Education Statistics, 2011 (NCES 2012-001), Chapter 2*. Washington, DC: U.S. Department of Education.

National FFA Organization. (1975). *FFA Advisors Handbook*. Alexandria, VA: National FFA Organization.

Phipps, L. J. (1965). *Handbook on Agricultural Education in Public Schools*. Danville, IL: Interstate Printers & Publishers.

Phipps, L. J. (1966). *Handbook on Agricultural Education in Public Schools* (2nd ed.). Danville, IL: Interstate Printers & Publishers.

Phipps, L. J. (1972). *Handbook on Agricultural Education in Public Schools* (3rd ed.). Danville, IL: Interstate Printers & Publishers.

Phipps, L. J. (1980). *Handbook on Agricultural Education in Public Schools* (4th ed.). Danville, IL: Interstate Printers & Publishers.

Phipps, L. J., & G. C. Cook. (1956). *Handbook on Teaching Vocational Agriculture* (6th ed.). Danville, IL: Interstate.

Phipps, L. J., & E. W. Osborne. (1988). *Handbook on Agricultural Education in Public Schools* (5th ed.). Danville, IL: Interstate Printers & Publishers.

Stahl, W. K. (1979). The U.S. and Native American education: A survey of federal legislation. *Journal of American Indian Education, 18*(3), 28–32.

Stevens, N. E. (1921). America's first agricultural school. *The Scientific Monthly, 13*(6), 531–540. doi: 10.2307/6538

Stick, D., & North Carolina, America's Four Hundredth Anniversary Committee. (1983). *Roanoke Island, the beginnings of English America*. Chapel Hill: Published for America's Four Hundredth Anniversary Committee by the University of North Carolina Press.

Stimson, R. W. (1919). *Vocational Agricultural Education by Home Projects*. New York: The Macmillan Company.

Stimson, R. W., & F. W. Lathrop. (1942). *History of Agricultural Education of Less Than College Grade in the United States: A Cooperative Project of Workers in Vocational Education on Agriculture and in Related Fields*. Washington: Federal Security Agency.

Struck, F. T., & J. C. Wright. (1945). *Vocational Education for a Changing World*. New York, London: J. Wiley; Chapman & Hall.

True, A. C. (1969). *A History of Agricultural Extension Work in the United States: 1785–1923*. New York: Arno Press.

True, R. H. (1939). *Sketch of the History of the Philadelphia Society for Promoting Agriculture*. Philadelphia: Philadelphia Society for Promoting Agriculture.

Urban, W. J., & J. L. Wagoner. (2000). *American Education: A History* (2nd ed.). Boston: McGraw-Hill.

U.S. Congress. (2006). *Carl D. Perkins Career and Technical Education Improvement Act of 2006*. Washington, DC: United States Government Printing Office.

U.S. Department of Agriculture. (2012). *History of American Agriculture*. Retrieved January 10, 2012.

U.S. Department of Education. (2005). *High School Transcript Study (HSTS), 2005*. Washington, DC:: Institute of Education Sciences, National Center for Education Statistics.

U.S. Department of Education, N. C. f. E. S. (2012). *Digest of Education Statistics, 2011 (NCES 2012–001), Chapter 2*. Washington, DC: U.S. Department of Education.

White, J. (1585). *Village of the Secotan in North Carolina*. A Watercolor in the British Museum, London.

Wright, S. J. (1949). The development of the Hampton-Tuskegee pattern of higher education. *Phylon, 10*(4), 9.

5

Organization and Structure of Agricultural Education

objectives

This chapter presents information about the organization and structure of agricultural education. It has the following objectives:

1. Describe the relationship of agricultural education to local school boards
2. Explain the levels of local, state, and federal administration
3. Relate local programs to the National FFA Organization
4. Discuss issues related to local autonomy

Since the federal government does not carry out secondary school agricultural education, the responsibility for doing so has been delegated to the various states. Each state has plans and procedures that meet with federal approval. These plans include how local school districts will go about delivering the education.

A key concept in American education is that local school authorities, particularly boards of education, are given considerable autonomy over the educational process. Some people, however, say that many state regulations must be met to receive funds from state government.

Often there is a great distance from the state capital to local programs of agricultural education. Fortunately, a workable structure is in place to assure quality agricultural education.

terms

autonomy
board of education
curriculum
elective course/program
high school
junior high school
middle school

policy
postsecondary program
principal
procedure
required course
school district
superintendent

AGRICULTURAL EDUCATION AND LOCAL SCHOOL BOARDS

Agricultural education is provided at the local level through state-approved programs in the nation's schools. Agricultural education programs in different schools include the three major program components—classroom and laboratory instruction, supervised agricultural experience, and student organization (FFA)—and the subject is taught in basically the same manner. However, the content of the instruction varies from school to school. The program should be designed to meet the needs of the local community.

State administration for agricultural education is provided through various state agencies and institutions. The most common lead agency for agricultural education programs is a state department of education. Federal administration is provided through the Office of Vocational and Adult Education (OVAE), U.S. Department of Education.

Today's agricultural education programs are focused on the science, business, and technology of the plants, animals, and natural resources systems. About 1 million students participate in agricultural education programs offered in grades 7 through adult throughout the fifty states and three U.S. territories.

School Districts

A *school district* is a geographical area under the supervision of a given school board. It may include several schools or attendance centers, or it may have only one school. Sizes of school districts vary considerably. A district may be congruent with the limits of a city or county or some other defined geographical area.

A central administrative office may be responsible for coordinating functions among the schools. The superintendent and several assistant administrators, as well as support staff, are usually part of this central office.

A *board of education* is a group that has the legal authority to take action and use resources in the operation of a school district. Boards of education are provided for in the laws of a state and are sometimes known as local school boards. Members of a board may be elected or appointed. The superintendent reports directly to the

5–2. A school district maintains an office for the local board of education.
(Courtesy, Education Images)

board. All other employees of the board report through the superintendent. Usually, a board delegates matters only to the superintendent. The superintendent may further delegate to staff members. Together, the board and the superintendent may establish guidelines for operation of the schools.

A local school board is a public body, and its meetings are open to the public. Teachers, students, parents, and others may attend school board meetings to keep abreast of issues discussed by the board. If a teacher or another individual wishes to present an item to the board, he or she must follow the board's policies for having the item placed on the board meeting agenda. It is wise for a teacher to inform the school principal if he or she intends to present information at a school board meeting.

A good agricultural education advisory committee can often be utilized to interface with the school board through both formal and informal communications channels. Advisory groups have no administrative or legislative authority and cannot establish policy. Their function is to increase understanding between the school and the community. A strong agriculture program has a local advisory committee to provide community input into the content and design of the local agriculture program. Good teachers working with a good local advisory committee are able to design and deliver high-quality programs that meet the needs of the local students and community. It is at the local level that teachers have the most opportunity for using their initiative and creativity in shaping the agricultural educational program to meet the needs of their students, school, and local community.

Policies and Procedures

A *policy* is a general guiding principle. Many policies are based on laws. For example, a state law may require a local board of education to have a policy on teacher certification. Once the board has established the policy, the superintendent prepares procedures to assure that it is properly implemented. Policies are made by individuals or groups that have the authority to do so, such as school boards. Policies are established to guide the management, procedures, and decision making especially of government bodies, such as school boards. In policy making, a local board may seek teacher and citizen input.

A *procedure* is the way a policy is to be implemented or carried out. Some procedures are routine or standard operating procedures. An example is purchasing supplies for a horticulture program. School districts have definite purchasing procedures that should be followed.

State approval is usually required to establish and maintain a local agricultural education program, but the local program operates at the pleasure of the local board of education. Local school board members represent the community in establishing policies for the operation of the local school system.

Local school boards must follow state and federal guidelines to ensure their schools are eligible to receive state and federal funds and to ensure their graduates are qualified to enter the workforce and/or advance to higher education programs. While much of a school's curriculum is mandated by state laws and regulations, there remains considerable local control of educational programs and school policies. For example, the local school board is responsible for approving and overseeing the

5–3. School districts publish student handbooks with information and expectations in line with school board policies. (A student teacher reviewing a handbook where she is student teaching.)
(Courtesy, Education Images)

school's budget and for hiring and terminating employment of administrators, teachers, and other school employees. The local board also establishes and/or approves most of the operational policies and procedures that students, teachers, and other school employees must follow.

Programs and Curriculums

The local board of education is responsible for assuring that all programs and curriculums meet the standards established by the state board of education.

The programs or courses in a school district are approved by the local board of education. The professional staff of the district is responsible for assuring the programs or courses are properly implemented so that students receive the appropriate education.

The **curriculum**, simply, is the list of courses offered in a school. Curriculum is sometimes defined more specifically as the learning experiences a student has under the direction of school personnel. Some disparity develops between what may be specified in the outline of a course and the learning experiences actually provided.

At the secondary level, some courses are required, and others are elective. **Required courses** are those needed to meet core requirements for graduation or admission to an institution of higher education. The programs or courses offered become the local curriculum. These offerings are ultimately determined by the local community, as expressed through local school board policy. **Elective courses/programs** are those that are not mandated by state policies or standards and that

5–4. A high school principal discusses student learning activities with a horticulture teacher at the school's land lab.
(Courtesy, Education Images)

students are not required to take to graduate from high school. These courses and programs, which include agricultural education, are either offered or not offered according to the will of the local board of education.

Administrative Staff

A school district may have several administrative positions. The role of individuals in these positions is to see that quality education is provided in an efficient manner.

A *superintendent* is the one who has executive oversight and charge. The local school board hires a superintendent as the chief executive for the school system, and the superintendent, in consultation with the board, provides for the preparation and administration of the school system budget and the hiring of other school administrators, including school principals. A state usually has policies requiring certain qualifications or levels of preparation for an individual to become a superintendent.

A *principal* is a person who has controlling authority or is in a leading position. The school principal is in charge of a local school site and responsible for leading and managing a particular school. Principals evaluate their school faculty and staff and make recommendations regarding hiring and firing employees at their school. They also administer policies and manage the budget for their school. The principal reports to the school system superintendent.

LOCAL, STATE, AND FEDERAL ORGANIZATION

Local, state, and federal levels are organized so that educational programs can be carried out. Local districts differ in their organization but have many similarities. The same is true of the states.

Local District Organization

Agricultural education at the local level may be offered in middle schools, high schools, and vocational centers and in special settings, such as academies and magnet schools. One or more of each may be found in a school district, depending on the size and student population of the district. Each site where students go to school may be known as an attendance center. A rural school district with sparse population may have one attendance center where all grades attend.

Agricultural education is sometimes offered at the middle school level, where the programs focus on exploring the broad subject of agriculture. A *middle school* is a school that commonly includes grades 6 through 8. Instruction at this level is focused on helping students understand the plants, animals, and natural resources systems and helping them make informed choices about their future. The first year of agricultural education is usually introductory in nature and covers FFA, supervised agricultural experience programs, an overview of the agricultural industry, and agricultural careers. The introductory course may be provided at the middle school, junior high school, or high school. In addition, some basic leadership and agricultural skills are often taught in introductory courses. The middle school instruction should be articulated with that of the high school the students will attend after completing middle school.

A *junior high school* is a school that includes grades 7 and 8 and sometimes grade 9. Agriculture instruction may be offered at the junior high school level. The curriculum is articulated with that of the high school the students will attend after completing junior high.

A *high school* is a school that includes grades 9 or 10 through 12. In some states, agricultural education is first offered at the junior high school level and/or high school level. Most high school courses are designed to develop specific skills in agriscience, agribusiness, technology, leadership, and human relations. Agricultural education is also offered beyond the high school level.

High school programs may be articulated with postsecondary programs. A *postsecondary program* is one that is sequenced following the high school level, such as a program at a community college or at a career and technical center. Postsecondary agricultural education programs are more specialized and provide students with advanced technical competencies to enter and progress in specific agricultural careers. Some states also provide agricultural education programs for adults in the community.

At the school site, the local agriculture department may be recognized as a separate department of the school, be part of a career and technical education department, be part of a science department, or be included in some other school organizational structure established under the leadership of the school principal.

Principals operate their individual schools under guidelines provided by the school board and administered by the superintendent and other central office administrators. In a small district, principals may report directly to the superintendent, while in a larger district, there may be assistant superintendents or directors in the chain of command between the principals and the superintendent.

The head of the agricultural education department usually reports to the school principal and/or to a career and technical education director. Principals are charged with the responsibility of evaluating, managing, and leading employees in their individual schools. Teachers should be evaluated periodically by a principal or an assistant principal and the results of these evaluations should be placed in the teachers' personnel files. Principals also make recommendations regarding the hiring and firing of their school's employees, including teachers.

Individual agriculture teachers may report to the head of the agricultural education department or, in a small school, directly to the school principal. The final decision to employ or dismiss an agriculture teacher or other school employee rests with the school board, which must follow its own policies and procedures in

5–5. A secondary agricultural education facility with a specialized aquaculture program.
(Courtesy, Education Images)

personnel matters. The board's decisions are usually guided by the recommendation of the principal and the superintendent. It is important for agriculture teachers to maintain positive relationships with all local school administrators and school board members.

State Organization

State policy for education is established by state legislatures and state boards of education. State education policies are administered through state departments of education, which generally have career and technical education divisions or groups.

State support and services for agricultural education programs are usually provided through the career and technical education division of the state department of education, with one individual designated as the lead or head supervisor, consultant, or coordinator of statewide agricultural education programs and activities. However, state structures for administering agricultural education vary. Some states delegate the authority for state administration of agricultural education to a state university, state department of agriculture, community college, or some other agency or institution.

Local programs receive support and direction from states in varying degrees. Some states have greater capacity to support local programs than do others. There are also differences from state to state in the amount of direction local schools receive regarding what is to be taught. Some states have required agricultural education curriculums that must be taught, while other states leave the decision about what to teach almost entirely up to the local schools. State funds generally provide a large portion of the financial support for local agricultural education programs.

As noted above, each state has an individual designated as the lead person for agricultural education programs. This individual may devote full time or part time to the administration of agricultural education programs and usually serves as advisor to the state FFA association.

Usually, one individual has overall responsibility for leading the entire agricultural education program in a state. Often another individual, under the leader, coordinates the statewide FFA component, working with the state's FFA officers and coordinating the FFA activities conducted in the state. Less often, a state has several agricultural education staff members, working under the leader, serving as consultants or supervisors for local agriculture programs and/or managing specific aspects of the state's agricultural education program. There is considerable variation among states in the location and structure of state leadership and in the amount of local support provided by state leaders. It is important for local teachers to know their state leaders personally and to understand how their state's leadership system works.

Federal Organization

Federal policies for education are established by the U.S. Congress and administered through the U.S. Department of Education. The federal government also provides some funding for agricultural education. States offer agricultural education under federal guidelines they receive through the U.S. Department of Education. The federal guidelines provide general direction to states and require submission of state plans for career and technical education. The plans describe each state's priorities for career and technical education, including how federal funds will be expended.

Federal administration of agricultural education is provided through the U.S. Department of Education. Steve Brown is the federal employee working with agricultural education and the FFA at the national level. Dr. Brown also serves as chair of the National FFA Board of Directors and as the National FFA Advisor. He is an employee of the Office of Vocational and Adult Education (OVAE), U.S. Department of Education, Washington, DC.

Since the 1970s, federal direction for agricultural education has decreased, as the guidelines have focused primarily on the broad field of career and technical education. With limited federal direction, state programs of agricultural education have become less consistent in their focus and delivery. To maintain some national consistency in program direction, national agricultural education leaders have collaborated with various stakeholders in establishing and refining a national vision for agricultural education, a mission statement with goals, and a strategic plan for implementing the vision and mission.

The National Council for Agricultural Education is the umbrella national organization for school-based agricultural education. The council's mission is to provide leadership and coordination to shape the future of school-based agricultural education. Representatives of twelve national organizations/entities serving agricultural education make up the council. They serve as catalysts for the development and implementation of the vision, mission, goals, national policies, and programs for agricultural education. The council recommends that states utilize the national vision, mission, goals, and strategic plan to guide the development of state and local visions, missions, goals, and strategic plans for keeping agricultural education abreast of agricultural and educational trends and issues.

In 2009, the National Council for Agricultural Education released the National Quality Program Standards for Secondary (Grades 9–12) Agricultural Education. These standards were developed to provide some consistency in delivering high-quality agricultural education programs nationwide.

LOCAL PROGRAMS AND THE NATIONAL FFA ORGANIZATION

A high-quality agricultural education program includes all three of the agricultural education program components. The integral nature of classroom instruction, supervised agricultural experience, and the FFA and the interrelationships between these three components ensure that resources directed at any one part of the agricultural education program affect the entire program.

The National FFA Constitution provides a common focus for agricultural education across the nation. With funding from members' dues, product sales, and the National FFA Foundation, the National FFA Organization has the financial and human resources to provide considerable leadership and direction for FFA chapters.

Through various partnerships and alliances, the National FFA also provides leadership for agricultural education. For example, the FFA partnered with the National Council for Agricultural Education in developing the Local Program Success and the Reinventing Agricultural Education for the Year 2020 initiatives that provided resources to improve and enhance local agricultural education programs.

Local FFA chapters are chartered by the state association, which in turn is chartered by the National FFA Organization. To remain in good standing with the state and national levels, each local chapter must have a constitution that complies with the state and national constitutions. Local chapters must also submit state and national membership dues, submit reports as requested by the state and national levels of the organization, and conduct activities that follow the ideals and purposes of the state and national levels. Chapters in good standing are eligible to participate in all state and national events and activities according to established guidelines.

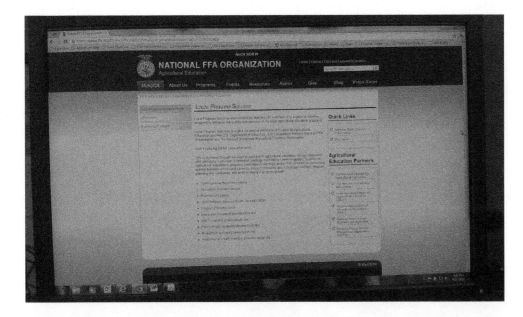

5–6. The FFA Web site contains many local program resources for teachers and students. (Courtesy, Education Images)

The National FFA Organization maintains a Web site that provides valuable information for teachers and students. Local teachers order many of their supplies for degrees and awards, chapter banquets, and FFA instruction; official dress uniforms; and various other items from the National FFA Organization. The site includes other resources including award applications, degree and proficiency award handbooks, the *Guide to Local Program Success*, and many other tools/materials needed by local agriculture teachers.

LOCAL AUTONOMY

Autonomy is the quality or state of being self-governing or self-directed. Local autonomy has been a part of the agricultural education program philosophy since the inception of the program. Teachers have always been encouraged to structure agricultural education programs around the needs and resources of their local communities.

As mentioned earlier in this chapter, considerable variation exists among states regarding the amount of local decision making authorized for agricultural education program management. Some states have numerous standards that programs are required to meet and consistently monitor local schools to ensure these standards are met. Other states provide general guidelines with little, if any, monitoring of how local schools provide agricultural education.

Strong teachers strive to involve the local community in establishing a vision and a mission for the local agricultural education program and in creating implementation strategies for achieving the vision and the mission. The National Council for Agricultural Education developed a *Community-Based Program Planning Guide* for assisting agriculture teachers in developing visions, missions, and strategic plans to guide their agriculture departments. Through the process outlined in the guide, representatives of all agricultural education stakeholders in a community are invited to become involved in the program, and their collective wisdom is captured to provide the vision and a "road map" for achieving the vision. This is a powerful strategy for determining what should be taught in the local agriculture program and for ensuring that the program meets the needs of the students enrolled. This process is outlined in Chapter "P" of *A Guide to Local Program Success* found on the National FFA Web site.

5–7. A sign indicates the facilities of an agriculture department and represents a level of autonomy for the program.
(Courtesy, Education Images)

Even in states that mandate statewide curriculums, local teachers can and should incorporate lessons, experiences, strategies, and other learning opportunities that address local community needs. This is essential for maintaining strong community support for agricultural education.

Good agriculture teachers thrive in schools where they are allowed to be innovative in determining the direction of local agriculture programs. They build on school and community strengths to ensure that high-quality instructional programs are designed for and delivered to the students. Local autonomy does not work well for weak teachers who often flounder without strong leadership, supervision, and guidance from state leaders. Strong local agriculture teachers are the most critical ingredient for maintaining high-quality agricultural education programs.

REVIEWING SUMMARY

Agricultural education is provided in the nation's schools through state-approved local programs. All agricultural education programs are expected to include the three major program components—classroom and laboratory instruction, supervised agricultural experience, and student organization (FFA)—and the subject is taught in much the same manner everywhere. The content of the instruction and the quality of the programs vary from school to school. The most important factor in providing a high-quality program is the agriculture teacher. An agriculture program should be designed to meet the needs of the local community, and teachers should actively seek community direction and support for the program.

State leadership for agricultural education is provided through a wide variety of sources and structures.

It is important that local teachers know the leaders and the leadership structure for their state. Federal administration for agricultural education is provided through the Office of Vocational and Adult Education, which is part of the U.S. Department of Education. Some national focus for agricultural education is also provided through the National Council for Agricultural Education, which is a federation of national organizations/entities that represent the agricultural education community. The National FFA Organization also provides leadership through numerous programs, events, and activities.

Local autonomy works well for good teachers, because they seek and utilize school and community input in planning and delivering the agricultural education program. This strategy generates interest, enthusiasm, and support throughout the community.

QUESTIONS FOR REVIEW AND DISCUSSION

1. Why are local school boards important to agricultural education?
2. Describe the three levels of agricultural education administration and explain why each level is important.
3. What does the National FFA Organization provide for local programs?
4. Is local autonomy an asset or a hindrance to maintaining high-quality local programs? Explain your answer.

ACTIVITIES

1. Use the Internet to investigate resources provided to local agriculture programs by the National FFA Organization.
2. Prepare a written report that describes the state administration structure in your state. Include how the state administration interacts with local agriculture programs, how leadership is provided for the total agricultural education program, and how leadership is provided for the FFA.
3. Prepare a written report that describes the autonomy of agricultural education programs in your state.
4. Arrange to attend a local board of education meeting. (School board meetings are open to the public.) Obtain a copy of the order of business and carefully observe the proceedings of the meeting. Prepare a report that includes who chaired the meeting, how items were presented, how decision-making processes were carried out, and what financial expenditures were approved.

CHAPTER BIBLIOGRAPHY

Caffarella, R. S. (2002). *Planning Programs for Adult Learners* (2nd ed.). San Francisco: Jossey-Bass.

Kahler, A. A., B. Morgan, G. E. Holmes, and C. E. Bundy. (1985). *Methods in Adult Education* (4th ed.). Danville, IL: The Interstate Printers & Publishers, Inc.

Lee, J. S. (2000). *Program Planning Guide for Agri-Science and Technology Education* (2nd ed.). Upper Saddle River, NJ: Pearson Prentice Hall Interstate.

Moore, G. E. (Ed.). (1998). A primer for agricultural education. *The Agricultural Education Magazine, 71*(3).

Moore, G. E. (Ed.). (2000). Theme: Reinventing agricultural education for the year 2020 (Entire Issue). *The Agricultural Education Magazine, 72*(4).

National Council for Agricultural Education. (2000). *The National Strategic Plan and Action Agenda for Agricultural Education*. Retrieved from the FFA website.

National FFA Organization. (2002). *A Guide to Local Program Success*. Retrieved from the FFA website.

Newcomb, L. H., J. D. McCracken, J. R. Warmbrod, and M. S. Whittington. (2004). *Methods of Teaching Agriculture* (3rd ed.). Upper Saddle River, NJ: Prentice Hall.

Phipps, L. J., E. W. Osborne, J. E. Dyer, and A. L. Ball. (2008). *Handbook on Agricultural Education in Public Schools* (6th ed.). Clifton Park, NY: Thomson Delmar Learning.

True, A. C. (1929). *A History of Agricultural Education in the United States, 1785–1925*. U.S. Department of Agriculture, Miscellaneous Publication No. 36. Washington, DC: GPO.

part two
Program Development and Management

6
Program Planning

Brian is a firm believer in the value of having a complete agricultural education program. But since his is a one-teacher program, what does "complete" mean? Do his food science students, who typically take only that one agriculture class, need supervised agricultural experience? Should they be expected to be active state and national FFA members? Should he be teaching exploratory agriculture to seventh and eighth graders?

Brian wants to answer these and other questions about his program. The university teacher educator in his state has suggested he use a program planning process. Brian is excited to begin.

Agriculture teachers often have many decisions to make. The nature of a program varies with the state and the local school district. A smart teacher always knows what is expected and attempts to meet the expectations.

objectives

This chapter explains the importance of planning a total agricultural education program. It has the following objectives:

1. Explain the meaning and importance of program planning

2. List three program components and describe the role of each

3. Explain procedures in program development, including the use of community needs assessments and technology-based program development aids

4. Describe evaluation procedures used with local programs

terms

community survey
connecting activity
contextual learning
intracurricular
Local Program Success
needs assessment

program components
program evaluation
program planning
technology-based development
work-based learning

6–1. Program planning is an important responsibility of agricultural educators. These two teachers are preparing a master calendar.
(Courtesy, Education Images)

WHAT IS PROGRAM PLANNING?

Every agriculture teacher plans, even one who seems to make decisions on a day-to-day or crisis-to-crisis basis. The old saying "If you fail to plan, then you are planning to fail" is most likely true. Successful agriculture teachers chart the course of their agricultural education programs, continuously monitor progress, and make course corrections as needed.

Program planning is, first and foremost, a process. The result is a written plan containing the agricultural education program vision and mission, short- and long-term goals and plans, and strategies for achieving these. Program planning is often said to be ". . . all of the activities needed to design and implement local-school agricultural education" (Lee, 2000, p. 13).

Developing the Program Plan

A good program plan is developed with considerable input from stakeholders. A stakeholder is an individual who has an interest or share in an enterprise or initiative. From one perspective, all taxpayers are stakeholders. In program development, individuals who represent a cross section of the community are desirable on the planning group. A good program plan also conforms to state standards and meets local expectations. This section covers important concepts in program planning.

Participants   Program planning includes participation of several groups and input from numerous sources. The agricultural education program operates within the local school and community; therefore, both must be represented in the program planning process. The local agricultural education advisory committee and citizen groups, as discussed in Chapter 7, are logical groups to participate. The local school board's policies, the state agricultural education guidelines, and the national agricultural education vision, mission, and goal statements should all be sources of input. Approval, input, and guidance from school administrators are essential in the program planning process.

The program planning process should involve people representing all groups affected by the local agricultural education program. Selecting these representatives will require the agriculture teacher to think broadly to ensure that a wide diversity of

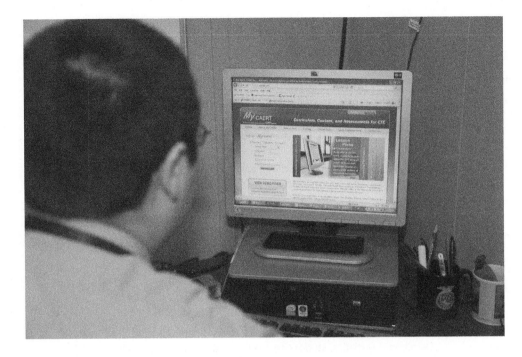

6–2. A teacher is using Web-based planning materials.
(Courtesy, Education Images)

agricultural education clients is included. Current students, parents of current students, and graduates of the local program should all be involved in the process. Representatives from all aspects of agricultural production and business in the community should be present. For example, full-time farms, part-time farms, and specialty farms will probably have different needs and viewpoints. The agriculture teacher must include all areas of agriculture, such as horticulture, landscaping, natural resources, small animal care, and others applicable to the local community.

Representation of the diversity within groups is important. Gender and ethnicity probably come to mind first, but there are also other areas of diversity. A wide range of ages should be included. If possible, both people who have been in the community for a short time and those whose families have been in the community for generations should be included. If the community includes both rural and developed areas, representatives should be selected from both. If the community is entirely rural or entirely suburban/urban, the agriculture teacher should consider bringing in someone from outside the community to represent the missing viewpoint.

Components   There are fourteen major components to a program plan. These are described in Table 6–1. Not all fourteen components will be applicable to every local agricultural education program. The components are to serve as a guide, and others may be added as needed. However, some components are vital for every agricultural education program.

The first component of a written program plan is a program, school, and community description. Narrative should be used for the description and observations. Demographics can be provided in tabular form. The purpose of this section is to provide the setting in which the agricultural education program operates.

Program planning components 2, 3, and 4 develop and support the program vision. Philosophies are practiced every day but take on greater meaning and awareness when articulated and written down. A vision that is debated and agreed upon brings groups together, provides focus and clarity, and stretches people to achieve a desired future. Advisory committees, partnerships, and volunteers are vital in connecting the program to the community.

Table 6–1
COMPONENTS OF A LOCAL AGRICULTURAL EDUCATION PROGRAM PLAN

Program Planning Component	Description
1. Program, school, and community description	Narrative or tables providing an overview of the setting
2. Personnel	Number of teachers, contract length(s), qualifications, professional growth plans
3. Program philosophy, vision, mission, objectives	Narrative and bullets on the direction of the program
4. Program management and planning	Administrative structure, advisory committee structure and tasks, community surveys and results, other community inputs, partnerships
5. Instructional planning and organization	Course of study descriptions, course frameworks, course outlines, teaching calendars
6. Student accounting and reports	Grading system, follow-up studies of graduates, other local and state reports
7. SAE coordination activities	Visitation/supervision plan, student record keeping, partnerships
8. FFA	Demographics, program of activities
9. Adult education program	Adult classes, Young Farmers chapter, partnerships, volunteer training
10. Recruitment and retention	Enrollment statistics, student-to-teacher ratio, marketing strategies
11. Safety training and practices	Safety instruction, inspection of facilities, safety practices
12. Summer schedule	SAE visitation/supervision, FFA activities, instructional activities
13. Physical needs and departmental budget	List, documentation, and timeline of physical needs in instructional materials, supplies and equipment, facilities, other; budget documentation
14. Long-range plan	5-year or longer goals

Program planning components 5, 6, 7, and 8 provide planning to the three components of the total agricultural education program: classroom and laboratory instruction, supervised agricultural experience (SAE), and FFA. Approximately 7 percent of agricultural education teachers have some adult or young farmer responsibility (Kantrovich, 2010). Even an agricultural education program that does not currently have an adult program may want to include program planning component 9 in its written plan.

Program planning components 10, 11, 12, and 13 provide support for the other components. School administrators are more accepting of changes and more willing to fund expenditures when the need for these has been documented through a program planning process. Recruitment and retention efforts are important, as agricultural education is an elective in most schools. A marketing plan should be followed to increase communications within the school and with the community.

Component 14, the development of a long-range plan consisting of goals that will take five or more years to accomplish, keeps the program focused on the future. These goals can also provide a starting point for the next program planning process.

Logistics   A written program plan is designed to guide all decisions regarding the agricultural education program, so it should be utilized on a continual basis. Annually, the agriculture teacher(s), in conjunction with the local advisory committee, should revisit the plan to check for progress, make needed adjustments, and set short-term goals and objectives for the next school year. Every three to five years, the entire plan should be reviewed and revised. Completing the program planning process is helpful before going through major program changes, such as adding new agriculture teachers, facilities, or curriculum areas.

The Process

The first step in developing a program plan is to assess the current situation in each of the fourteen components described in the previous section. What is the program currently doing in each of the components? How well is the program currently meeting standards and needs in each of the components? If a previous program plan was in use, what are the accomplishments, what goals and objectives have been met, and what components need improvement? The National Quality Program Standards Online Assessment (National Council for Agricultural Education, 2009b) is a useful tool for assessing an agricultural education program.

Next, the agriculture teacher convenes meetings with the groups described in the "Participants" section of this chapter. Each meeting must be organized, have a specific purpose, and be conducted in a timely fashion. Most people are willing to give of their time to worthy causes but do not tolerate what they perceive as wasting their time. It is recommended that each meeting cover three or fewer topics and last one to two hours. This ensures that appropriate time is devoted to each topic. If possible, the meeting order of business and background materials should be sent to each participant before the meeting. Providing refreshments at all meetings helps with attendance and is a way of showing thanks to the participants.

The first meeting should be devoted to brainstorming and visioning. What changes have occurred and are occurring in agriculture and education? What knowledge, skills, and dispositions will agricultural education graduates need for agricultural occupations in the future? What should this agricultural education program look like and do in the future? What resources does this agricultural education program need to do the best job of preparing students? Subsequent meetings should be held to develop the vision and mission statement, goals and objectives, and implementation strategies.

Once the plan is developed and written, it should be implemented. The written plan now provides a guide for making programmatic decisions. The final step in the program planning process is to conduct periodic program evaluations. These evaluations feed the next cycle of revising the program plan.

COMPONENTS OF A TOTAL AGRICULTURAL EDUCATION PROGRAM

Agricultural education involves much more than content learned without context. Agricultural education has three main **program components**: classroom and laboratory instruction, SAE, and FFA. Each of these components is critical for students

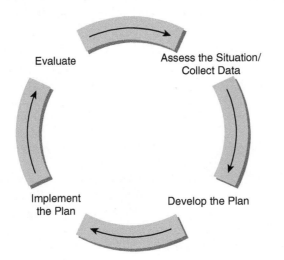

6–3. The program planning process is a cycle.

Evaluate

Assess the Situation/
Collect Data

Implement
the Plan

Develop the Plan

6–4. A model of the agricultural education program can be pictured as three interlocking circles.

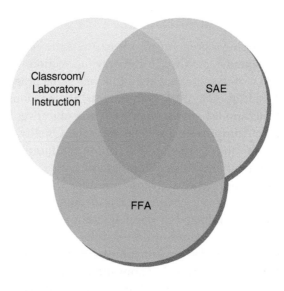

to receive full educational benefit. In addition, agricultural education programs may conduct agricultural literacy efforts in elementary grades, exploratory courses in middle school, and adult education.

Classroom and Laboratory Instruction

Classroom and laboratory instruction is the foundation for everything else that occurs in the agricultural education program. Instruction can also be called contextual learning. In traditional instruction, content is taught as bits of information, such as $2 + 2 = 4$.

Contextual learning is putting the instruction within a perspective to which it is easy for students to relate. Effective agricultural education instruction teaches content using the context of the plants, animals, and natural resources systems. Chapter 8 describes the process used for determining what content should be taught. Chapter 13 describes the process used for determining how the content should be taught.

Public schools are increasingly held accountable for their students' meeting state and national standards typically in academic subjects, such as English, math,

6–5. This teacher is demonstrating a scientific principle using the context of agriculture.

(Courtesy, Education Images)

and science. Agricultural education instruction can include English, math, and science concepts within the context of agriculture. For example, students learn math concepts of fractions, decimals, and calibrations within the context of agricultural power and technology. Also, within agricultural power and technology, they learn physics concepts, such as the characteristics of fluids under pressure.

Supervised Agricultural Experience

Supervised agricultural experience (SAE), or work-based learning, is that part of agricultural education that allows students to practice in a workplace what they have learned in the classroom or laboratory. Ideally, all agricultural education students would have SAE that directly relates to what is being learned in the classroom and that prepares the students for occupations in which they are interested. Chapter 22 describes the importance and types of SAE.

Work-based learning is a component of agricultural education that sets it apart from most other subjects. Students are able to explore areas of interest and then develop skills to a much greater depth than is possible within the regular school classroom. Tasks performed and problems encountered in SAE can be used in the classroom to provide real-world examples of concepts being learned. Exploratory SAE can also be used to develop agricultural literacy.

FFA

The career and technical student organization in agricultural education is FFA. FFA provides connecting activities. A *connecting activity* is one that establishes relationships between school and life. FFA is the student development component of agricultural education. It connects classroom learning to life in the areas of leadership development, personal growth, and career success. Chapter 23 describes the structure and activities of FFA.

FFA is an *intracurricular* component of agricultural education. This means that it is an integral part of the program, not an extracurricular club. Class period instructional time may sometimes be used for FFA instruction. At least on the local level, all agricultural education students must be involved in the connecting activities that FFA provides.

6–6. Supervised agricultural experience provides students with work-based learning.
(Courtesy, Education Images)

6–7. For more than eighty years, FFA has been an integral part of agricultural education. These members are preparing for an official FFA event.

(Courtesy, Education Images)

Other Components of a Total Agricultural Education Program

Agricultural literacy is knowledge and understanding of the plants, animals, and natural resources systems, from production through consumption. It encompasses the idea that an agriculturally literate person can evaluate and communicate basic information about agriculture (Frick, Kahler, and Miller, 1991). Most agricultural education literacy efforts are focused at the elementary school level. These activities include Partners in Active Learning Support (PALS), a peer mentoring program; Food for America, an agricultural literacy educational program; and other limited-term efforts. Chapter 15 describes school and general-population agricultural literacy in greater detail.

Middle school agricultural education is a part of approximately 20 percent of U.S. agricultural education programs. In most instances, this instruction is on an exploratory or introductory level. Depending on the state, agricultural education students in grades 7 and 8 can be a part of FFA. Chapter 16 further describes middle school agricultural education.

Adult agricultural education is a part of approximately 10 percent of U.S. agricultural education programs. This instruction may be as adult classes or Young Farmer organization programs. A few states still have full-time adult agricultural education teachers. Chapter 18 further describes adult agricultural education.

Hughes and Barrick (1993) published a revised model for agricultural education in secondary schools. It placed the agricultural education program within the context of the local school and community. Additionally, it restructured the three-circle model into classroom and laboratory instruction, application, employment and/or additional education, and career. This clarified that all aspects of FFA and SAE are related to classroom/laboratory instruction and that a career is the desired student outcome.

PROCEDURES IN PROGRAM DEVELOPMENT

Input from a number of sources is used in local program development. This includes program success materials as well as information collected about the local community.

Updated Model of Agricultural Education

(Hughes, M., & Barrick, R.K. (1993). A model for agricultural education in public schools. *Journal of Agricultural Education, 34*(3), 60.)

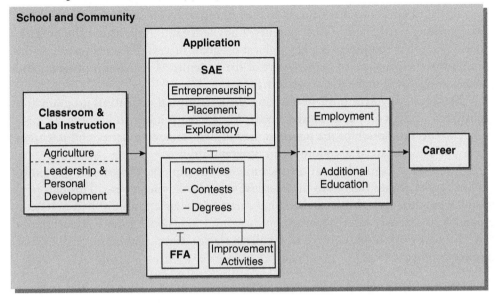

6–8. The Hughes and Barrick model for agricultural education in secondary schools updated the three-circle model.

(*Source:* Hughes & Barrick, 1993, p. 60. Used with permission.)

State Standards and Curriculum Guides

Many states have established standards and provided curriculum guides for use in developing local programs. The standards and guides are known by different names such as frameworks, blueprints, and performance standards. Some states have established uniform achievement testing programs based on the content and expectations in the state-established materials.

In states where such materials have been developed, teachers in local programs are expected (and in some cases required) to use them in planning and to meet specified minimum standards as may be assessed on statewide achievement tests. Failure of a teacher to address these standards often results in student test scores below the desired level.

The state-provided guides are often on a course-by-course basis. For example, such guides have been established for introductory agriscience, horticulture, landscaping, forestry, biotechnology, and aquaculture, among other subjects. Some states may provide incentive funding to local schools that use the guides and have students who achieve certain levels of performance.

In 2009, the National Council for Agricultural Education published content standards for the Agriculture, Food, and Natural Resources (AFNR) career cluster. These are organized into eight pathways: agribusiness systems, animal systems, biotechnology systems, environmental service systems, food products and processing systems, natural resources systems, plant systems, and power structure and technical systems. Many states have adopted or modified these pathways. In addition, curricular materials are being mapped to these standards.

Local Program Success

Local Program Success (LPS) is a national initiative designed to enhance the quality and success of local agricultural education programs (National FFA Organization, 2012). A national task force verified that quality and success on the local level depends on seven keys: classroom/laboratory instruction, SAE, FFA, partnerships, marketing, professional growth, and program planning. Program planning is the focus of this chapter. Partnerships, marketing, and professional growth are discussed next. The National FFA Organization maintains Local Program Success resources at its website.

The agriculture teacher cannot do everything or be an expert in every subject area. Forming partnerships with parents, knowledgeable individuals, businesses, groups, and other volunteers is a key to local program success. Partnerships allow the community the privilege of being a part of the agricultural education program and provide resources to make the program stronger.

Marketing involves both publicizing the agricultural education program and identifying student and community needs. Marketing is a key to keeping the program current. Promotion of the program, recruitment of new students, and retention of current students are vital to long-term program survival.

The fields of agriculture and education change rapidly. Discoveries and innovations render things that were the best last year obsolete next year. Through professional growth, agriculture teachers stay current in their knowledge and practices, expand their abilities, and renew their enthusiasm. Professional growth opportunities exist through workshops, conferences, college classes, travel, seminars, reading, and numerous other sources.

Community Needs Assessments

A hallmark of agricultural education is that the content and activities of the program reflect the agricultural industry of the local community. This provides accountability to taxpayers and curriculum relevance to careers available. Periodically, the agriculture teacher should reevaluate the agricultural education program in light of community needs and characteristics.

A *needs assessment* is a program planning process used to collect and analyze community data. Sources of data include reports available from state and national statistical services. The U.S. Census of Agriculture provides detailed information on a county-by-county basis for all agricultural products. State and national labor departments (the official names of which vary) provide statistics on employment and employers. The local chamber of commerce, the Extension Service office, the visitors' bureau, and even the telephone yellow pages are all sources of information.

Most likely, however, complete information will not be available from these sources. The agriculture teacher will have to conduct a community survey to obtain a total local picture that is accurate. A *community survey* is a process to collect comprehensive and specific local information from various areas of the agricultural industry. The survey may include written questionnaires, telephone or face-to-face interviews, and review of documents. Information collected includes types and numbers of agricultural industries and employers, types of occupations, descriptions of competencies performed, and current number of employees and expected future demand.

The agriculture teacher, along with the advisory committee, analyzes the collected data and then evaluates the local agricultural education program. What knowledge, skills, and dispositions do students need to be employable in the local agricultural industry? What modifications need to be made to the agricultural education curriculum? What opportunities exist for student supervised experience? What FFA personal development activities, Career Development Events, and leadership events match local community needs and the agricultural education curriculum?

Technology-Based Program Development

Technological tools are increasingly used to assist in the program development process. *Technology-based development* is the use of curriculum databases, related matrixes, and technology-based access to assist in planning the agricultural education

Community Needs Survey

Name of business _____

Person completing form _____

Title _____

Business address _____

_____

Telephone _____ Fax _____

E-mail _____ Web site _____

Agricultural products or business description _____

_____

_____

Job Title	Competencies	Education Required	No. of Employees	No. Needed Next 5 Yrs.

6–9. A sample form for collecting community information.

program. This allows standardization within a state as well as local flexibility. Technology-based development provides the local agriculture teacher the power to customize curriculum to meet local community needs. What only a few years ago would have taken days or even weeks to accomplish can now be done in one day or less.

With curriculum databases, whether accessed through the Internet, CD-ROM, or some other means, the agriculture teacher can search for and organize curriculum based on content, objectives, academic standards, or other variables. The teacher can personalize and customize easily yet still present a professional-quality product. The curriculum is adaptable to the variety of audiovisual display technologies and methods of instruction.

Matrixes provide documentation that the curriculum meets certain standards. They can be organized to present information based on agricultural education standards, academic standards, or other variables. In conjunction with other evaluation data, matrixes help agriculture teachers be accountable to outside audiences.

Technology-based development allows the agriculture teacher to try "what if" scenarios easily. For example, suppose the local agricultural education program is considering adding food science to its course offerings. With technology-based development, the agriculture teacher can plan the course of study electronically and quickly add or delete subject areas. The teacher could then prepare documentation demonstrating what academic standards the food science course would cover in addition to the agricultural education standards. Then, the teacher could prepare a professional-looking packet to present to the advisory committee, school administrators, the school board, or others.

PROGRAMMATIC EVALUATION PROCEDURES

A *program evaluation* is an assessment of progress in achieving stated goals and outcomes of the program plan. It is a process that should be done annually to facilitate program planning for the next school year. The results of the program evaluation should be shared with the local advisory committee, school administrators, and state agricultural education leadership as appropriate. News releases and other publicity should be used to inform the general public of program growth and successes.

Program evaluation follows many of the same steps as development of the written program plan. Data need to be collected and input sought from key groups. Results need to be compared against the goals, objectives, and outcomes set forth in the program plan. Recommendations for changes and improvement must be developed and should include action steps and implementation strategies.

Program evaluation is a vital part of the program planning process, not an afterthought. Although formal complete program evaluation can realistically be done only once a year, components of the agricultural education program should be evaluated on a continual, less formal basis. An activity should be assessed during and immediately after, with notes taken for commendations, suggestions, and things to do differently the next time. The agriculture teacher should be consistently asking if the activity is effective in its outcomes and if it is important to the agricultural education program vision, mission, and goals.

What Should Be Evaluated?

Each of the agricultural education program components (instruction, SAE, and FFA) should be evaluated. Data relating to the fourteen program plan components should be included. Figures 6–9, 6–10, and 6–11 are sample forms indicating what data should be collected.

Making Interpretations

Once information has been collected, the data must be tabulated and analyzed. A summary of the findings is then prepared. Some findings will be numerical and can be summarized with frequencies, means, ranges, and other descriptive statistics. Other findings will be qualitative and can be summarized through patterns, themes, and quotes. Afterward, synthesis of the findings is needed to draw conclusions and prepare recommendations for improving the local agricultural education program. As the data are synthesized, these questions may be helpful in making interpretations. What trends emerged? What priorities were identified? What were lesser priorities that could be eliminated or scaled back to allow resources to be better utilized? What is a realistic timeline for implementing changes? What resources will be needed? Who is responsible for implementing the recommendations and what accountability measures will be put in place? These are important responsibilities. Teachers may get the assistance of school district evaluation specialists as well as advisory committee members in this process. Administrators and school board members may wish to be informed. Local media can be used to inform citizens of the worthwhile findings.

Instruction Evaluation Survey

School year _____ – _____

Course name _____

What are the best characteristics of this course? _____

_____

_____

_____

_____

What are suggestions for improving this course? _____

_____

_____

_____

_____

How has this course prepared students for careers or further education in agriculture? _____

_____

_____

_____

_____

For the next questions, circle the number that best matches your opinion on the statement.
1 = Strongly Disagree, 2 = Disagree, 3 = Agree, 4 = Strongly Agree

1 2 3 4   This course has challenged me to think.

1 2 3 4   This course is interesting.

1 2 3 4   I would take another agriculture course.

1 2 3 4   My agriculture teacher is one of the best teachers in the school.

1 2 3 4   My agriculture teacher has a positive attitude about teaching and learning.

1 2 3 4   My agriculture teacher presents a professional image.

1 2 3 4   My agriculture teacher used a variety of teaching methods (lectures, demonstrations, experiments, field trips, resource persons, project-based learning, etc.)

6–10. A sample form for evaluating instruction. (This form should be completed by each student.)

Supervised Experience Evaluation Survey

School year _____–_____

Number of students in agricultural education _____

Number of students with an SAE _____

Number of students with an entrepreneurship SAE _____

Number of students with a placement SAE _____

Number of students with an experimental SAE _____

Number of students with an exploratory SAE _____

Number of paid hours worked by students in SAE _____

Number of unpaid hours worked by students in SAE _____

Total gross dollars earned by students in SAE _____

Number of proficiency awards given on local level _____

Number of proficiency awards submitted to state _____

Number of State FFA Degrees received last year _____

Number of American FFA Degrees received last year _____

Percent of graduates entering the workforce in area of their SAE _____

Percent of graduates pursuing further education in area of their SAE _____

Ask students these two questions. Summarize results and attach.

What is the best part of the SAE program at _____ Agriculture Department?

What needs to be improved about the SAE program at _____ Agriculture Department?

Ask supervisors/owners at SAE sites these two questions. Summarize results and attach.

What is the best part of the SAE program at _____ Agriculture Department?

What needs to be improved about the SAE program at _____ Agriculture Department?

6–11. A sample form for evaluating supervised experience. (This form should be completed by the agriculture teacher and should include a summary of responses from students and supervisors/owners at SAE sites.)

FFA Evaluation Survey

School year _____–_____

Number of students in agricultural education _____

Number of FFA members (Include out-of-school.) _____

Number of students whose highest degree is Greenhand _____

Number of students with a placement SAE _____

Number of students whose highest degree is Chapter FFA _____

Number of students whose highest degree is State FFA _____

Number of students whose highest degree is American FFA _____

Number of proficiency awards given on local level _____

Number of proficiency awards submitted to state _____

Number of State FFA Degrees received last year _____

Number of American FFA Degrees received last year _____

Percent of eligible graduates maintaining FFA membership _____

Percent of FFA members participating in at least one local FFA activity
(other than meetings) _____

Percent of FFA members participating in at least one FFA activity above
the local level _____

Percent of FFA members participating in a CDE _____

Percent of FFA members participating in a leadership development event
(workshop, camp, etc.) _____

Ask students these two questions. Summarize results and attach.

What is the best part of the FFA program at _____ Agriculture Department?

What needs to be improved about the FFA program at _____ Agriculture Department?

NOTE: The National Chapter Award Application can also be used as an evaluation and
documentation tool. Both Form I and Form II should be completed.

NOTE: Additional information may be provided as appropriate. Some states have FFA reporting forms that can be used.

6–12. A sample form for evaluating FFA. (This form should be completed by the agriculture teacher and should include a summary of responses from students.)

REVIEWING SUMMARY

Program planning is essential to the success of the local agricultural education program. Although program planning takes time up front, it will greatly save the agriculture teacher time as it is implemented. Program planning is not done in a vacuum by the agriculture teacher alone but instead involves input and participation from several school and community groups.

Every local agricultural education program consists of three components. Classroom and laboratory instruction is the foundation of the program. Without strong instruction, the program cannot succeed. Supervised agricultural experience is the work-based learning part of an agricultural education program. Students take what is learned in the classroom and apply it in the workplace. They also bring their workplace experiences back into the classroom. FFA is the intracurricular student organization that provides connecting activities between instruction and the workplace. FFA is also the primary student development component of agricultural education.

Local Program Success (LPS) is designed to assist agriculture teachers in having successful total agricultural education programs that meet local community needs. LPS provides tools and strategies that can save agriculture teachers time and effort in the program planning process.

Program evaluation is vital to the success of a local agricultural education program. A systematic evaluation should be conducted annually. The results should be shared with the school, community, and general public. To be effective, program evaluation must have an impact on the program plan.

QUESTIONS FOR REVIEW AND DISCUSSION

1. Why is program planning important to the local agriculture teacher?
2. Who should be involved in developing the local program plan?
3. What are the fourteen major components of a program plan?
4. What are the steps in program planning?
5. Why is it important that all agricultural education students be involved on the local level in the three main program components?
6. What are other components that may be included in the local agricultural education program?
7. What are the seven keys identified in LPS that are critical to achieving local program quality and success?
8. What is included in a community survey?
9. How does program evaluation lead to local program success?
10. What should be included in a local program evaluation?

ACTIVITIES

1. Investigate the program planning process used when starting a new agricultural education program at a high school. Prepare a report on your findings.
2. Explore the Local Program Success resources on the National FFA Organization Web site. Write a report about the impact of Local Program Success (LPS) on agricultural education in your state.
3. Using the *Journal of Agricultural Education*, *The Agricultural Education Magazine*, and the proceedings of the National Agricultural Education Research Conference, write a report on research about program planning within agricultural education. Current and past issues of these journals can be found in a university library. Selected past issues can be found online at American Association for Agricultural Education or National Association of Agricultural Educators websites.

CHAPTER BIBLIOGRAPHY

Association for Career and Technical Education (ACTE). (2012). *ACTE Web Site.* Accessed on November 5, 2012.

Binkley, H. R., and R. W. Tulloch. (1981). *Teaching Vocational Agriculture/Agribusiness.* Danville, IL: The Interstate Printers & Publishers, Inc.

Caffarella, R. S. (2002). *Planning Programs for Adult Learners* (2nd ed.). San Francisco: Jossey-Bass.

Frick, M. J., A. A. Kahler, and W. W. Miller. (1991). A definition and the concepts of agricultural literacy. *Journal of Agricultural Education, 32*(2), 49–57.

Frick, M., and S. Stump. (1991). *Handbook for Program Planning in Indiana Agricultural Science and Business Programs.* Unpublished manuscript, Purdue University.

Hughes, M., and R. K. Barrick, (1993). A model for agricultural education in public schools. *Journal of Agricultural Education, 34*(3), 59–67.

Indiana Department of Education. (1999). *Teacher/ Local Team Self-study of Standards and Quality Indicators for Agriscience and Business Program Improvement.* Indianapolis: Author.

Kahler, A. A., B. Morgan, G. E. Holmes, and C. E. Bundy. (1985). *Methods in Adult Education* (4th ed.). Danville, IL: The Interstate Printers & Publishers, Inc.

Kantrovich, A. J. (2010, October). *The 36th Volume of a National Study of the Supply and Demand for Teachers of Agricultural Education 2006–2009* (online). Accessed on November 5, 2012.

Lee, J. S. (2000). *Program Planning Guide for AgriScience and Technology Education* (2nd ed.). Upper Saddle River, NJ: Pearson Prentice Hall Interstate.

Moore, G. E. (n.d.). *Licensure in Education for Agricultural Professionals.* Accessed on November 30, 2011.

Moore, G. E. (Ed.). (1998). Theme: A primer for agricultural education (Entire Issue). *The Agricultural Education Magazine, 71*(3).

Moore, G. E. (Ed.). (2000). Theme: Reinventing agricultural education for the year 2020 (Entire Issue). *The Agricultural Education Magazine, 72*(4).

National Association of Agricultural Educators (NAAE). (2012). *NAAE Web Site.* Accessed on November 5, 2012.

National Council for Agricultural Education. (2009a). *National Agriculture, Food and Natural Resources (AFNR) Career Cluster Content Standards.* Alexandria, VA: Author.

National Council for Agricultural Education. (2009b). *National Quality Program Standards for Secondary (Grades 9–12) Agricultural Education.* Alexandria, VA: Author.

National FFA Organization. (2003). *Local Program Resource Guide: 2003–2004* [CD-ROM]. Indianapolis: Author.

National FFA Organization. (2011). *Official FFA Manual.* Indianapolis: Author.

National FFA Organization. (2012). *Local Program Success* (online). Accessed on November 5, 2012.

Newcomb, L. H., J. D. McCracken, J. R. Warmbrod, and M. S. Whittington. (2004). *Methods of Teaching Agriculture* (3rd ed.). Upper Saddle River, NJ: Prentice Hall.

Phipps, L. J., and E. W. Osborne. (1988). *Handbook on Agricultural Education in Public Schools* (5th ed.). Danville, IL: The Interstate Printers & Publishers, Inc.

True, A. C. (1929). *A History of Agricultural Education in the United States, 1785–1925.* U.S. Department of Agriculture, Miscellaneous Publication No. 36. Washington, DC: GPO.

7

Advisory and Citizen Groups

The first order of business at the Spruce Pine Agricultural Education Advisory Committee meeting was to review the courses offered by the agricultural education program. Bob Wells was the new teacher at Spruce Pine and had just completed his first year. He had followed a long-term career teacher, who had taught at Spruce Pine School for thirty-five years. For much of the last ten years, the curriculum had centered on production agriculture and agricultural mechanics.

Committee chairperson John Lee opened the meeting: "Thank you all for coming tonight. Let's get started. Bob Wells has some ideas for revising the curriculum here at Spruce Pine."

Bob moved to the front of the room where a list of courses was projected on a video screen. "These are the courses we have offered at Spruce Pine for the last ten to fifteen years. In that time period, the agricultural industry has changed considerably. There are more livestock operations in the area. With the opening of Alpha Greenhouses here in town, there has been huge growth in landscape design and horticulture."

Bob then projected a series of slides that portrayed trends in the local labor market, and careers in agriculture. He also provided the committee with the results of a career survey of graduates of the program, and the results of a student survey on career choice in agriculture.

Bob continued, "Based upon these data, I propose that we discontinue Agriculture I and II, and offer Horticulture and Animal Science beginning next school year."

Committee member Joan Jones spoke up, "This sounds like a good idea, and you've certainly done a lot of homework on the issue, but do we have the resources to do this? You don't have a greenhouse for teaching horticulture."

Bob replied, "You're right, Joan, we don't have a greenhouse just yet, but I completed an inventory of the instructional equipment we have, and we can offer both new courses with what we have. Our principal has agreed to let us build a greenhouse next to the agricultural science classroom, but because of funding cuts, we have to supply the greenhouse ourselves."

terms

advisory committee
craft committee
FFA Alumni
Herbert M. Hamlin

layperson
mission statement
program advisory committee

7–1. *Advisory committee members are meeting in an agricultural education classroom.*
(Courtesy, Education Images)

Committee member Terry Mooring countered, "I think I have the answer. My neighbor Frank Barnes has the frames for two greenhouses on his farm. I'm sure he would be willing to donate them."

John Lee spoke, "Yep, and I think we can get Alpha Greenhouses to supply the covers for the greenhouses, especially if we start sending them students to work for them as part of the SAE program."

A couple of others in the group nodded in agreement. Then, Joan spoke again, saying, "Getting the new horticulture and animal sciences courses in place here at Spruce Pine is exactly in line with where agriculture is going." Discussion continued for several minutes. More of the members began to agree with the new course plan. By the time the meeting ended, the committee had decided to request the curriculum revision and get started on the new agricultural science program at Spruce Pine High School.

CITIZEN PARTICIPATION IN SCHOOLS

Citizen participation is an essential component in any school. Schools are both a reflection and product of their community. Citizens, through their government, arrange for a system of schooling that provides for the development of effective citizens who contribute to society. Citizens expect schools to help youth develop social and cognitive skills that lead them to responsible citizenship and a successful vocation. One of the inherent characteristics of a highly evolved society is that its members are able to communicate and reason together in such a way as to increase understanding and learn from one another (Darling-Hammond & Bransford, 2005). In this sense, the school is an extension of the community. As Dewey explained, "What the best and wisest parent wants for his own child, that must the community want for all of its children" (Dewey, 2001). As such, the community has the responsibility to be involved in the development and improvement of educational programs.

Citizen Participation in Agricultural Education

One of the most important characteristics of a local agricultural education program is the interaction between the program and the community served by the school. Agriculture teachers depend upon the agriculture and business segments of the community to provide support in many ways, such as supervised experience for students. If the community supports a local agricultural education program, it will have an improved supply of workers entering the workforce upon graduation from high school and postsecondary school.

Citizens support agricultural education through tax dollars, through volunteerism, and in other ways. Their purpose is to provide high-quality educational experiences for their children. This transactional relationship requires that agricultural education teachers involve citizens in educational programs and encourage local citizens to support their agricultural education program proactively.

Among the strongest proponents of using citizen advisory groups in agricultural education was **Herbert M. Hamlin**. While a professor of agricultural education at the University of Illinois, and later at North Carolina State University, Hamlin wrote extensively about input into educational policy formulation by lay groups. He felt that decisions about local agricultural education programs should only be made with citizen input. One of Hamlin's best-known books is *Public School Education in Agriculture: A Guide to Policy and Policy-Making* (1962). He reminds us that, "In characteristic American fashion, public agricultural education originated in communities. State initiative followed. National action came later"(Hamlin, 1962, p. 5). The agricultural education program should mirror the wishes of the community in which it resides.

The Importance of Citizen Participation in Learning

Students enter the doorway of the school with their ideas about how the world works. Many of these ideas are the result of their personal experiences and the experiences of their family and friends. In some cases, students walk into the agricultural education classroom with misconceptions based upon personal or family experiences. These misconceptions may be the result of insufficient prior knowledge or inaccurate prior knowledge. Misconceptions distort new knowledge, and make it difficult for the instructor to teach new and emerging concepts in agriculture. It is the teacher's responsibility to clear up these misconceptions, but the community can be an important source of assistance in this endeavor. Advisory groups can provide the support and validation a teacher needs when making innovative changes to the curriculum.

Meaning and Kinds of Advisory Groups

An **advisory committee** is a group of laypersons who provide advice on educational programs. A **layperson** is an individual who is not an agricultural educator or other professional or certificated person employed by the school. An advisory committee may also be identified as an advisory group, a citizen advisory group, an advisory committee, a lay advisory committee, or a citizens' committee. Regardless of the exact wording of the committee's name, the intent is the same—to provide advice to decision makers on the quality and direction of the local agricultural education program.

An advisory group does not have the power to make decisions, but its input can be used by a legally empowered board in making decisions or in a less authoritative way as a source of advice for an agriculture teacher or administrator. Advisory groups should be representative of the community served by the school. It is best if an advisory group is formally organized and has a lay chair, who directs the affairs of the group.

An agricultural education program may have several advisory groups. A **program advisory committee** may serve to provide advice for the overall program. It may have several subcommittees that serve various needs or areas. These

7–2. A committe of advisory group members is at work.
(Courtesy, Education Images)

subcommittees are in the specialty areas within agricultural education. They are sometimes known as craft committees. A **craft committee** is an advisory group that is specific to a particular area, such as horticulture or forestry. The focus of a craft committee is on a technical area being served by the instruction. A craft committee may be known by the name of the area in which it provides advice—for example, the biotechnology advisory committee.

Barriers to Citizen Participation in Agricultural Education

Getting people involved sometimes poses a challenge. A number of barriers inhibit the necessary relationship between agricultural education programs and schools. Citizen participation in schools may be discouraged somewhat by the tightening of security requirements for school visitors. A school usually requires campus visitors to report to the school office or face possible trespassing charges. Visitors are required to wear special visitor badges and in some cases be escorted to the agricultural education facilities. In light of efforts to limit access to school campuses, are citizens welcome on school campuses? Regardless of the perceived barriers, advisory committees provide a valuable and essential service for agricultural education programs. The best agricultural education programs have useful and viable advisory groups to help them carry out effective education. Although school security has become more rigid in some cases, that does not preclude the involvement of parents and others who have a genuine interest in the success of schools. The push to increase educational accountability in schools has increased the need for community involvement instead of diminishing it.

This is the age of accountability in schools. Many agricultural education instructors are finding themselves teaching a standardized curriculum that was produced for them by a state board of education. Some teachers do not see a need for advisory committees in agricultural education once the standard curriculum is adopted. Where is the need for advisory committees if the subject matter being taught is mandated by the state curriculum? The curriculum may be the result of a state mandate, but the teacher has flexibility in how that curriculum is taught. An advisory committee can help the teacher and school administration fit the curriculum to the unique needs of the community.

The number of high school students continuing on to postsecondary education has increased significantly in the last decade, and many agricultural education students will continue their education beyond high school at some form of post-secondary institution (National Center for Education Statistics, 2011). These students are graduating from high school and leaving their home communities for college, and most never return to the local workforce. If high schools are becoming more and more like prep schools for postsecondary institutions, then is there a need for vocational education? Is there a need for advisory committees in this situation? As more and more students seek postsecondary education after high school graduation, the agricultural education program must consider broadening the scope of its mission. An advisory committee can provide valuable support and assistance to the school so that the best decisions are made and agricultural education continues to serve the changing needs of youth.

ROLE OF ADVISORY COMMITTEES

What exactly does an advisory group do? The two basic functions of an advisory group are to assist professional agricultural educators in program development and to help in evaluating the local agricultural education program. These are carried out based on realistic, long-range goals (Martin, 1994). Too often, an advisory committee is used to carry out a predetermined agenda for the agricultural education program. An advisory committee should work with the faculty and the school administration to set the agenda for agricultural education in the school.

Program Development

The advisory committee for an agricultural education program has a major role in helping develop the program so that it meets most effectively the needs and interests of the community served. Of course, all the input from the committee must be adapted to state-mandated standards and curriculum blueprints. Further, the local school board has the authority over the agricultural education program through the appropriate administrative channels of the school district.

7–3. Advisory groups need orders of business to assure that important areas are covered in meetings.
(Courtesy, Education Images)

Examples of areas in program development where an advisory committee (including subcommittees) can be useful are given in Figure 7–4.

Career Development
Determining needs for pre employment instruction.
Guiding the career planning process.
Providing resources for career orientation.
Providing professional development opportunities for students and faculty.
Providing placement services for graduates and students with SE programs.
Conducting job analyses in specialized areas to determine what skills are needed by graduates in these areas.
Providing instructors with information on industry innovation and trends.
Keeping instructors informed of current events in agricultural education.
Providing job shadowing and other career observation experiences.

Curriculum Development
Determining the courses to be offered and the course sequence.
Finding resources persons and others sources of support in the community.
Assisting with projects or activities of the agricultural education program.
Locating sources of teaching materials and equipment.
Making recommendations on physical facilities, equipment, and supplies needed for the curriculum.

Overall Program Development
Identifying and addressing the needs of disadvantaged students.
Conducting cost/benefit studies.
Conducting future planning of the agricultural education program.
Recruiting students.
Providing learning resources for instructional purposes in classrooms.
Developing partnerships with other agencies working with agricultural education.
Communicating with educational and political leaders in the community.
Identifying the equipment, tools, and supplies needed.
Identifying facility modifications needed to achieve curriculum goals.

Program Evaluation
Evaluating the effectiveness of CTE program(s).
Utilizing assessments to measure whether or not instructional competencies are being met.
Reviewing courses of study for content relevance and accuracy.
Reviewing textbooks and other instructional materials as to their suitability for classroom use.

7–4. An advisory committee can provide assistance for many areas of program development.

(*Source:* Adapted from *Advisory Groups: Advisory Councils and Committees* by the Iowa Department of Education, Bureau of Career and Technical Education Services, 2012).

Program Evaluation

The advisory committee plays an important role in program evaluation. Program evaluation is the process of assessing the extent to which a program is meeting the objectives established for it. It answers the question: Is the program effectively and efficiently doing what it should be doing? An agriculture teacher is usually evaluated by a school official solely on the basis of teaching skills. The school principal or assistant principal visits a class in progress and completes a formative or summative assessment of the instructor's ability to teach. Often, there is not any additional review of the program except in cases where an accrediting agency visits the school and performs a periodic evaluation. The advisory committee can evaluate the agricultural education program in a timely manner and provide specific recommendations for the school administration to consider.

Specific items that a well-established advisory committee can evaluate include the following:

- The validity of the curriculum to economic need within the community and adjacent communities
- The quality and availability of instructional materials
- Long-range plans for curriculum and instruction
- The quality and sophistication of facilities and equipment as measured against the current state of agricultural technology
- The overall perception of the agricultural education program in the community
- The effectiveness of supervised experience in meeting the vocational needs of youth

Once the advisory committee has completed an evaluation of the agricultural education program, it can recommend changes in the curriculum, instruction, and the physical plant that address deficiencies or magnify strengths in the program (Martin, 1994).

Other Functions

The advisory committee has a unique view of the present economic situation in a local community. Members of the committee know where local needs are, based upon their interaction with others in the community, and can recommend suitable locations for students to have supervised experience in agriculture. In addition to finding potential sites for supervised experience, advisory committee members can make recommendations to the agricultural education faculty on policy and procedures. Committee members can also identify problem areas in the supervised experience program and head off trouble before it escalates. The advisory group can also serve as an independent voice for agricultural education in promoting and defending supervised experience to school officials (Martin, 1994).

Becoming Proactively Involved in the Politics of Education   Public education in the United States was created in the political arena and goes about its operation subject to political pressures and decisions. The advisory committee can work to

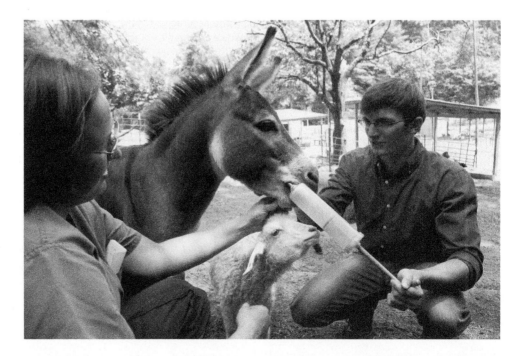

7–5. Placement supervised experience provides the opportunity to learn skills not otherwise readily available.
(Courtesy, Education Images)

establish healthy support for agricultural education among school and community leaders. By providing accurate information about the mission and goals of the local agricultural education program to school administrators, the school board, and civic leaders, the advisory committee encourages continued and increased funding of the program.

The advisory committee can help "tell the story" of agricultural education in the local community. The local community is more likely to support agricultural education when it understands its mission and its purpose. A high-quality advisory committee ensures that agricultural education is visible and respected in the school and the community.

ORGANIZING AND USING AN ADVISORY GROUP

Advisory committees can do good things for agricultural education, but rarely can they establish themselves and get to work without leadership and guidance from the agriculture teacher. Developing an advisory committee is a relatively easy process; however, it requires more than just calling a meeting and asking for volunteers. Setting up the effective advisory group requires careful thought and attention to detail.

Before beginning the process of establishing an advisory committee, the agriculture teacher should carefully analyze the community. Every community has its unique characteristics, and the advisory committee should be representative of these characteristics. Otherwise, the agriculture teacher runs the risk of not getting accurate advice and input on the future direction of the agricultural education program. A community survey provides valuable information to assist the teacher in selecting the right members for the advisory group as well as essential data for making good decisions. Figure 7–6 lists the components of a good community survey.

Six Steps to a Productive Advisory Group

Step 1: Develop a Mission and a Purpose   Advisory committees serve a multitude of roles and purposes. The agricultural education teacher should develop an initial mission and operating guidelines for the committee. One of the first questions often asked by school administrators and potential advisory committee members is, "What will the advisory committee do?" This question can be answered easily if the

Demographic Information About the Community	Economic Development Plans	Agricultural Industry Information	Local Education Trends and Information
▪ Ethnicity ▪ Educational level ▪ Home ownership ▪ Poverty levels ▪ Population trends	▪ Number and types of businesses ▪ New and emerging industries ▪ Unemployment rates ▪ Labor Force ▪ Manufacturing Layoffs ▪ Percent Working Age ▪ Population Percent Working in Professional and Business Services	▪ Types of agribusiness ▪ Farm ownership and type ▪ Natural resources ▪ Total Agricultural Receipts ▪ Average Farm Size ▪ Total Crop Receipts ▪ Total Livestock Receipts	▪ Graduation and dropout rates ▪ Enrollment trends ▪ School budgets ▪ Per Student Expenditures K–12 ▪ Percent With High School Diploma

7–6. Information to gather about the area served by a school when establishing an advisory committee.

appropriate groundwork has been laid. Although the advisory committee can amend its mission to meet the changing needs of the agricultural education program, the agriculture teacher should provide a basis for establishing the committee.

Step 2: Gain Support of School Administrators and the Board of Education   The second step in creating an advisory committee is to meet with the appropriate school administrators and inform them of the purpose of the advisory committee and the need for the group. School administrators can provide advice on the selection of committee members, assist in getting the advisory group officially recognized by the school system, and provide administrative and clerical assistance where needed. It is often a good idea to get approval of the board of education in establishing such a committee.

Step: 3: Choose the Members   After securing administrative support for an advisory group, the next step is to choose the members. One of the most common mistakes an agriculture teacher can make in creating this committee is choosing favorite parents or supporters to serve on it. That virtually ensures that the committee will not address the difficult issues that arise in the agricultural education program. The committee might be reluctant to disagree on the issues. On the other hand, a committee staffed with skeptical people who do not trust each other will be paralyzed by continual disagreement. A good advisory committee should include people who

- Are genuinely interested in the success of the agricultural education program
- Are from diverse backgrounds and have varied interests
- Have significant experience in agriculture
- Understand their purpose on the committee
- Are committed to participating actively on the committee
- Have the time and resources necessary to serve
- Understand their role on the advisory committee

Foster (2003) recommends avoiding the inclusion of too many parents on the committee. A committee heavily staffed with parents gives the impression that it is little more than a booster club for the agricultural education program. Include a representative from the local FFA Alumni chapter. Business and industry representatives and others who have a substantial knowledge of all segments of agriculture in the local community should be part of the committee. Include a school administrator as an ex officio member to forge an effective relationship between the committee and the school system.

It is very important that the advisory committee represent the diverse aspects of the community. An advisory committee that has only one or two segments of the community represented on it will be hindered by its narrow focus. Figure 7–7 shows the representation to include when selecting members.

Step 4: Choose or Elect a Chair and Other Officers   The chair plays a very important role in the success of the advisory committee. The chair should be someone with excellent leadership skills and the ability to build a consensus on issues. The chair serves as the liaison between the committee and the agriculture teacher and school administrators. Some responsibilities of the advisory group chair include developing agendas for meetings, following up on committee decisions to ensure that they are acted upon, and attending school meetings to represent the advisory committee. In some cases, the chair may be asked to serve on school-wide advisory committees.

Traditionally, the agriculture teacher has often served as the secretary to the advisory group. Since the agriculture teacher has access to clerical equipment and

• School administrator (ex officio)
• Agricultural education teacher (ex officio)
• FFA Alumni chapter representative
• FFA chapter president
• Agriculture industry representatives
• Parents

records and has a mechanism for contacting committee members and hosting meetings, the job of secretary seems like a natural fit. However, the advisory committee should select someone else as secretary, because agriculture teachers should focus their attention, time, and effort on delivering high-quality instruction. The secretary corresponds with members and processes the paperwork of the committee. The secretary also prepares drafts of the advisory committee's minutes and corresponds with school administrators on matters related to the advisory group.

Step 5: Organize the First Meeting   The first meeting of the advisory committee will set the tone and pace of the committee's work in the agricultural education program. The first meeting should firmly set the mission and purpose of the committee and establish members as owners in the committee's work. Figure 7–8 shows an example of the first meeting agenda of an advisory committee. The first meeting should involve school administrators, who help establish the importance and validity of the committee. A good way to involve the committee quickly in its duties would be to make plans for a program evaluation.

Step 6: Establish Operating Procedures for the Committee   For the committee to function successfully, some procedures need to be established governing its activities. Sometimes a committee develops a constitution to serve as a guide, but usually a committee does not need anything this elaborate. A simple list of guidelines may be all that is needed for the committee to carry out its mission. Some items that need to be included in the guidelines are as follows:

- Time, location, and general dates of meetings. A committee should probably not meet more than four times nor less than two times annually.
- Rotation schedule for members of the committee.
- Process for selecting new members to serve on the committee.
- Process for selecting the chair and other officers.
- General rules for handling business. Most committees adopt parliamentary procedure as the method of choice.

Members of the advisory committee should be selected to serve three-year terms. A system can be established so that only a few members rotate off the committee every year. In this manner, the institutional memory of the committee is preserved.

Program Evaluation by the Advisory Committee

It is essential that the advisory committee begin its work with an in-depth examination of the agricultural education program. Just as a physician begins every visit with a physical assessment, so should the committee check the "blood pressure" and the "temperature" of the agricultural education program. The advisory committee needs

7–8. Sample order of business for the first meeting of an advisory committee.

Agenda

Advisory Committee

Spruce Pine High School Agricultural Education Program

January 28

2800 Hall Building

6:30 PM

Reception	Hosted by the FFA chapter officers
Welcome and Committee Mission	Ms. Sharon Dean, Assistant Principal
Recognition of the advisory Committee Chair	Mr. Bob Wells, Agriculture Teacher
New Business	Ms. Erin Jones
Program Evaluation Plan	
Next Meeting Date	
Tour of Agriculture Facility	Mr. Bob Wells
Adjournment	

---

data in order to make good decisions. The worst thing that can happen for the advisory committee is to meet and make arbitrary recommendations about curriculum, planning, and goals. Sometimes such arbitrary recommendations are inconsistent with reliable data. To guard against arbitrary decision making, the advisory committee should perform a program evaluation that answers the following questions (Martin, 1994):

- Does the curriculum adequately reflect the career needs of students?
- Are agricultural education courses sufficiently challenging and relevant to the best practices in the industry?
- Do all students have the opportunity to take agricultural education courses, provided the courses meet the students' educational and vocational goals?

7–9. Former and current students can provide useful input to an advisory group.
(Shutterstock © Monkey Business Images)

SAMPLE MISSION AND GOALS OF AN ADVISORY COMMITTEE

Mission

The mission of the Oakwood High School Agricultural Education Program Advisory Committee is to give citizens a voice in the development of the total instructional program that includes both classroom instruction and supervised agricultural experiences in agricultural sciences and technology, agricultural engineering, and agricultural business management. In addition, the committee works to develop the leadership capabilities of youth through the activities and programs of the Oakwood High School FFA Chapter. We accomplish this by advising, advocating, and making recommendations for the enhancement, availability, and quality of this total instructional program so that youth served by the program are prepared to enter the workforce or to proceed into postsecondary education.

Goals

In developing goals, the Oakwood High School Agricultural Education Program Advisory Committee considered goals that would be essential in facilitating accomplishment of the school's mission:

- Provide advice and recommendations for the instructional program and the improvement of school and program facilities.
- Advocate for the program and school.
- Assist the agriculture teachers and school administrators in developing and improving the program.

- Does the agricultural education program administer a supervised experience program, and do all agricultural education students have valid supervised experience?
- Do graduates have viable job opportunities upon completion of the agricultural education program?
- Are facilities and instructional equipment current with technology in the agricultural industry?
- Do adequate resources exist for carrying out the instructional program?
- Does the agricultural education program have a well-prepared and highly skilled and motivated agriculture teacher?
- Does the agricultural education program have an active FFA program, and do all students have the opportunity to join the FFA chapter?
- Is there an active FFA Alumni group affiliated with the agricultural education program?

In some cases, this work will have been largely done through a school accreditation program. Most schools are required to be periodically evaluated by an accrediting agency for quality assurance purposes. The questions above may have been answered by the reports of the accrediting agency. However, if the accreditation process has not answered some or all of the questions, then the advisory committee should take the task upon itself to find the answers.

General Suggestions on Advisory Committees

When organizing an advisory committee for the agricultural education program, there are some recommendations for its operation that might make things run more smoothly. Advisory committee membership should be free to address the important issues in a hospitable environment where all opinions are welcomed and discussed.

For agricultural education programs to grow, difficult problems must be solved with new and innovative ideas. New ideas are often met with resistance because "That's not the way we've always done it." The agriculture teacher must be sensitive to the need to maintain a hospitable environment and must help the committee move through the difficult issues with everyone's anger and pride in check.

Rewarding Members   As members of the advisory group complete their terms of service, make certain that they are adequately rewarded. It is recommended that each retiring member receive a thank-you letter from the agricultural education program and the school administration and that the FFA chapter consider awarding the member the Honorary FFA Degree or other similar honor (Doerfort, 2003).

Making Use of Resource Persons and Data   It may be necessary from time to time for the advisory committee to bring in resource people to provide information about issues important to the agricultural education program. For instance, in determining the effectiveness of supervised experience programs, the county economic development director may be brought in to talk about trends in the industry (Doerfort, 2003). Without viable data, the committee is essentially guessing about the direction the agricultural education program should take. The agriculture teacher should provide reports and resources of interest to the advisory committee to validate its decisions.

Develop a Rotational Plan for Terms of Service   The work in an agricultural education program is ongoing. It is never complete. There is something to do all the time to adapt and improve the program. Steps must be taken to prevent those individuals who donate their time, effort, and expertise to the advisory committee from becoming overburdened with duties. If the advisory committee is well organized and has a rotational plan for terms of service, then it is unlikely that any one individual will be overworked. A three-year term of service is recommended for every committee member. At the end of three years, it is time for the person to retire from the committee and allow someone else to bring some new ideas to the table. The individual may serve again in the future, but terms should not be consecutive.

Keep Committee Members Productive   It is often said that the busiest people make the best workers because of their ability to get things done. That is not always the case. Sometimes the busy people are so very busy because they are disorganized, are unable or unwilling to delegate, and procrastinate in their work.

Select individuals who are adept at carrying out plans and projects to completion. They do not have time to waste, nor do they like having their time wasted in meaningless meetings and poorly planned activities. If task-oriented committee members are given substantial work to do, they will have ownership in the agricultural education program and be its champions for many years to come.

FFA ALUMNI

One of the most productive methods for developing community support for an agricultural education program is involving the FFA Alumni. Formed in 1971, the **FFA Alumni** is an adult group within the National FFA Organization that supports agricultural education by raising funds, promoting community awareness, and providing resources needed by the programs. In 2011, there were approximately 52,555 alumni members in local affiliates of FFA chapters in the United States.

Contrary to the way most alumni organizations work, membership in the FFA Alumni is open to anyone who has a sincere interest in supporting FFA. Because the National FFA Alumni Association is a subordinate entity of the National FFA Organization, its mission is very similar to that of FFA.

> The Mission of the National FFA Alumni Association is to secure the promise of FFA and agricultural education by creating an environment where people and communities can develop their potential for premier leadership, personal growth and career success.

7–10. Mission statement of the National FFA Alumni Association.
(*Source:* National FFA Organization 2012a).

Mission Statement

A ***mission statement*** is a written statement of the purpose of an organization. It reflects the best efforts of an organization to go about achieving its purpose. Figure 7–10 presents the mission statement of the National FFA Alumni Association.

Beyond the mission, the National FFA Alumni Association has endeavored to operate by a governing vision. This vision further explains the FFA Alumni's mission and provides a foundation for present and future activities of the organization.

Organizational Structure of the FFA Alumni

The most active component of the FFA Alumni organization is the local alumni affiliate. The affiliate works closely with the local FFA chapter to conduct joint activities and plan independent activities that support the local FFA chapter.

The affiliates within a state form the state alumni association, which provides overall leadership and direction. The state association elects officers to serve according to the alumni constitution in that state. The state associations are chartered by the National FFA Alumni Association, which is headquartered in the National FFA Center in Indianapolis, Indiana.

Membership in the FFA Alumni is open to anyone with an interest in supporting FFA. The local FFA Alumni affiliate can create community awareness of FFA activities. As a civic organization, it is a natural advertising agency for the activities of the FFA chapter. Alumni members also provide valuable support of supervised experience because of their close relationship to the agricultural education program. Some alumni members are excellent teachers and trainers and provide instructional resources and ideas for improving instruction and for improving FFA chapter programs. Alumni members can provide opportunities for teachers to learn more about the subject matter they teach. Alumni groups are even a natural audience for adult education in agriculture.

Alumni members can accomplish many things in support of the FFA chapter and still be able to hold separate activities that appeal to the older generations found in the FFA Alumni. Examples of alumni activities include the following:

- Fund-raisers for the local FFA chapter
- State and national alumni conventions
- Big Brother/Big Sister programs
- Coaching and judging of Career Development Events
- Substitute teaching in the agricultural education program
- Celebrations, parades, and holiday activities

> The Vision of the National FFA Alumni Association is a world where people and communities can grow and develop to their fullest potential.

7–11. Vision for the National FFA Alumni Association.
(*Source:* National FFA Organization, 2012b).

Initiating an FFA Alumni Chapter

The first step is to select a nucleus of highly motivated and interested individuals. Choose one person to lead the effort to create an alumni chapter, and set up an organizational committee to share in the work. Members of this committee should be knowledgeable of the agricultural education program and FFA and should have effective leadership and organizational skills. Call an organizational meeting and have the FFA members advertise it to parents and the local community. An FFA chapter can create an interest in alumni membership through the following:

- An "Each one, reach one" campaign. Each member should bring one prospective alumni member to a meeting.
- Personal contacts. The members and the FFA advisor can contact individuals and ask them to join.
- Mailings and telephone calls. The agricultural education teacher can send letters to parents and other supporters of the program.
- Newspaper announcements. The local newspaper can run an advertisement about the upcoming FFA Alumni meeting.

At the meeting, provide an agenda, a draft constitution, and information about the benefits of FFA Alumni membership. Once a constitution has been agreed upon by the membership, the FFA Alumni chapter can then elect officers, set a budget, and begin developing a plan of activities. After the officers are elected, the organizational committee will cease to exist, and the officers will begin their service.

Recruiting and Retaining Members

The FFA Alumni is a social organization by nature. Including things in the program of activities that are relevant, fun, and exciting for members is imperative. FFA Alumni members need a cause in which they can become involved and to which they can make a useful contribution. Respecting the needs and wishes of individuals is important. For instance, the FFA Alumni should meet regularly but not too often. If the FFA Alumni meets too frequently, the members might become "burned out" and inactive. Politics, like some religious issues, can divide the FFA Alumni as well as the community. Avoid involvement in political activities through the FFA Alumni chapter.

The FFA Alumni can help the agricultural education teacher and the advisory committee carry out their mission (Agnew & Jumper, 2003). By engaging the support of the community through the alumni, an agricultural education program has the human resources necessary to accomplish great things for young people.

REVIEWING SUMMARY

The purpose of advisory committees is to assist the local school administration and instructor in their efforts to plan, develop, evaluate, and keep contemporary the agricultural education program (Iowa Department of Education, 2012). Such groups should reflect the nature of the community served by the school and be formally organized to carry out their work more effectively. The groups must be aware of their roles and that they do not make policy, as that is a function of the local board of education. Citizens' groups can assist in program development and evaluation, supervised experience, and the gaining of overall support for agricultural education.

Thorough knowledge of a community is needed before initiating an advisory committee. A survey can be made if the needed information is not otherwise available. Several important steps should be followed in establishing and operating an advisory committee:

- Develop a mission and a purpose.
- Gain support of school administrators and the board of education.

- Choose the members.
- Choose or elect a chair and other officers.
- Organize the first meeting.
- Establish operating procedures for the committee.

Teachers are responsible for getting an advisory group organized and helping it progress. This requires keeping members focused on their roles and providing

recognition for their service. A teacher is an ex officio member of an advisory group.

FFA Alumni groups can also provide support for the local agricultural education program. The National FFA Alumni Association is affiliated with the National FFA Organization. Members are not required to be former FFA members but should be committed to supporting the local FFA chapter.

QUESTIONS FOR REVIEW AND DISCUSSION

1. What are the steps to establishing an advisory committee in an agricultural education program?
2. What are some activities of the FFA Alumni that can support the mission of the local agricultural education program?
3. Why are the FFA and FFA Alumni missions similar?
4. Which organization would be better for parents who wish to support the agricultural education program—the FFA Alumni or the advisory committee? Explain your answer.

ACTIVITIES

1. Contact a local agricultural education program and ask to attend the next advisory committee meeting or FFA Alumni meeting. Talk to the alumni president or the advisory board chair about his or her leadership role. Interview the agricultural education teacher for suggestions on how to establish a successful advisory committee or FFA Alumni chapter.
2. Investigate the mission and programs of the National FFA Alumni Association. Begin your study with the National FFA Web site. You may

also gather information from the state supervisor of agricultural education in your state. Prepare a report on your findings.
3. Investigate the beliefs of Herbert M. Hamlin about citizen input into agricultural education programming. Refer to one or more of his publications, which include *Public School Education in Agriculture: A Guide to Policy and Policy-Making* (1962) and *The Public and Its Education* (1955). These books should be available from the library at your college or university.

CHAPTER BIBLIOGRAPHY

Agnew, D., and P. Jumper. (2003, July/August). Managing successful alumni relations. *The Agricultural Education Magazine, 76*(1), 8–9.

Darling-Hammond, L., and J. Bransford. (2005). *Preparing Teachers for a Changing World: What Teachers Should Learn and Be Able to Do* (1st ed.). San Francisco: Jossey-Bass.

Dewey, J. (2001). *The School and Society & The Child and the Curriculum*. Mineola, NY: Dover Publications.

Doerfort, D. L. (2003, November/December). Discussing the future with advisory committees. *The Agricultural Education Magazine, 76*(3), 20–21.

Foster, D. D. (2003, July/August). Revelations of a first year teacher. *The Agricultural Education Magazine, 76*(1), 16–17.

Hamlin, H. M. (1955). *The Public and Its Education*. Danville, IL: The Interstate Printers & Publishers, Inc.

Hamlin, H. M. (1962). *Public School Education in Agriculture: A Guide to Policy and Policy-Making*. Danville, IL: The Interstate Printers & Publishers, Inc.

Iowa Department of Education. (2012). *Advisory Groups: Advisory Councils and Committees*. Des Moines: Iowa Department of Education.

Martin, R. A. (1994). Advisory committees: Questions in search of answers. Accessed on December 10, 2003.

National Center for Educational Statistics. (2011). The Condition of Education 2011. United States Department of Education. Retrieved from their website.

National FFA Organization. (2012a). *Mission Statement of the National FFA Alumni Association*. Indianapolis, IN: Author.

National FFA Organization. (2012b). *Vision Statement of the National FFA Alumni Association*. Indianapolis, IN: Author.

8
Curriculum Development

Jay Jorgenson, then a new college graduate, began as teacher of agriculture at George Washington High School in 1982. He set about to design a curriculum that was in line with community needs and his interests. His students were almost all males, and most lived on farms or had families who worked in agriculture. Except for a few adjustments from time to time, including new course names, he teaches much the same content as he did when he began teaching.

Meanwhile, the nearby city has sprawled increasingly toward George Washington High School. Farmland has been bought and converted to residential areas, malls, parks, and athletic fields. New ordinances have been enacted that make it nearly impossible to keep livestock and poultry. Enrollment in the agriculture program has gone down to only a couple dozen students. All this has caused Mr. Jorgenson concern. He has checked with the state retirement system and may end his teaching career.

Assume you were following Mr. Jorgenson as the new agriculture teacher. What would you do to modernize the curriculum, recruit students, and, in general, breathe life into the agriculture program?

terms

academic subjects	scope
accountability	sequence
articulation	standard
block scheduling	teaching calendar
Carnegie unit	teaching schedule
course of study	traditional scheduling
curriculum guide	trimester scheduling
integration	

8–1. Teachers are working cooperatively on a curriculum project.
(Courtesy, Education Images)

CURRICULUM MEANING AND PROCESSES

Who determines what courses are offered in a local agricultural education program? How does the agriculture teacher decide what content to teach in each course? How are the scope and sequence of content within a course determined? Curriculum development addresses these questions.

Meaning of Curriculum

The term *curriculum* is used to describe several concepts. To some, curriculum is what is taught in a single course. To others, it is the sequence of courses taught within a particular area, such as agricultural education. Those with a broad view see curriculum as encompassing everything related to "learning activities and experiences that a student has under the direction of the school" (Finch & Crunkilton, 1999, p. 11).

For purposes of this chapter, *curriculum* is those learning activities and experiences that students have within agricultural education. This definition encompasses courses, content within courses, supervised agricultural experience (SAE), and FFA. Although students may learn much about agriculture from a trip to see grandparents or a vacation in Orlando, Florida, these experiences are informal and not included within this definition of curriculum.

Curriculum development primarily looks at content, depth of coverage, sequencing of content, standards, and related issues (Finch & Crunkilton, 1999). Curriculum is a broad overview of what students should learn. Lesson planning looks at specifics and how content is taught. It is discussed in Chapter 13.

Curriculum Processes

Decisions regarding the curriculum should not be made by the agriculture teacher alone. Just as program planning, discussed in Chapter 6, involves the school and the community, so should curriculum development. The agricultural education advisory committee can provide community feedback regarding the agricultural education curriculum. School administrators and other teachers can provide valuable input, especially in the articulation of agricultural education with other subjects within the school.

Articulation is the way in which educational activities are organized and connected with other educational activities (Lee, 2000). The agricultural education curriculum will have some degree of articulation between courses. The introductory course should prepare students for the more specialized courses at the next level. Concepts learned in a horticulture class, for example, should prepare students for applications in a landscaping class. Management practices learned in an animal sciences course can be used by students in their entrepreneurship supervised agricultural experience.

As a part of the overall school curriculum, agricultural education should have articulation with other subjects. For example, content in soil science should be articulated with content learned in chemistry and geology. The scientific concepts learned in chemistry can help students better understand the processes discussed in agriculture. If students have studied soils in agriculture first, they benefit in chemistry by being able to connect the abstract concepts and principles to real-world situations within agriculture. Students benefit by learning similar content in different settings, thereby taking advantage of the learning principles discussed in Chapter 12.

STANDARDS AND CURRICULUM

The agriculture teacher should not teach only what he or she wants to teach. National, state, and local educational entities require that certain standards be met. A *standard* is the expectation of what students should know and be able to do after completing a given area of instruction. The teacher is responsible for using his or her professional knowledge and experience to determine how to meet the standard.

Some standards are voluntary; others are required. Required standards are typically enforced by student standardized testing, funding, or accreditation. These standards have been developed with input from educators, business and industry, and the general public. They also may be benchmarked against content covered in standardized tests. Finally, they are officially adopted by the applicable governing body.

8–2. Standards may be available in print, CD-ROM, or online formats.
(Courtesy, Education Images)

Academic Standards

Academic subjects are subjects required of all students and include English/language arts, mathematics, science, and social studies. The curriculum in these subjects is mostly prescribed and fairly consistent among all schools within a state. The Common Core State Standards Initiative has developed standards in English language arts and mathematics that have been adopted by forty-five states (Common Core State Standards Initiative, 2012). Increasingly, subject areas outside of these four are being expected to assist students in learning and meeting academic standards. Agricultural education is uniquely situated to address standards in all four areas.

Through the study of the food, fiber, plants, animals, and natural resources systems, agricultural education addresses standards in all areas of science. Examples and problems used in the classroom can come from the real world and the workplace of agriculture. Computational math is used almost every day in agriculture. However, the content of algebra, geometry, and trigonometry is also found throughout agricultural areas of study. English/language arts are a part of every classroom, but the agricultural education components of SAE and FFA provide additional opportunities to learn and practice these skills. Finally, agriculture has been and is a vital part of our nation's history, geography, culture, politics, and economics.

If agricultural education is to assist students in achieving academic standards, articulation and integration must occur on a conscious and consistent basis. Articulation, as discussed earlier, works to have students learn similar concepts in different courses at an appropriate time and in an appropriate sequence.

Integration is the process of combining academic curriculum with career and technical education curriculum so that learning is more relevant and meaningful to students (Finch & Crunkilton, 1999; National FFA Organization, 2012). Whereas articulation allows students to build on prior or concurrent learning, integration is designed to eliminate the rigid distinction between academic and career and

8–3. Standards are developed and published by various organizations and agencies, such as the National Science Education Standards by the National Research Council. Agricultural educators often use academic standards to assure that the curriculum is appropriately integrated.
(Courtesy, Education Images)

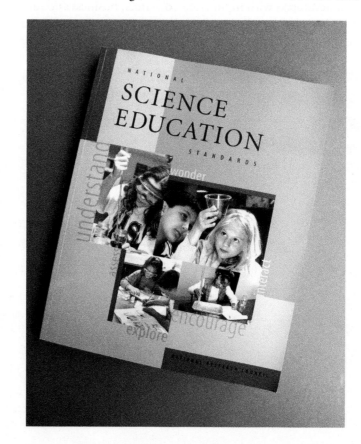

technical education. Integration is discussed in more detail in a later section of this chapter.

Agricultural Education Standards

Most states develop and provide curriculum guides. A ***curriculum guide*** is a written description of courses or areas/units of instruction, along with suggested lesson objectives. Curriculum guides can also include scope-and-sequence outlines, suggested teaching–learning activities, content outlines, and lesson resources. In most cases, these do not carry the weight or authority of standards, thus allowing agriculture teachers great flexibility in developing local curriculum.

Increasingly, agricultural education is becoming like other subject areas and including standards with the other components of curriculum guides. States may tie funding to documentation that standards are being used to design local agricultural education courses. Agricultural education programs may be required to show that a minimum percentage, maybe 70 to 80 percent, of state standards is being covered. This allows a local program some flexibility in meeting local community agricultural needs. Other states administer end-of-course testing based on the state agricultural education standards. Depending on the state requirements, only students who pass the test receive credit for the agriculture course.

The National Council for Agricultural Education has developed content standards for each of the eight career pathways in the Agriculture, Food, and Natural Resources career cluster (career clusters are discussed in Chapter 9). Each pathway has one broad standard, multiple performance elements that state what content should be provided, and performance indicators that are measurable at three levels. These are designed to guide the development or revision of state and local content standards (Pentony & Hall, 2008). Standards for the various states are often available online through the Web site of the state education agency.

The Curriculum for Agricultural Science Education (CASE) is a sequence of courses designed to increase the rigor and relevance of agricultural education instruction (National Council for Agricultural Education, 2008). CASE is a National Council for Agricultural Education project that began in consultation with Project Lead the Way, Inc. (for more information on PLTW, see their website.). CASE includes foundations courses in animals and plants as well as several specialization courses. Each course is designed with units and lessons that contain activities, projects, and problems, along with support materials. Each lesson is connected to LifeKnowledge, FFA, and SAE. In order to obtain and use a CASE course, teachers must participate in an intensive training on the curriculum. More information on CASE can be found on its Web site.

CLASS SCHEDULING PROCEDURES

Credit for high school courses is typically based on the Carnegie unit system. A ***Carnegie unit*** has traditionally been the amount of credit a student is awarded for a course that meets one instructional hour per day for 180 school days (36 weeks). An instructional hour may range from 45 to 60 minutes of actual clock time.

Carnegie units are important for determining high school graduation and college entrance requirements. So, a course that meets for 50 minutes per day for one semester (18 weeks) would be worth one half Carnegie unit. On the other hand, a course that meets for two 50-minute periods per day for the entire 180-day school year would be worth two Carnegie units.

In ***traditional scheduling***, high schools are organized with six or seven one-instructional-hour class periods for five days per week for the entire school year.

8–4. Block scheduling provides more time for hands-on learning.
(Courtesy, Education Images)

Under this system, students earn six or seven Carnegie units per school year. If classes are on a semester basis, students can take up to fourteen semester classes. Within the last fifteen years, other organizational systems have emerged, with block scheduling and trimester scheduling being the two most prevalent.

Block Scheduling

Block scheduling is a class scheduling system in which students take fewer classes each semester than in a traditional system but in which each class period occupies a larger block of time. Each class period may be 85 to 100 minutes. Students take four classes each day and earn eight Carnegie units per school year. If classes are on a semester basis, students can take up to sixteen semester classes.

This system has several advantages over a traditional schedule. Students take in fewer classes, so they can be more focused in their thinking and studying. Block scheduling allows more time for laboratories, in-class field trips, and projects. There is the potential for greater efficiency; half as much time needs to be spent on housekeeping duties at the beginning and ending of classes since there are half as many class periods. Pedagogically, one class period can be broken into three to five instructional variations. Potentially, this allows the teacher to demonstrate a concept, break it into its component parts, have students do a laboratory experiment, and debrief all in the same class period. Teachers have to prepare for fewer classes, as they will teach three or four class periods per day rather than five to seven. With four classes per day instead of seven, the chances for integration and articulation occurring are greater.

Block scheduling also has some disadvantages. The total amount of time for instructional purposes may be less than in a traditional scheduling system if the class meeting time is not doubled. Although teachers have fewer classes to prepare for, they must be better prepared for each class since it meets for a longer time. Students who miss a class period because of illness, a field trip, or some other situation miss the equivalent of two instructional periods and can fall behind very quickly. Finally, teachers who do not modify their instructional techniques to fit the longer time span lose student attention due to boredom, must deal with greater misbehavior problems, and experience reduced student learning.

There are two common forms of block scheduling: 4×4 and 4×2. A 4×4 block schedule has four class periods each day, with all classes meeting every day. An instructional year is completed in one semester. A 4×2 block schedule, often called an A/B schedule, has four class periods each day, but a different set of classes is offered on alternate days. Classes meet the entire instructional year. A modified block scheduling system combines block scheduling with traditional scheduling. For example, Monday and Wednesday may be A schedule, Tuesday and Thursday may be B schedule, and Friday all eight classes meet.

Trimester Scheduling

Trimester scheduling is another alternative class scheduling system that divides the school year into three 12-week trimesters. Each class period may be 70 to 80 minutes in length. Students take five classes each day during the trimester. Classes taken for two trimesters equal one Carnegie unit each, so students can earn seven to seven and one half Carnegie units per school year. Classes taken for one trimester are the equivalent of one semester, so students can take up to fifteen semester classes. A trimester scheduling system is designed to include the advantages of a block system without some of the disadvantages.

Alternative Scheduling Systems and Agricultural Education

With its rich history of hands-on learning, agricultural education has adapted well to alternative scheduling systems. Concepts can be taught at the beginning of a class period and then put into action the same class period through experiments or laboratory projects. Agriculture teachers who once had to break up laboratories and projects into multiple segments and days can now do them in one class period. Agricultural land laboratories and nearby facilities can be utilized more easily, more readily, and more often. In addition, some agricultural education programs have experienced increased enrollments, because students have fifteen or sixteen semester slots to fill as opposed to twelve to fourteen under a traditional scheduling system.

Alternative scheduling systems have presented more challenges for SAE and FFA than for classroom instruction. It is possible under a block 4×4 system for an FFA member to have an agricultural education class the first semester of the sophomore year and not have one again until the second semester of the junior year. In a trimester system, it is likely that students will have an agricultural education class only two of the three trimesters. This causes problems of continuity and contact with students. Students may be less likely to participate in FFA and SAE if they are not in an agricultural education class at the time. Agriculture teachers may teach twice as many students under an alternative scheduling system. This puts strains on a teacher's ability to make SAE visits and to involve all FFA members in FFA activities.

Recent research has shown that many of these difficulties can be overcome with planning and ingenuity. Moore, Kirby, and Becton (1997) recommended that better and more frequent communications regarding SAE and FFA be used to help keep students engaged and participating. Other suggestions include having two sets of officers (one in the fall and another in the spring), continually collecting FFA dues, and repeating Career Development Events in both the fall and the spring.

ACCOUNTABILITY

Increasing pressure is felt from parents, the private sector, and government agencies for schools, programs, and teachers to be accountable for student learning. *Accountability* means that the school, program, or teacher is held answerable or responsible for student learning and achievement. Most people agree that the primary role of schools is education, so holding schools accountable for this makes sense. However, what are not agreed upon are how to measure student learning and achievement and what the standard for accountability should be. In many states, standardized tests in the areas of English/language arts, mathematics, science, and social studies (core academic areas) have become the instruments for measuring accountability.

The No Child Left Behind Act of 2001 had accountability as the centerpiece of the legislation. A significant accountability measure in the law is mandatory academic testing of all students. Schools are to be held accountable for closing the achievement gaps between students of different racial, ethnic, and economic groups. Schools will have to demonstrate that they are staffed by teachers qualified to teach in their subject areas.

Accountability is the job of all teachers, not just English, math, science, and social studies teachers. Agricultural education teachers are also accountable for student learning in the four core academic areas. This means that agriculture teachers may have to give more writing assignments and more closely grade those assignments for grammar, spelling, punctuation, and sentence structure. They may have to provide a certain amount of time per week for silent reading, either in literature or agricultural publications. The use of math concepts in agriculture may become a more explicit part of lessons. Scientific concepts and principles will need to be explained more thoroughly, and experiments debriefed rather than just conducted for the "gee whiz" effect. When teaching about international trade, the agriculture teacher may need to bring in the social studies concepts of government and politics. These changes should not only result in improved learning in the core academic areas but should enhance learning in agricultural education as well.

PREPARING COURSES OF STUDY

Curriculum development is an art requiring application of principles to unique programs and local situations. A part of curriculum development is the preparation of courses of study. A ***course of study*** is a written guide that aids in teaching a particular subject or course. It typically contains objectives, an outline of content, a list of learning activities, means of assessment, and suggested instructional materials. Content includes known facts and subject matter; however, it is continually updated as research creates new knowledge and discoveries.

Some guiding principles of curriculum planning as is done in preparing courses of study are presented in Figure 8–5. These principles will help in determining the scope and sequence of the agricultural education curriculum. Scope and sequence are two important concepts in preparing a course of study.

Scope is the extent, depth, or range of the curriculum. Will courses be taught for grades 6 through 12 or grades 10 through 12? Will five class periods be devoted to a unit of instruction, or fifteen? Will emphasis be placed on a wide breadth of topics, or will fewer topics be taught in greater depth?

Sequence is the order or progression of the material covered. Sequence in agriculture may be determined in any of several ways, such as from simple to complex or by production cycle (crop planting to harvest). Will an introductory course be required before students can take an advanced course? Are general courses required before specialized courses? How will units be arranged within a course? Will content proceed from known to unknown, from simple to complex, from whole to parts to whole again? Will units be arranged according to seasons of the year?

Curriculum Planning Principles in Agricultural Education

1. The agriculture teacher is responsible for developing the curriculum for the local program.

2. The curriculum should reflect the present and future importance of agricultural situations and resources in the community and larger geographic area. Input from the local advisory committee and others should be used.

3. The curriculum should be planned in advance but be flexible to allow adaptability to student situations and conditions.

4. Teaching and learning activities should be planned as sequenced and systematic units.

5. Appropriate state and national standards must be included. The objectives and content that are most significant to the local situation and/or mandated by a state curriculum guide should be included.

6. Content, along with teaching and learning activities, should be planned based on student developmental levels, student learning modes, and principles of learning.

7. A balanced program of instruction should include agricultural literacy courses, agricultural career preparation courses, specialized courses, supervised experience, and FFA. Depending on the program and local situation, middle school exploratory courses and adult agricultural education are included.

8. Each year (or semester if on alternative scheduling or using semester courses) is articulated with other courses guided by state and/or national career pathways.

8–5. *Curriculum decisions should be guided by planning principles.*
(Adapted from Lee, 2000)

Curriculum Models

When determining the scope and sequence of courses in agricultural education, a curriculum model is a helpful tool. A curriculum model is a graphical representation of the scope and sequence of agricultural education courses. An example is shown in Figure 8–6.

The middle school courses may be taught in a six-, nine-, twelve-, or eighteen-week time frame. For example, a sixth-grade AgriScience Discovery course may be taught on a six-week rotation, with every sixth grader in the middle school taking this course. An eighth-grade AgriScience and Technology Explorations course may be taught for an entire eighteen-week semester and be available to interested students as an elective. The goal of middle school agricultural education is to increase agricultural literacy and motivate students to consider taking future high school agricultural education courses.

Students would enroll in yearlong introductory courses in the ninth and tenth grades. Advanced and specialized courses would be taken in grades 11 and 12 on either a semester or full-year basis.

Selecting and Sequencing Content

To select course content, the teacher should use the various sources of input, along with state curriculum guides. If a state curriculum guide mandates content, then that content must be included in the course. Any local-option content should be based on community needs and endorsed by the local advisory committee. Content articulated and integrated with academic subject areas should also be considered. Finally, articulation between agricultural education courses and their content must also be taken into account.

8–6. Curriculum model showing the scope and sequence of agricultural education for middle school through high school.

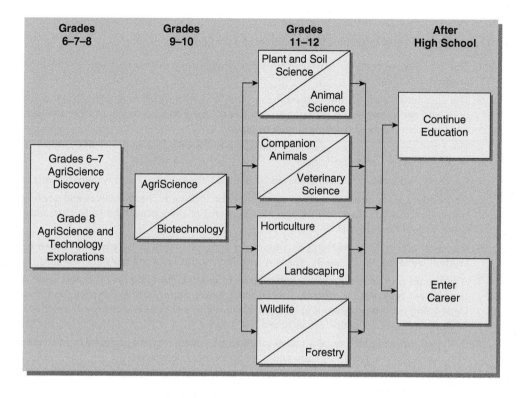

The agriculture teacher is responsible for the scope and sequence of the content taught in agricultural education courses. Determining the scope of the content for a course should involve answering these questions:

1. What is the purpose of the course? An introductory or exploratory course should introduce students to a variety of topics and content but may not go into much depth on any one topic. An advanced-level or career preparation course, on the other hand, may cover a limited number of topics but go into a greater depth of content so that students completing the course are extremely proficient in the area.

2. What is the level of standards in the curriculum guide? Content to be learned on the analysis, synthesis, and evaluation levels (see Chapter 13 for an explanation of these terms) may require more class periods than content to be learned on the knowledge level.

3. What is the background of the students? What prerequisites have the students already taken? Learning totally new content requires more class periods than learning content that expands previous learning or relates to previous learning. If the students are unfamiliar with basic terms, concepts, and practices, then class time will need to be devoted to learning these before moving on to more complex content.

4. What content that relates to this course is covered in other classes? Content covered in other classes, whether other agricultural education classes or not, deserves special consideration. Is it critical content that should be repeated and reinforced in this course? Is it content that needs to be covered only once and therefore should be left out of this course? Is it content that provides a foundation for further learning in this course?

Determining the sequence of the content for a course should involve answering these questions:

1. What content that relates to this course is covered in other agricultural education and academic courses? If it is complementary, then the agricultural education instruction in this course should occur close in time to the instruction in the other courses.

2. What prerequisite knowledge do students need for learning this content? In general, content should be sequenced from the known to the unknown. Prerequisite knowledge, by definition, should be sequenced first.

3. What is the order of complexity of the content? In general, content should be sequenced from simple to complex.

4. Is there a season or other time of year that is most logical for content to be learned? In agriculture, content should be sequenced so that it occurs just before or concurrently with the agricultural practice. For example, in a horticultural science course, drawing landscaping designs might be sequenced first. This would be followed by selecting plants and landscaping materials, which would come before site preparation, which would come before landscaping the site.

5. What are the interests and what is the readiness level of the students? Younger students typically are more interested in small animals than in livestock. If small-animal care is sequenced first, this may build interest and motivation for learning about livestock next.

6. What conditions or special activities occur throughout the school year that may influence content sequencing? Weather may be a deciding factor. For example, in a horticultural science course, drawing landscaping designs might be done in the winter or rainy season when no outside work can be accomplished. School activities, such as science fairs, homecoming, or prom, may influence what time of the year certain content is taught. The nature and timing of student supervised experience should be taken into consideration when sequencing content. Finally, FFA Career Development Events and other activities may provide input into the sequencing of course content.

Developing a Teaching Schedule and a Teaching Calendar

Just as a curriculum model shows the scope and sequence of agricultural education courses, a *teaching schedule* shows the scope and sequence of content within an individual agricultural education course. A basic teaching schedule is shown in Figure 8–7. Broad units of instruction are listed in the order in which they would be taught during the semester or school year. A suggested number of class periods to spend on each unit is also given. Other information that can be on a teaching schedule includes grade level, length of course, and citations for textbooks used in the course. Some school corporations require that standards and assessments also be included.

A *teaching calendar* is a specialized type of teaching schedule that provides additional details on the time when units of instruction will be taught. The teaching calendar may be on a yearly, semester, monthly, or weekly basis, depending on the specificity needed. If known, testing days, holidays, and other non-instructional activities should be placed on the teaching calendar. The teaching calendar provides advantages over the basic teaching schedule. With a teaching calendar, an agriculture teacher can order supplies and materials, schedule field trips, and involve resource persons in a more timely and effective manner. A teaching calendar also serves to keep the

Teaching Schedule: Plant and Soil Science (Grade 10 or 11; 18–36 weeks or 90–180 hours) Note: Major units are listed, followed by suggested hours for instruction.	Suggested Hours for Instruction
Orienting to Class and FFA	2–4
Using Plant and Soil Science in Agriculture/Horticulture	4–8
Sustaining Plant and Soil Productivity	4–8
Exploring Plants as Living Organisms	5–10
Exploring Plant Reproduction	5–10
Exploring Plant Growth	5–10
Determining Cultural Practices in Producing Plants	6–12
Exploring Soil Materials and Formation	4–8
Practicing Land Classification and Use	3–6
Determining Soil Fertility and Management	4–8
Using Irrigation	3–6
Using Soilless Plant Production	4–8
Using Integrated Pest Management	5–10
Managing Insect and Nematode Populations	5–10
Managing Plant Diseases	4–8
Managing Weeds	5–10
Exploring Plants and Environmental Factors	3–6
Producing Crops (grain; sugar and oil; fiber; forage and turf; vegetable, fruit, and nut; ornamental; and forestry crops)	13–26
Planning Supervised Experience	3–6
Assembly Programs, Testing, etc.	3–6
Textbook: Biondo, R. J., and Lee, J. S. (2003). Plant & Soil Science and Technology (2nd ed.). Upper Saddle River, NJ: Pearson Prentice Hall Interstate.	

8–7. This basic teaching schedule for a plant and soil science class shows the order in which units are taught and the number of class periods devoted to each unit of instruction.

agriculture teacher on-track for wiser use of instructional time. Finally, it serves as a public relations tool with administrators, the local advisory committee, and others.

INTEGRATING ACADEMICS AND AGRICULTURAL EDUCATION

Earlier sections of this chapter have discussed the role standards and accountability play in education today. Also discussed was how articulation and integration can help students learn more effectively. This section discusses levels of integrating academics and career and technical education (Finch & Crunkilton, 1999).

The first level of integration involves putting more academic content into career and technical education courses. Academic courses may include agricultural topics in examples and problems. With this level of integration, academic and agricultural education teachers may or may not be discussing curricular issues with each other.

The next level of integration involves formally bringing academic and career and technical education teachers together to devise ways to enhance the curriculum. Strategies are devised to infuse academics into agricultural education and agriculture into academic subjects.

The third level of integration involves aligning the two curriculums. Teachers meet regularly and plan content and instructional activities to supplement what is happening in the other area. Both curriculums may be modified to better facilitate integration.

The highest level of integration eliminates the hard distinction between academic and career and technical education. Content may be organized around career clusters. In this model, the content of each subject is taught using the career cluster as a theme and motivating factor. Academies and magnet schools demonstrate this level of integration.

REVIEWING SUMMARY

The agricultural education curriculum includes courses, content within courses, SAE, and FFA activities. The curriculum may be specified by a state agency, be totally under local control, or be at some point on a continuum between the two. Regardless, curriculum development should occur with input from the school and the local community. The local agricultural education advisory committee should be involved in curriculum decisions.

Increasingly, agricultural education is bound by standards. These may be standards for English/language arts, mathematics, science, and social studies that all areas must assist students in meeting. There may also be specific standards for agricultural education. Teachers and agricultural education programs are accountable for students achieving to required levels on standards.

Schools have a variety of class scheduling systems. The three most popular scheduling systems are traditional, block, and trimester. Each system has advantages and disadvantages for students and teachers. Changing from one class scheduling system to another typically requires the agriculture teacher to revise curriculum plans and evaluate teaching strategies.

Agricultural education teachers are responsible for determining the scope and sequence of their courses. Teachers should follow curriculum planning principles when developing curriculum models and teaching schedules. Although teaching schedules should be detailed, they should leave room for flexibility and student interests.

The integration of academics and career and technical education is important in meeting standards and being accountable for students' educational experiences. Schools may desire to integrate at various levels, ranging from the infusion of some content to the development of highly integrated curriculums involving career clusters, academies, or magnet schools.

QUESTIONS FOR REVIEW AND DISCUSSION

1. What is the curriculum for a local agricultural education program?
2. What are the sources of input for curriculum development in agricultural education?
3. What are examples of articulation between agricultural education courses?
4. Why should agricultural education be articulated with other subjects?
5. What purpose do standards serve in agricultural education?
6. How are curriculum guides used in your state?
7. What are the advantages and disadvantages of each scheduling system for agricultural education?
8. How are courses of study developed on the local level?
9. Why are scope and sequence important in student learning?
10. What are the levels of integration between academic and career and technical education?

ACTIVITIES

1. Investigate the degree of integration of academics and career and technical education at a selected school. You may want to interview the principal, the agriculture teacher, and one or more academic teachers.
2. Develop a teaching calendar for an agricultural education course. Remember to include time for testing, special days, and holidays. Address scope and sequence in the teaching calendar and have a strong rationale for your scope and sequence.

3. Using the *Journal of Agricultural Education, The Agricultural Education Magazine*, and the proceedings of the National Agricultural Education Research Conference, write a report on block scheduling research in agricultural education. Current and past issues of these journals can be found in the university library.

Selected past issues can be found online at American Association for Agricultural Education or National Association of Agricultural Educators websites.

4. Research the foundational literature regarding curriculum and instruction. Make sure to include a review of Ralph W. Tyler's work.

CHAPTER BIBLIOGRAPHY

Binkley, H. R., and R. W. Tulloch. (1981). *Teaching Vocational Agriculture/Agribusiness*. Danville, IL: The Interstate Printers & Publishers, Inc.

Caffarella, R. S. (2002). *Planning Programs for Adult Learners* (2nd ed.). San Francisco: Jossey-Bass.

Canales, J., J. Frey, C. Walker, S. F. Walker, S. Weiss, and A. West. (Eds.). (2002). *No State Left Behind: The Challenges and Opportunities of ESEA 2001*. Denver: Education Commission of the States. (ERIC Document Reproduction No. ED468096).

Common Core State Standards Initiative. (2012). *Common Standards Web Site*. Accessed on October 3, 2012, at 6.

Finch, C. R., and J. R. Crunkilton. (1999). *Curriculum Development in Vocational and Technical Education: Planning, Content, and Implementation* (5th ed.). Needham Heights, MA: Allyn and Bacon.

Frick, M., and S. Stump. (1991). *Handbook for Program Planning in Indiana Agricultural Science and Business Programs*. Unpublished manuscript, Purdue University.

Indiana Department of Education. (1999). *Teacher/Local Team Self-study of Standards and Quality Indicators for Agriscience and Business Program Improvement*. Indianapolis: Author.

Kahler, A. A., B. Morgan, G. E. Holmes, and C. E. Bundy. (1985). *Methods in Adult Education* (4th ed.). Danville, IL: The Interstate Printers & Publishers, Inc.

Lee, J. S. (2000). *Program Planning Guide for Agri-Science and Technology Education* (2nd ed.). Upper Saddle River, NJ: Pearson Prentice Hall Interstate.

Lynn, R. L., E. L. Baker, and D. W. Betabenner. (2002). *Accountability Systems: Implications of Requirements of the No Child Left Behind Act of 2001* (CSE Technical Report 567). Los Angeles: Center for the Study of Evaluation, UCLA. (ERIC Document Reproduction No. ED467440).

Moore, G., B. Kirby, and L. K. Becton. (1997). Block scheduling's impact on instruction, FFA, and SAE in agricultural education. *Journal of Agricultural Education, 38*(4), 1–10.

National Association of Agricultural Educators (NAAE). (2011). *NAAE Web Site*. Accessed on November 30, 2011.

National Council for Agricultural Education. (2008). *Curriculum for Agricultural Science Education*. Accessed on June 3, 2008.

National Council for Agricultural Education. (2009). *National Agriculture, Food and Natural Resources (AFNR) Career Cluster Content Standards*. Alexandria, VA: Author.

National FFA Organization. (2003). *Local Program Resource Guide: 2003–2004* [CD-ROM]. Indianapolis: Author.

National FFA Organization. (2012). *Local Program Success* (online). Accessed on November 5, 2012.

Newcomb, L. H., J. D. McCracken, J. R. Warmbrod, and M. S. Whittington. (2004). *Methods of Teaching Agriculture* (3rd ed.). Upper Saddle River, NJ: Prentice Hall.

Pentony, D., and D. Hall. (2008). ANFR National Content Standards: Project Roundtable Sessions [PowerPoint presentation]. Accessed on June 3, 2008.

Phipps, L. J., and E. W. Osborne. (1988). *Handbook on Agricultural Education in Public Schools* (5th ed.). Danville, IL: The Interstate Printers & Publishers, Inc.

True, A. C. (1929). *A History of Agricultural Education in the United States, 1785–1925*. U.S. Department of Agriculture, Miscellaneous Publication No. 36. Washington, DC: GPO.

U.S. Department of Education. (1996). National Science Education Standards. Washington, DC: National Academy Press.

U.S. Department of Education. (2001). Back to School, Moving Forward: What "No Child Left Behind" Means for America's Educators. Washington, DC: Author. (ERIC Document Reproduction Service No. ED466790).

9

Student Enrollment and Advisement

Rosco Francisco is the agriculture teacher at Etowah High School. He is a popular teacher and always has good enrollment in his classes. Once students take one of his classes, they return for others. He carefully plans what he is to teach and has an abundance of instructional resources. Most years his classes rapidly reach peak enrollment, and students are turned away. Last year Mr. Francisco was named teacher of the year in the school district.

Ronnie Lee is the agriculture teacher at a high school near Etowah. His classes are small, and he has to struggle each year to have enough students to offer the classes. Those who take one class usually fail to return for a second class. Mr. Lee requests very little money from the school administrators for his program. At a recent board of education meeting, the superintendent mentioned that money for teacher salaries could be saved by eliminating the agriculture classes.

What makes the difference between these two programs? If you were advising Mr. Lee, what would you say?

objectives

This chapter explains the importance of student enrollment and advisement to the total agricultural education program. It has the following objectives:

1. Explain the meaning and importance of student enrollment
2. Discuss student recruitment and identify useful strategies
3. Discuss student retention and identify useful strategies
4. Describe the role of guidance counselors and school administrators in student enrollment
5. Discuss barriers to recruitment and retention
6. Describe the meaning and importance of student advisement
7. Describe the role of career guidance
8. Describe the role of personal guidance

terms

career cluster
career guidance
enrollment barrier
feeder school
free elective
guidance
promotional material
recruitment
retention
student advisement
student enrollment

9–1. A school counselor discusses class scheduling with students.
(Courtesy, Education Images)

STUDENT ENROLLMENT

Secondary agricultural education courses are elective courses. An elective course is a course that is not specifically required for high school graduation. In some states, agricultural education courses count as free electives. A *free elective* is a course that does not fulfill any requirements. In these instances, agricultural education competes with all other electives for students. In other states, agricultural education courses may fulfill an open requirement or restricted elective. For example, three science courses may be required for graduation, with biology and chemistry specified to be taken. The third course can be selected from a list of choices that include agriscience courses. Another example is a sequence of agricultural education courses used to fulfill a career pathway requirement.

Student enrollment is the total number of students in the agricultural education program. This is important, as it justifies the number of agriculture teachers employed as well as gives an indication of the impact the agricultural education program has within the school. How do students enroll in agricultural education? What factors influence students' decisions about whether to enroll? Why do students re-enroll during subsequent years? These questions will be addressed in the following sections.

Because of its elective status, students must be recruited into agricultural education. *Recruitment* is the process of seeking and soliciting students to enroll in agricultural education courses. There are six key variables in the successful recruitment of students into an agricultural education program. As shown in Figure 9–2, the agriculture program, the recruitment program, student characteristics, parents, school support, and community support all influence whether a student enrolls in an agricultural education course (Myers, Dyer, & Breja, 2003).

STRATEGIES FOR RECRUITING AND ENROLLING STUDENTS

Student enrollment in agricultural education comes from three sources: (1) students who enroll as a result of the recruitment program, (2) students who as a result of retention re-enroll in agricultural education, and (3) students who enroll for reasons

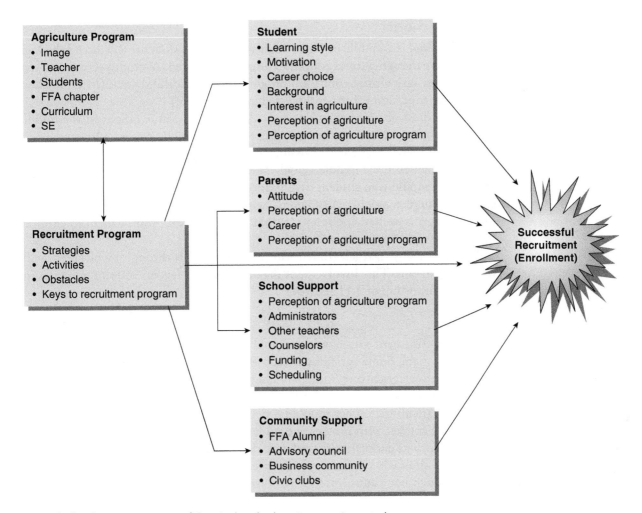

9–2. The key factors in a successful agricultural education recruitment plan.
(*Source:* Myers, Dyer, & Breja, 2003, p. 97. Used with permission.)

outside of their control, also known as disavowance reasons. The first two sources will be discussed separately. Disavowance reasons such as "there were no other courses to take" or "I thought this course would be easy" are typically not considered when developing recruitment and retention strategies.

Hoover and Scanlon (1991), using literature (Quey, 1971) on group dynamics, stated that students take a class such as agricultural education for three reasons:

- Students take an agricultural education class if they like the curriculum or activities in the class.
- Students enroll in agricultural education if they like the people, such as the teacher and their peers, in the class.
- Students enroll in agricultural education if they are able to satisfy a need, such as competition or career preparation, through the class.

Students who have knowledge, observations, and information relating to agricultural education courses may have a greater tendency to enroll in those courses (Fishbein & Ajzen, 1975). Several strategies consider this and are effective for student recruitment. These include feeder school contact, agriculture teacher–student contact, FFA chapter events, publications and other promotional materials, curriculum, parents and other support groups, and recruitment events (Myers, Dyer, & Breja, 2003). These will be discussed next.

Feeder Schools

A *feeder school* is a middle or junior high school whose students will attend the high school where the agricultural education program is located. A strong recruitment tool is to have an agricultural education program in the middle school. (Middle school agricultural education is further discussed in Chapter 16.)

Regardless of whether or not agricultural education is taught in the middle school, there are other ways the agriculture teacher should connect with students in grades 5 through 8. Whenever possible, agriculture students/FFA members should conduct activities in the middle school. These may include demonstrations in science classes, presentations to student organizations, or the hosting of a middle school science fair. Middle school FFA programs, with developmentally appropriate activities, can get students excited about agricultural education and prompt them to enroll in high school agricultural education.

Another way to reach students in this age group is through involvement with the 4-H program. This involvement may be as a club leader, in joint activities, or through sponsorship of 4-H activities.

Contact

Another recruitment strategy is personal contact with prospective students and visibility to them. Some agriculture teachers make home visits to eighth graders who are prospective agriculture students. They may also use telephone calls, cards, or face-to-face meetings as recruitment tools.

The agriculture teacher should be visible throughout the school. Important for establishing contact with prospective students is attendance at sporting events, class activities (such as dances), and academic competitions (such as science fairs), along with daily visibility in the hallways and cafeteria.

Although not a substitute for personal contact, a well-constructed agricultural education program Web site is beneficial. Students who are exploring classes to take may use the Web site during this process. While final responsibility rests with the agriculture teacher, it may be helpful for a high school student to develop and maintain the Web site, possibly as their SAE. Depending on the school's use policy, social media may also be used for recruitment.

9–3. Personal contact is an effective recruitment strategy.

(Courtesy, Education Images)

FFA

FFA can be a tremendous recruitment tool. The activities associated with FFA are highly visible and can fulfill student needs for achievement, recognition, and self-improvement.

Some of the most visible are Career Development Events, leadership development events and activities, and out-of-school opportunities, such as trips and scholarships. A caution is that whatever the image students have of FFA, it is typically transferred to agricultural education classes. For example, if students perceive FFA to be for rural white males, then the agriculture teacher will have a difficult time recruiting into agricultural education students who do not fit this image. To alleviate this problem, the agriculture teacher should strive to involve all students in FFA and have an FFA program that appeals to a diversity of student backgrounds.

Publications and Other Promotional Materials

Publications and other promotional materials are what most people think of when they hear about recruitment. ***Promotional material*** is a product or approach used to gain desired responses from people. It is somewhat like advertising.

Publications and promotional materials include brochures, bulletin boards, newspaper articles and ads, radio and television spots, posters, CD-ROMs and Web pages, social media sites, videos, school announcements, roadside signs, and school display cases. Promotional materials also include giveaways, such as calendars, USB drives, and other items with the agricultural education program name on them. The success of these is difficult to gauge. They definitely raise awareness of the program name and provide valuable information regarding the agricultural education program. In fact, other than by word of mouth, promotional materials are how most people become aware of something. However, these are rarely cited as the number one reason for a student taking an agricultural education class.

Curriculum

Ultimately students and parents must be satisfied with the curriculum for enrollment to take place. What in the agricultural education curriculum adds value for the student?

9–4. Student enrollment should reflect the diversity of the school.
(Courtesy, Education Images)

The agricultural education curriculum must be current and address local community priorities. Ideally it should be career-focused, even if it leads to additional education at the postsecondary level. It is a strong selling point if agricultural education courses count as science, communication, or economics credit.

For recruitment of the broadest range of students, at least part of the curriculum must have an agriscience focus that appeals to students without a farming background. This includes broadening FFA activities and opportunities in SAE (Myers, Breja, & Dyer, 2004).

Parents and Others

Parents and other support groups can serve as voices in the community for agricultural education recruitment. For this reason, they should be involved in the agricultural education program as advisory committee members, Young Farmer or FFA Alumni members, chaperones for trips, and possible classroom resource persons. It is important that they feel appreciated for their efforts. Many agriculture teachers give special recognition to volunteers and supporters at appreciation breakfasts or FFA chapter banquets.

Events

Recruitment events include open houses, agricultural awareness days, National FFA Week activities, orientation nights, and eighth-grade assemblies. These are usually large events and must be planned well in advance of the dates of occurrence. They serve an awareness function similar to publications but hold the potential for personal contact that publications cannot give. Several questions must be answered when planning a recruitment event. Who is the audience? What needs does the audience have? What is the message the audience should get from the event? Who will be the speakers? What activities will be conducted? How will success be gauged? Typically, the agriculture teacher does not plan a recruitment event alone. Depending on the activity, FFA chapter committee(s), the advisory committee, and/or volunteers/supporters should assist in planning and implementing the event.

RETENTION

Retention is gaining repeated enrollment of a student in agricultural education classes. Several practices support students re-enrolling in the agricultural education program. Many of the practices used in recruitment aid in student retention. Once students are in agricultural education courses, they need to see themselves fitting into the image of agricultural education and FFA at their school. Because of this, agriculture teachers need to be constantly aware of the image that is portrayed and compare that against what they want to be reality. What is the focus of the curriculum? Is it current, and does it meet community/student needs? Is the agricultural education program/FFA inclusive and open to a diversity of student backgrounds? Are students of various backgrounds accepted and encouraged to maintain their uniqueness? Does the teacher work to reduce cliques and exclusionary practices?

Just as with recruitment, personal contact and word of mouth are very effective for retention. Students want to know that they belong and that the agriculture teacher cares about them. With FFA and SAE, the agriculture teacher has two vehicles to get to know students better and to affect them on a personal level.

Retention is important for programmatic and student reasons. Obviously, from a programmatic perspective, student retention means increased enrollments. It also means that re-enrolled students can serve as ambassadors for the program and recruit new students. Finally, retention makes program planning easier because the agriculture teacher knows there will be a cohort of students who need specific agriculture courses.

Retention is also good from a student perspective. Supervised agricultural experience is designed to build from one year to the next. Students who continue in agricultural education courses expand their SAE and gain valuable career skills. Continuers also have the benefit of the personal growth, leadership development, and career success activities available to them through involvement in FFA. In addition to this, they have the opportunity for scholarships, awards, recognition, travel, and other benefits of FFA. Finally, students who continue in agricultural education courses increase their agricultural literacy and build their skills and knowledge for agricultural careers.

ROLE OF GUIDANCE AND ADMINISTRATION

Guidance counselors and school administrators are critical for student enrollment in agricultural education. These are the people who meet with students to register them for classes, so they have a great influence on what classes students take. Counselors and administrators who are knowledgeable about the value of the agricultural education program are more inclined to advise students to enroll in its courses. They are also cognizant that agricultural education has benefits for all students, including college-bound, non-rural, and even those without specific career interests in agriculture.

Counselors and administrators are typically involved in designing the master course schedule. This schedule charts where each course in the school will be taught during the school day and how many sections of a course will be offered. Where an agricultural education course is placed within the daily schedule may make the difference between students taking the course and not being able to take it. This is especially true in a small-enrollment school where section offerings of required courses may be limited. An agricultural education course placed at the same time as the only section of a required course for a particular grade level eliminates a group of students from taking agriculture.

9–5. Guidance counselors can be valuable allies in recruiting students into the agricultural education program.
(Courtesy, Education Images)

BARRIERS TO RECRUITMENT AND RETENTION

Agricultural education researchers have identified several barriers to successful recruitment and retention of students. An **enrollment barrier** is a perception or other element in the school or community environment that promotes non-enrollment in agricultural education. Some barriers may be beyond the immediate control of the agriculture teacher.

Breja, Ball, and Dyer (2000) identified schedule-related problems as some of the greatest barriers to student enrollment. College-bound students who have schedule conflicts between courses required for college admission and agricultural education courses will not be able to take the agriculture courses. If increased graduation requirements reduce the number of available slots for elective courses, then enrollment in agricultural education suffers. Finally, if guidance counselors and school administrators have negative perceptions of the agricultural education program, they will be less likely to encourage students to enroll.

Student Perceptions of Image

Image can also be a barrier to enrollment. This includes both the image of agriculture and of agricultural education. Some students, particularly students from ethnic minority groups, may view agricultural careers as low-paying menial labor with no opportunities for them. Others may view agriculture as just production agriculture. It takes effort on the agriculture teacher's part to overcome both of these stereotypes.

As mentioned earlier, students in some schools view agricultural education as being only for rural white males. This image of agricultural education is inaccurate. According to the National FFA Organization (2012), FFA membership is about two thirds nonfarm, two fifths female, and one fourth ethnic minority.

Program Quality

Another barrier may be the quality of the agricultural education program and teacher. Eliminating this barrier is within the control of the agriculture teacher. Why is the program of poor quality? Does the curriculum need updating? Are the facilities inadequate? Does the program focus fit community and student needs? Why is the agriculture teacher not performing up to standards? Does the teacher need technical retraining? Is there a mismatch between teacher expertise and program emphasis? Is the teacher receiving appropriate parental and administrative support? Are there personal reasons for the teacher performing below standard? These questions should be addressed thoughtfully.

Student Expectation Barriers

Other identified barriers include a mismatch between student interests and the focus of the agricultural education course and a mismatch between student career goals and agricultural education. These barriers may be difficult to overcome and may be legitimate reasons for students not to enroll in agricultural education courses.

Student Advisement

Agriculture teachers are in a unique position to offer their students meaningful advice. **Student advisement** is the process of offering assistance to students regarding their career and educational decisions or actions. As stated earlier in this chapter, students may take agricultural education courses for multiple years. This allows the agriculture teacher to know students better academically and personally. In the role of FFA advisor, the agriculture teacher gets to know students even more on academic,

9–6. An agriculture teacher should be trusted by students and provide guidance on career and certain personal issues as a part of instruction.
(Courtesy, Education Images)

personal, and social levels. Finally, through supervised experience, the agriculture teacher meets students' parents or employers and develops a deeper understanding of the students' career goals.

Advisement is not telling students what they should do or making decisions for them. Rather, it is providing information, asking thought-provoking questions, and recommending possible solutions. However, before any of these, the foremost requirement is listening. Students want to be heard before they can hear. Sometimes students only want a sounding board and are not expecting the agriculture teacher to give advice.

Agriculture teachers are often asked to give guidance and counseling. *Guidance* is the process of providing assistance to help people make wise decisions regarding choices or changes (Phipps & Osborne, 1988). One technique within guidance is counseling. Counseling typically involves an individual, face-to-face meeting with the purpose of addressing a specific situation. A counselor is professionally trained and licensed to provide proper guidance to students. Depending on school size and location, a school may have one or more counselors, nurses, psychologists, psychiatrists, social workers, and law enforcement officials who are trained to provide various types of counseling.

Because of the relationships developed with students, an agriculture teacher provides guidance and acts as a counselor in many instances. This is appropriate as long as the teacher is acting as a listening and caring individual who knows when to refer a student to others. However, most agriculture teachers are not trained as counselors, so they should limit the scope of the counseling they provide. For example, students who are depressed or suicidal require the careful guidance that only a trained counselor can provide. Other issues that are best referred to trained counselors include drug or alcohol use, school violence, child abuse, sexual activity, and bereavement.

Agriculture teachers provide guidance in two general areas. One is assisting students to choose a career path and plan the preparation necessary to achieve that goal. The other is assisting students in numerous personal issues common to almost all teenagers. Both of these are discussed in the following sections.

9–7. A counselor reviews scholarships with students.
(Courtesy, Education Images)

CAREER GUIDANCE

Career guidance is the process of helping people proceed through career development stages. It involves providing information regarding careers and helping students understand their interests, abilities, preferences, and values so they can make wise decisions regarding career planning. Career guidance traces its modern roots to Frank Parsons, who in 1909 wrote of the three factors influencing a person's wise choice of a vocation (Brown, Brooks, & Associates, 1990). In choosing vocations, people should (1) understand themselves and their abilities, interests, and circumstances, (2) have knowledge of requirements, advantages and disadvantages, and opportunities of various career paths, and (3) have sound reasoning regarding the relationship between these two (Parsons, 1909).

Vocational Life Stages

Super et al. (1957) theorized that people go through vocational life stages. These stages are further divided into substages. Although Super assigned an age range to each stage, it is understood that there are transitional phases between stages and that people proceed through the stages at differing rates. The five stages are growth (birth–age 14), exploration (15–24), establishment (25–44), maintenance (45–64), and decline (age 65 and on). Middle school and high school students are in the first two stages.

The growth stage has three substages. During the fantasy substage (4–10), children are involved in role-playing. In the interest substage (11–12), children's likes and dislikes influence what they believe they will do as careers. At the capacity substage (13–14), children focus more on their abilities and the requirements of jobs/careers. It is important throughout the growth stage to provide opportunities for students to role-play, explore their interests, and discover job requirements.

In the exploration stage, students examine themselves and try different roles. They explore jobs/careers through school activities, leisure activities, and part-time employment. The exploration stage also has three substages. The tentative substage (15–17) includes most high school students. Students examine their interests, abilities, values, and needs in light of their circumstances. They make tentative career decisions based on many factors and experiences. They try out these decisions through fantasy; discussions with peers, significant persons, and others; courses in school; clubs and activities; work; and other experiences. The transition substage (18–21) involves more consideration toward reality. Students either enter the job market or pursue postsecondary education and training. In the trial substage (22–24), people typically decide on a field, obtain a beginning job, and try it out for their life's work.

Throughout the exploration stage, students continue to need chances to learn about and try out careers. They also need guidance in matching their interests and abilities with the requirements of the careers. When young adults choose careers that are either too demanding or not demanding enough for their abilities, this causes mismatches that result in frustrations. The agriculture teacher, through classroom and laboratory instruction, FFA, and SAE, can help students explore various careers and find ones that fit their interests and abilities.

School to Career is a career guidance initiative to assist students in making a successful transition from school to the world of work (National FFA Organization, 2005). School to Career has three components: contextual learning, work-based learning, and connecting activities. In agricultural education, classroom and laboratory instruction is the contextual learning component. Within the context of the plants, animals, and natural resources systems, students explore requirements of careers and learn the requisite knowledge for entry-level positions or further education and training. The work-based learning component is SAE. Here students gain skills necessary in the workplace and explore career opportunities in a hands-on manner. FFA through Career Development Events, proficiency awards, and other programs is the connecting activity between school and careers.

Career Clusters

The number of careers available for students to enter is staggering. It is also important to note that careers are changing constantly. This means that continual retraining and lifelong learning will be essential for students as they progress through their working lives. An effort to assist students in this task is career clusters.

A *career cluster* is a broad grouping of occupations to help students decide on and prepare for further education and careers (NASDCTEc, 2012). Career clusters help organize curriculum so students learn both academic and technical skills to be better prepared for the world of work (Ruffing, 2006). All possible occupations are included in one of sixteen career clusters. For example, one of the clusters is Agriculture, Food, and Natural Resources. The career cluster concept is an organizational tool for student learning. It fits in well with the Local Program Success (LPS) initiative within agricultural education. Career clusters link classroom instruction with career-based learning. They also outline pathways for students to obtain additional education and training needed for the occupations within a cluster. Finally, as recommended in LPS, the career cluster concept encourages collaboration and partnerships between schools, business and industry, and the community.

Career Guidance Tools

A school guidance counselor can be an ally to the agriculture teacher in advising students regarding their career development. School guidance counselors are trained professionals who can work with students in group or individual settings. They can also advise the agriculture teacher on specific resources and tools available to assist students. The agriculture teacher must keep the guidance counselors updated on the agricultural education curriculum, FFA and SAE activities, and career opportunities within the Agriculture, Food, and Natural Resources career cluster.

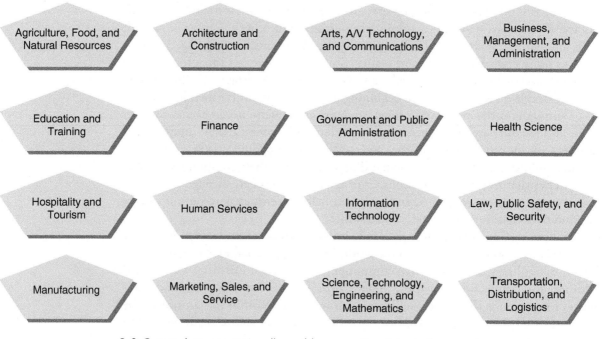

9–8. Career clusters organize all possible occupations into similar groupings.
(NASDCTEc, 2012)

There are numerous career guidance tools available to guidance counselors and agriculture teachers. These include software programs, the *Occupational Outlook Handbook*, aptitude tests, and interest surveys. The school guidance department will have many of these tools and can arrange for administering others. The main advantage of these tools is that they allow the student to narrow his or her focus down to those occupations that meet needs, interests, and abilities. Many of these tools also describe working conditions with pay ranges, training/education required, and employment outlook.

Career development interventions in secondary schools fall into four categories (Dykeman et al., 2001). Work-based interventions include apprenticeships, service learning, job shadowing, and job placement. Advising interventions include academic planning and counseling, conferences, tutoring, career library, and various assessments and testing. Introductory interventions include career days and fairs, career field trips, and short-term lessons on development. Curriculum-based interventions include longer-term interventions, such as career information infused into the school curriculum, and various career-based models, such as magnet schools and career academies. Career and technical education courses, such as agricultural education, fall into this fourth category.

The Agriculture Teacher in Career Guidance

The agriculture teacher is in a unique position to assist students in career guidance. The agriculture teacher is one of the few school personnel who gets to follow students from possibly the sixth grade through post–high school graduation. This gives the agriculture teacher a broad and deep perspective of students' experiences, academic abilities, interests, and skills. The agriculture teacher can assist students in career guidance as SAE supervisor, FFA advisor, provider of information, and career counselor. Many agriculture teachers arrange career fairs specifically in the career cluster of Agriculture, Food, and Natural Resources. This allows students to focus on one career cluster when obtaining information and other resources. Agriculture teachers also host guest speakers as resource persons from a number of occupational areas. In this way, students are exposed to people from the community representing various occupational areas.

Personal Guidance

As mentioned earlier, students may experience certain personal issues that the agriculture teacher must refer to trained counselors. However, the agriculture teacher is qualified to counsel in numerous other situations. A student must feel trust and comfort in sharing with the agriculture teacher. Listening on the part of the agriculture teacher is often more important than talking and offering advice. Some of the best personal guidance is given when the agriculture teacher listens, asks questions, and affirms that the student has been heard.

Students will share information with their agriculture teacher about relationships, self-concept, family, financial concerns, and other personal issues. This information must be treated as confidential. The teacher must not gossip or get in the middle of disputes between students. Doing so would lead to mistrust, accusations of favoritism, and a lessening of teacher authority.

For privacy reasons, personal guidance should be given in an office or other setting where a student's peers cannot hear. The student may become emotional and could become embarrassed if heard by peers. In all situations, the agriculture teacher should strive to maintain the dignity of the student.

The agriculture teacher should exercise caution when meeting individually with students. The conference needs to be arranged in such a way as to remove any hint of impropriety. For example, the agriculture teacher should never meet alone with a student before or after school. During the day, it is best to meet in an office with an inside window or open door.

REVIEWING SUMMARY

Recruitment and retention are important components of an agricultural education program plan. Successful recruitment and retention strategies depend on a number of variables. It is imperative that the agriculture teacher maintain a quality program.

Students enroll in a class because of (1) curricular factors, (2) people factors, and (3) need factors. In addition, providing information and observation opportunities will increase the likelihood that students will enroll. Successful recruitment strategies are built around these basic reasons for enrolling.

Student retention is an important issue for an agricultural education program. It is desirable that students who take one agricultural education course subsequently enroll in additional courses. This has benefits for both the agricultural education program and the students.

Guidance counselors and school administrators affect enrollment in agricultural education on two fronts. First, they are the point of contact for students when decisions are made regarding what courses to take. A positive recommendation from a guidance counselor could encourage a student to take an agricultural education course. Conversely, a negative comment or no information at all is a detriment to enrollment. Second, counselors and administrators put together the master schedule. It is important that they understand how agricultural education courses are sequenced and the target class grade for each course.

Among the barriers to successful recruitment and retention of students are some that are out of the agriculture teacher's control. These are ones that are schedule-related or related to high school graduation or college admission requirements. Other barriers can be overcome with planning and dedication.

Agriculture teachers often provide advisement to students. Students share with their agriculture teachers because they trust the teachers. Agriculture teachers provide guidance to students in two areas: career and personal. Career guidance involves assisting a student in choosing a career and gaining the preparation necessary to enter the career. Personal guidance involves all the issues that are a part of the student's life. The agriculture teacher may not be qualified to deal with some of these issues. In that case, the student should be referred to a professional counselor.

QUESTIONS FOR REVIEW AND DISCUSSION

1. Should every student be required to take at least one agricultural education course sometime in either middle school or high school? Why or why not?

2. What are the variables in the successful recruitment of students into agricultural education programs?

3. Why do students enroll in an agricultural education course?

4. What are several strategies for effective student recruitment?

5. Why is a middle school agricultural education program an effective recruitment strategy?

6. Why do certain students respond to particular recruitment strategies while others do not?

7. Why is retention so important for agricultural education programs?

8. What are the benefits of retention for the students?

9. What are the major barriers for student enrollment in agricultural education?

10. Why do students frequently come to their agriculture teacher for guidance?

ACTIVITIES

1. Investigate the confidentiality laws of your state and how these are practiced in the schools. Include sharing of student information, such as grades. Also, investigate requirements regarding teachers reporting child abuse and other illegal activity to authorities.

2. Develop a recruitment and retention strategy for an agricultural education program in a local high school. Include an electronic presentation that could be used in a middle school to promote enrollment. Give an oral report in class on what you develop.

3. Using the *Journal of Agricultural Education*, *The Agricultural Education Magazine*, and the proceedings of the National Agricultural Education Research Conference, write a report on career guidance research in agricultural education. Current and past issues of these journals can be found in the university library. Selected past issues can be found online at the American Association for Agricultural Education and the National Association of Agricultural Educators Web sites.

4. Design a brochure that would be useful in student recruitment for a high school agricultural education program. Write the content, develop line art or photographs, and lay out the brochure.

CHAPTER BIBLIOGRAPHY

Anderson, C., and L. C. Rampp. (Eds.). (1993). *Vocational Education in the 1990s, II: A Sourcebook for Strategies, Methods, and Materials.* Ann Arbor, MI: Prakken Publications.

Binkley, H. R., and R. W. Tulloch. (1981). *Teaching Vocational Agriculture/Agribusiness.* Danville, IL: The Interstate Printers & Publishers, Inc.

Breja, L. M., A. L. Ball, and J. E. Dyer. (2000). Problems in student retention: A Delphi study of agriculture teacher perceptions. *Proceedings of the 27th Annual National Agricultural Education Research Conference,* San Diego, CA, 27, 502–513.

Brown, D., L. Brooks, and Associates. (1990). *Career Choice and Development: Applying Contemporary Theories into Practice* (2nd ed.). San Francisco: Jossey-Bass.

Dykeman, C., M. Ingram, C. Wood, S. Charles, S. M. Chen, and E. Herr. (2001). The taxonomy of career development interventions that occur in America's secondary schools. (ERIC Digest No. CG-01-04). Accessed on November 14, 2003.

Fishbein, M., and I. Ajzen. (1975). *Beliefs, Attitudes, Intentions, and Behaviors.* Reading, MA: Addison-Wesley.

Frick, M., and S. Stump. (1991). *Handbook for Program Planning in Indiana Agricultural Science and Business Programs.* Unpublished manuscript, Purdue University.

Gliem, R. R., and J. A. Gliem. (2000). Factors that encouraged, discouraged, and would encourage students in secondary agricultural education programs to join the FFA. *Proceedings of the 27th Annual National Agricultural Education Research Conference,* San Diego, CA, 27, 251–263.

Hoover, T. S., and D. C. Scanlon. (1991). Recruitment practices—A national survey of agricultural educators. *Journal of Agricultural Education, 32*(3), 29–34.

Indiana Department of Education. (1999). *Teacher/ Local Team Self-study of Standards and Quality Indicators for Agriscience and Business Program Improvement.* Indianapolis: Author.

International Association for Educational and Vocational Guidance (IAEVG). (n.d.). *Strategies for Vocational Guidance in the Twenty-first Century.* Accessed on November 14, 2003.

Jordaan, J. P., and M. B. Heyde. (1979). *Vocational Maturity During the High School Years.* New York: Teachers College Press, Columbia University.

Kahler, A. A., B. Morgan, G. E. Holmes, and C. E. Bundy. (1985). *Methods in Adult Education* (4th ed.). Danville, IL: The Interstate Printers & Publishers, Inc.

Maddy-Bernstein, C. (2000). Career development issues affecting secondary schools. The highlight zone: Research @ work No. 1. Accessed on November 14, 2003.

Myers, B. E., L. M. Breja, and J. E. Dyer. (2004). Solutions to recruitment issues of high school agricultural education programs. *Journal of Agricultural Education, 45*(4), 12–21.

Myers, B. E., J. E. Dyer, and L. M. Breja. (2003). Recruitment strategies and activities used by agriculture teachers. *Journal of Agricultural Education, 44*(4), 94–105.

National Association of Agricultural Educators (NAAE). (2003). NAAE Web Site. Accessed on August 22, 2003.

National Association of State Directors of Career Technical Education Consortium (NASDCTEc). (2012). *Career Clusters at a Glance.* Accessed on November 7, 2012.

National FFA Organization. (2005). *Local Program Resource Guide: 2005–2006* [CD-ROM]. Indianapolis: Author.

National FFA Organization. (2011). *Local Program Success* (online). Accessed on November 30, 2011.

National FFA Organization. (2012). *FFA Statistics.* Accessed on April 26, 2012.

Newcomb, L. H., J. D. McCracken, J. R. Warmbrod, and M. S. Whittington. (2004). *Methods of Teaching Agriculture* (3rd ed.). Upper Saddle River, NJ: Prentice Hall.

Parsons, F. (1909). *Choosing a Vocation.* Boston: Houghton Mifflin.

Phipps, L. J., and E. W. Osborne. (1988). *Handbook on Agricultural Education in Public Schools* (5th ed.). Danville, IL: The Interstate Printers & Publishers, Inc.

Quey, R. (1971). Functions and dynamics of work groups. *American Psychologist, 26*(10), 1081.

Ruffing, K. (2006). *The History of Career Clusters* (online). Accessed on August 14, 2008.

Super, D., J. Crites, R. Hummel, H. Moser, P. Overstreet, and C. Warnath. (1957). *Vocational Development: A Framework for Research.* New York: Teachers College, Columbia University.

Sutphin, H. D., and M. Newsom-Stewart. (1995). Student's rationale for selection of agriculturally related courses in high school by gender and ethnicity. *Journal of Agricultural Education, 36*(2), 54–61.

10

Classroom and Laboratory Facilities

objectives

This chapter explains the importance of facilities and equipment in teaching and learning. It has the following objectives:

1. Describe the role of facilities in teaching
2. Identify the kinds of facilities needed for agricultural education
3. Explain the organization and maintenance of facilities
4. Develop a plan for the continual updating of facilities and equipment
5. Maintain a complete inventory of equipment and supplies
6. Practice safety in instructional environments

Glenda Anderson is teaching in a high school where student numbers are rapidly increasing. The original school building, including the agriculture facility, was constructed forty years ago. Temporary classrooms are used for some classes. The board of education has obtained land for building a new, larger, modern school facility. The voters in the school district have approved the sale of bonds to finance a new building.

Planning for the building is underway by the superintendent, principal, and others. Ms. Anderson is delighted that a new agriculture facility will be included. She has been invited to a meeting at the superintendent's office with the architect so that she can share her ideas. Now, she is pondering how to prepare, what to take, and what to say. She is even wondering about the nature of the classroom and laboratory facility that is needed.

For a teacher to have input into facility design is unusual. Most teachers are at schools that have been established, and they work with what exists. To offer ideas on a facility is a wonderful opportunity, but the teacher must be prepared.

terms

agriscience laboratory
aquaculture laboratory
classroom
facility
inventory

laboratory
learning stop
safety
storage facility

10–1. A principal and an agriculture teacher are reviewing a proposed facility plan.
(Courtesy, Education Images)

ROLE OF FACILITIES IN TEACHING AND LEARNING

Agricultural education is a facilities-intensive program. A ***facility*** is a building, a specialized area within a building, an outside area, or a large item that is not moveable or that is attached. Although students are capable of learning in any environment, facilities make a difference in the quality, breadth, and depth of instruction. It is very difficult to demonstrate, experiment with, and practice concepts without the proper facilities.

Facilities and Program

Facilities must be matched to the type and size of program desired. The program plan, as discussed in Chapter 6, should guide the facilities planning process. Ideally, a program is planned first and the facilities are established afterward so that the program can be efficiently and effectively implemented.

Here are a few questions to consider: What is the program emphasis? What are the program goals? What are the needs of the community? What agricultural education courses will be offered? What is the expected student enrollment? How many agriculture teachers will the facilities need to support?

The Teacher's Role

When a new school building or new agricultural education facilities are being planned, the agriculture teacher must inform planners, decision makers, and architects of the needs of the agricultural education program. Along with the local advisory committee and the state agricultural education specialist (state supervisor), the agriculture teacher should give input on items of which the others may not be aware. Information to be provided includes type and size of facilities needed; layout of rooms and equipment; special electrical, ventilation, or plumbing needs; and agricultural considerations, such as optimal location for a greenhouse. Input given before construction begins will save time and money and will enhance teaching effectiveness for the life of the facilities.

Building codes, state and local standards, and safety and fire codes must be met when any facility is being constructed. In addition, facilities must be accessible to persons with disabilities. In some states, facilities must meet state education standards if state funds are to be available for construction and equipment.

10–2. The C. E. Vickery Agriculture Center at Sandra Day O'Connor High School in Texas is a large complex of laboratory and supervised experience facilities.

(Courtesy, Education Images)

KINDS OF FACILITIES NEEDED FOR AGRICULTURAL EDUCATION

Agricultural education facilities are typically in two major areas: classroom and laboratory. Classrooms often have many similar qualities. Laboratories vary widely based on the instruction to be provided and the needs of the local community served by the school. Storage and work areas are parts of each.

Facility Location

Agricultural education facilities are typically found in one of three locations:

- **In a separate building**—A school may have the agricultural education facilities as a separate building or complex close to the main school building. This arrangement has the advantages of providing space for the program to expand, removing noise and odors from the main school building, and providing program visibility to the community.
- **Within the main building**—A school may locate the facilities within the main school building, usually within a wing housing other career and technical education programs. This arrangement has the advantages of efficiently utilizing the building infrastructure, involving agricultural education students and teachers in the total school, and providing for sharing of space between programs.
- **Within other departments**—A school may sometimes place the agricultural education facilities within other school areas, such as the science wing. This has the advantage of providing for sharing of space between programs, viewing the agricultural program as an integral part of the school, and recruiting students who might never see the program otherwise.

A teacher who has input into design should promote a location that enhances instruction. Access is very important with some instructional areas, such as farm machinery, livestock, and horticulture. Location can also enhance student participation by convenience of access as related to locations of other classes at a school facility.

10–3. An agriculture classroom with individual student desks at an Illinois high school.
(Courtesy, Education Images)

Categories of Facilities

There are four broad categories of agricultural education facilities: classroom, laboratory, storage, and office/teacher workroom. Within each of these categories can be specialized areas for different agricultural emphases.

Classroom   A *classroom* is a facility designed for group instruction. It typically has individual student desks or tables and chairs. It is important for the agricultural education program to have its own classroom(s). Experiments may need to be left in place for several days. The room may contain specialized equipment, such as GPS units, microscopes, aquariums, or food processing machines. FFA activities, student projects, and supervised experience records all require special room arrangements or storage places within the classroom.

An agricultural education classroom should have a minimum of 33 square feet of floor space per student. Ideally, the largest agricultural education classroom should have 45 square feet of floor space per student to allow for special activities. This translates into 825 to 1,125 square feet of floor space for a classroom designed for twenty-five students. Tables and chairs are preferable to tablet desks because they provide space and flexibility for experiments, student presentations, and group projects.

At least one wall of the classroom should be equipped with a writing surface (chalkboard or whiteboard). Other walls of the classroom can be equipped with bulletin boards and storage cabinets. Sometimes an agricultural education classroom is equipped with a row of Internet-connected computers along one wall. These computers should be networked to a printer and other peripherals. A projection screen should be located in the front or in a corner of the classroom. Use the guidelines in Chapter 13 to determine proper screen size for the room. Natural light through windows is desirable in the classroom, but blinds or other means must be available to control light level for audiovisual use. If applicable, stations with water, electricity, and gas are important additions to the classroom.

Audiovisual equipment is an important component of the agricultural education classroom. Depending on the school's resources, the classroom may have dedicated equipment or shared equipment may need to be scheduled through the school's media center. Programs that have dedicated audiovisual equipment tend to use

10–4. Part of a bench-type lab facility for agriscience at a California high school. (Courtesy, Education Images)

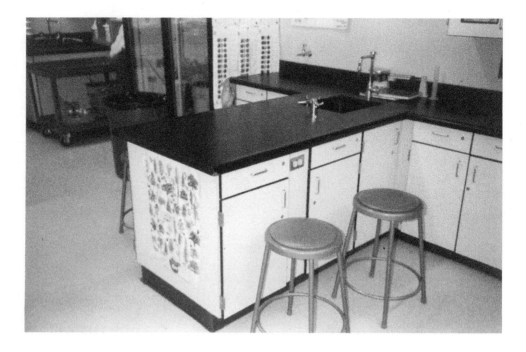

that equipment more often. Ideally, the classroom should have a dedicated interactive whiteboard. This should be paired with a digital video projector and Internet-connected computer capable of playing Blu-ray and DVDs. In some programs, a VCR player, overhead projector, and a slide projector may be useful.

Laboratory   Some classrooms may have built-in laboratory facilities. Possible examples are those for food processing, small-animal care, computer technology, hydroponics, or aquaculture. However, most laboratory facilities are separate rooms or buildings. With either arrangement, an agricultural education facility should have restrooms, sinks for cleaning up, and a water fountain. For all facilities applicable National Electrical Code (NEC) requirements, National Fire Protection Association (NFPA) codes and standards, and Occupational Safety and Health Administration (OSHA) laws and regulations must be met.

Agricultural education programs require several types of laboratory facilities. A *laboratory* is an area for individual or group student experiments, projects, or practice in agriscience, food science, aquaculture, agricultural mechanics, horticulture, plant and soil science, animal science, and natural resources. Each of these will be discussed in further detail.

Agriscience.   An *agriscience laboratory* is a facility used in teaching the science and math principles and concepts associated with agriculture. This type of facility should have 160 square feet of floor space per student. For a twenty-five-student class, this would mean 4,000 square feet. This amount of space allows for bench-type experiments, demonstrations, and projects. The agriscience laboratory may need water, electricity, and gas at each workspace. Ventilation and floor drainage must be provided.

Equipment storage is a necessity. Microscopes, measurement devices, and other equipment need storage cabinets. Depending on the community, the laboratory will need cages for small animals, aquaculture tanks, hydroponics units, grow lights, soil-erosion experiment tables, and other agriscience-related equipment.

Food Science.   Food science laboratories are becoming increasingly popular. One type of food science laboratory includes the equipment and facilities for

10–5. Exterior of an enclosed aquaculture facility in West Virginia.
(Courtesy, Education Images)

processing meat and other foods. The other type of facility provides for experiments, cooking, and preparation of food products. State guidelines should be followed when designing and equipping food science laboratories. Food safety, sanitation, and processing regulations vary by state and locality and must be followed.

Aquaculture.   An *aquaculture laboratory* is a facility used for providing learning experiences in fish farming and related areas. The laboratory may be indoors, using tanks or vats, or outdoors, using ponds or raceways.

Agricultural Mechanics.   An agricultural mechanics laboratory should have 150 to 200 square feet of floor space per student. For a twenty-five-student class, this would mean 3,750 to 5,000 square feet. The type of agricultural mechanics program needed by the community will determine the design of the facility. Possible types include woodworking and construction, metals and welding, large engines and hydraulics, small engines, electricity, plumbing, concrete, and paint and bodywork. Each type requires a different layout and different equipment and tools. Adequate storage for projects, tools, and supplies is a necessity. An agricultural mechanics laboratory must be designed with student safety in mind. Building and fire codes must be adhered to strictly. Student health considerations are important when designing agricultural mechanics facilities. Dust, welding fumes, volatile organic compounds, exhaust fumes, and other contaminants must be properly handled.

Phipps and Osborne (1988) provided useful guidelines for designing an agricultural mechanics laboratory. The central portion of the laboratory should be clear to allow space for large projects. An overhead door at least 24 feet wide and 18 feet high will permit tractors and such to be brought into the facility. A regular-sized entrance door should be provided so people can enter and exit without having to open the overhead door. The ceiling should be 20 to 24 feet high. Concrete floors provide for numerous uses of the space. The electrical wiring should provide for both 110- and 220-volt current. Depending on the equipment needs, three-phase current will also need to be provided. An outside, fenced-in patio can provide both work and storage space. Inside and outside water service, including appropriate floor drains, is needed for many projects.

10–6. Ducts help remove welding fumes from a laboratory area with welding booths.
(Courtesy, Education Images)

Horticulture. Horticultural laboratories include greenhouses, headhouses for preparation and sales, indoor or outdoor landscaping areas, floriculture areas, and aquaculture/hydroponics areas. Note: Depending on purchasing regulations, greenhouses are sometimes designated as equipment rather than facilities.

A primary greenhouse should be well constructed of glass or hard plastic and have 60 square feet per student, but no less than 1,600 square feet total. The headhouse should have 30 square feet per student, but no less than 1,150 square feet total. The agriculture teacher must decide on bench arrangement, walkways, watering system, and temperature regulation system. Cost is almost always a limiting factor, but in all decisions, functionality as a teaching facility must take priority.

Floriculture can be taught using a classroom or the headhouse but requires some special equipment. Various-sized coolers and display cases are necessary to maintain the flowers and plants. Storage cabinets are important for the care and maintenance of tools and equipment.

Landscaping facilities can be both indoors and outdoors. Indoor facilities can be used for water garden displays, displays of landscaping materials, and construction of items such as gazebos. Ideally, an outdoor landscaping laboratory will be at least 1 acre in size. The area can be sectioned into smaller plots as needed. For many agricultural education programs, the school grounds or a town park becomes the outdoor landscaping laboratory. With this arrangement, care must be taken that tasks and projects conducted are for educational purposes.

Plant and Soil Science. Land laboratories may be located on the school grounds or a distance from the school. These may range in size from as little as 1 acre to 100 acres or more. Larger land laboratories may serve as moneymakers for the agricultural education program, as supervised experience for students, or as demonstration plots for agricultural producers in the community. Some programs own the planting, tillage, and harvesting equipment themselves, while others contract with community farmers for these services. Land laboratories are most useful when used for year-round agricultural education instruction. Best utilization requires summer employment for the agriculture teacher and extended class periods, such as in block scheduling, during the school year.

10–7. An animal laboratory is a highly useful feature at this California school.
(Courtesy, Education Images)

Animal Science.   Animal science laboratories range in size from what is needed for housing small animals indoors, to barns for animals, to several acres for livestock grazing. Animal facilities must follow health and safety guidelines in addition to humane animal-care guidelines. Provisions must be made for food and water, waste disposal, odor control, and disposition of the animals at the end of instruction. Note that when housing animals in the classroom or indoor laboratory, any state or local indoor air quality regulations must be followed.

Natural Resources.   Natural resources laboratories include Christmas tree farms, wildlife areas, ponds, nature areas with trails, and timber production areas. These laboratories are excellent for teaming with other high school teachers and with middle school and elementary school teachers. These laboratories, more so than others, require signage and learning stops. A *learning stop* is a location in a lab where students stop and read information, do a task, or observe some phenomenon. Land laboratory facilities should be designed using a master plan in order to build in as many learning experiences as possible.

Storage   Agricultural education programs need an assortment of tools, equipment, textbooks, audiovisuals, resource materials, and consumables. When not in use, these items must be properly stored. A *storage facility* is a secure location where materials and equipment can be inventoried, organized, and protected from loss and damage. Storage facilities in agricultural education include separate rooms, as well as cabinets, bookcases and racks, lockers, and various types of shelving.

The agricultural education classroom should have storage for textbooks and other resources, FFA materials, student supervised experience record books, and student notebooks. If the classroom is dedicated to a specific function, such as agriscience, then storage for frequently used tools, equipment, and consumables should also be provided in the classroom.

A separate storage room is best for tools and equipment that are valuable, used infrequently, or bulky. A guideline is to have 10 to 16 square feet of storage space per student. Wall and cabinet storage should have tool silhouettes so the agriculture teacher can easily see if any tools are missing.

Office/Teacher Workroom Dedicated office space allows the agriculture teacher to be organized, professional, and responsive to student and community needs. Each agriculture teacher should have a minimum of 120 square feet of office space. Teachers can share office areas when a school has more than one agriculture teacher. The office can be combined with a workroom or a conference room.

In a teacher office, each agriculture teacher should have, at a minimum, a desk, a desk chair, a worktable, and a filing cabinet. The office should have at least one telephone with an outside line. A telephone on each desk is preferred. An agriculture teacher is required to communicate with employers, adult agricultural education students, suppliers, and others. This justifies an agriculture teacher having an outside line when other teachers may not. An Internet-connected computer with a printer is important for the agricultural education teacher office. Increasingly, curricular, FFA, and supervised experience resources are located on the Internet or require Internet communication.

ORGANIZING AND MAINTAINING FACILITIES

Facilities should be properly organized and maintained to assure their efficient use and long life. A neat facility represents the agriculture teacher(s) and program well with the publics who see it.

Organization

Agricultural education facilities should be organized for effective and efficient instruction. In addition, they should be attractive and professional in appearance.

Agricultural education facilities may be used by the community more often than other school instructional facilities and can be good public relations tools.

The state supervisor of agricultural education in each state will likely have sample facility plans for various curriculum offerings. Safety is a major concern in establishing facilities. Plan doors, windows, ventilation systems, and other features to promote student and teacher safety.

Maintenance

Equipment and tools must receive both daily and periodic maintenance. Daily tasks, such as cleaning, checking for wear and tear, and storing properly, help maintain appearance and functionality. Periodic maintenance schedules should be followed for replacing parts, lubricating, making adjustments, and cleaning thoroughly. Well-maintained equipment and tools will provide years of quality service.

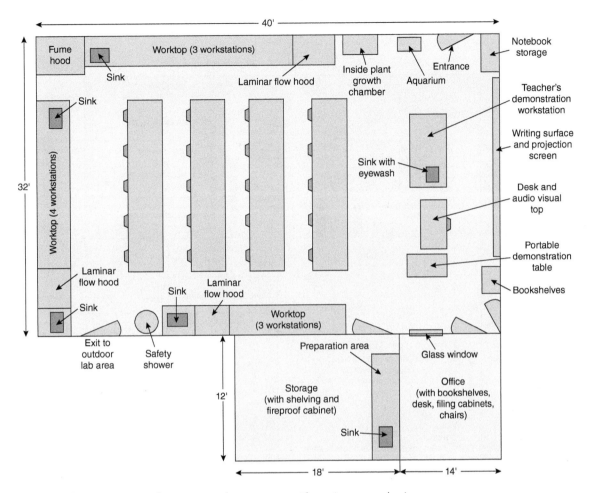

10–8. Facility arrangement for a one-teacher program with a science emphasis.

UPDATING FACILITIES AND EQUIPMENT

A replacement and updating plan should be developed for the facility and equipment. This will extend the life of the facility and equipment while preventing them from becoming completely outdated. Local, state, and federal funding sources will have various requirements for documenting the need for new or replacement facilities and equipment. The agriculture teacher should develop one-, three-, and five-year plans.

The one-year plan should include items needed immediately and included in the budget. It is good to have additional items in the one-year plan in case more funds become available during or at the end of the school year. Safety equipment, such as eye, hearing, and body protection, should be replaced annually. The three-year plan should include replacement items as well as items needed for new curricular thrusts. The five-year plan should be a capital improvements plan that will give school administrators and the school board time to budget for larger-expense items.

INVENTORYING EQUIPMENT AND SUPPLIES

An *inventory* is a complete list (by kind and number) of equipment, tools, and supplies. An inventory should be taken as soon as possible upon initial employment and yearly thereafter. The results should be communicated to

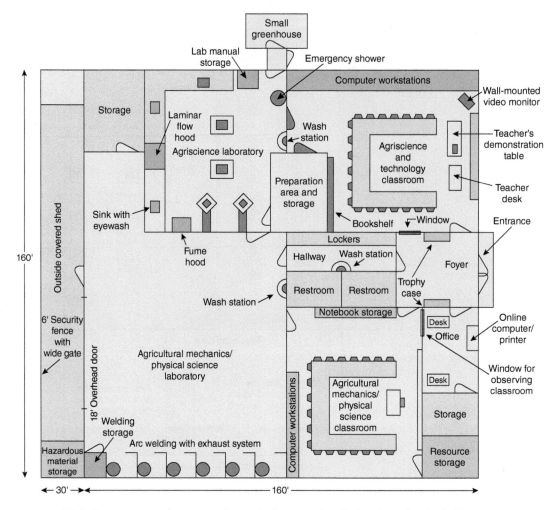

10–9. Arrangement of a two-teacher agriscience and agricultural mechanics facility.

administrators. A sample inventory form is shown in Figure 10–11. The teacher can use this form when walking around viewing the items. Then, the information can be entered into a computer database created in a software program such as Microsoft Excel or Access. Note that in addition to counting the number of items, documenting their condition and usefulness is also necessary. Items that are in less than desirable condition should be repaired or replaced. Consideration should be given to getting rid of items that were not used the previous year and are not anticipated to be used the next year. Not only do these take up space that could be used for other purposes, but they contribute to an environment of clutter and disorganization.

SAFETY IN INSTRUCTIONAL ENVIRONMENTS

All agricultural facilities require that safety be practiced. *Safety* is the reduction of risk and of the likelihood of personal injury. Many items used in the agricultural education curriculum can cause injury if misused or if safety is not practiced. Chemicals, power tools, heavy objects, pinch points, electricity, motorized vehicles, hot objects, and animals all have the potential to cause injury or harm to students.

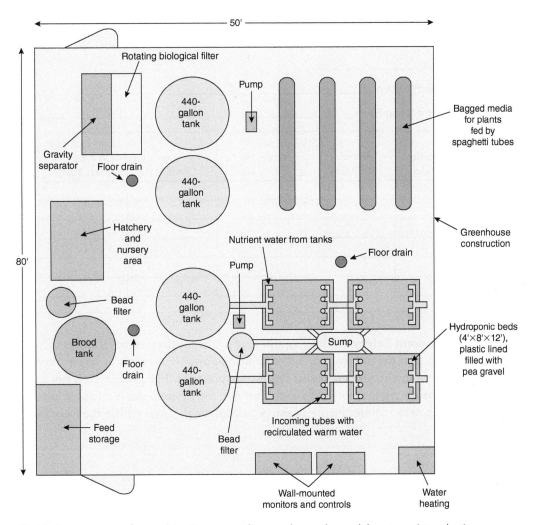

10–10. Arrangement of a combination aquaculture and greenhouse laboratory. A teacher's office, storage room, classroom, and headhouse are provided separately from this main facility.

Safety is a vital concern for the agriculture teacher. Instructional time must be devoted to safety so students both know and practice safe habits. Safety equipment must be provided in working condition, and its use strictly enforced. Finally, the agriculture teacher must be a role model in the use of safe practices.

Agriculture teachers may be held liable for actions they may or may not take within the classroom and laboratory (Lee, 2000). An agriculture teacher is expected to do what a reasonable, prudent person would do. The teacher should provide instruction and practice in the safe use of equipment, tools, and supplies. Documentation must be kept that shows students have passed a safety knowledge test and have demonstrated safe operation and practices. Safety guards and equipment must be installed, and students must understand their use. Tools and equipment must be properly maintained or taken out of use until repaired. Finally, tools and equipment should be used for designed purposes. Other uses may place students at risk of harm or injury.

Practices for Teachers

Teachers are responsible for the safety of their students. Many teacher actions and practices contribute to a safe instructional environment. Lee (2000) has summarized these into ten essential safety practices for teachers (Table 10–1).

10–11. An inventory form for equipment, tools, and supplies.

Inventory Form

School Year _____–_____

Agriculture Program _____

Person Completing Form _____

Item Name and Brand	Number of Items	Model No./ Serial No.	Condition	Used Last Year or Will Be Used Next Year?

Power Equipment Safety Power equipment presents safety considerations for the eyes, hearing, and the extremities. Whether using power hand tools or larger stationary power equipment, students must follow safe practices. Eye protection is not only important for the person operating a tool but also for those in the immediate area. A good practice is to require the use of eye protection at all times within the laboratory

Table 10–1
SAFETY PRACTICES IN AGRICULTURAL EDUCATION

Practice	Explanation
Plan and practice.	Teachers should plan learning activities and try out what students will be doing.
Instruct students.	Teachers should instruct students in potential dangers to avoid and safe practices to follow. Documentation is critical.
Supervise the learning environment.	Teachers should always be present with their students and be attentive to what is happening in the learning environment.
Keep up the facilities.	Teachers should keep laboratory facilities clean, in good repair, and free of potential hazards. Broken equipment, dangerous chemicals, and safety devices should always be dealt with properly.
Be a good example.	Teachers should set a good example by dressing and grooming properly and wearing proper protection. They should never violate safety procedures.
Use safety cleanup materials.	Teachers should have the appropriate cleanup materials readily available in case of an accident.
Know and follow school policies.	Teachers should know and follow school policies. The hazardous waste officer should be asked to review the laboratory and teaching content.
Obtain professional educator liability insurance.	All educators need to have insurance coverage that protects them while they are conducting their professional duties. However, insurance does not provide protection if the teacher is negligent.
Be a good housekeeper.	Keeping the classroom and the laboratory in good condition helps avoid some hazards.
Post emergency numbers.	Teachers should post emergency telephone numbers near all telephones.

space. This eliminates students' forgetting to put their eye protection on and the need to define the dividing line for when eye protection must be worn.

Hearing protection is important for the student operating a power tool or power equipment. It may also be required for those within the immediate area. Teenagers may not realize the extent to which exposure to loud noises can cause damage, so it is critical that the agriculture teacher be a role model and enforce use of hearing protection.

Often overlooked is the need to keep long hair from contacting moving surfaces. Long hair should be pinned up or put under a hat.

Other safety considerations around power equipment include proper clothing and footwear and removal of jewelry. Students should also be aware that slick-bottomed shoes are not appropriate for the laboratory. There may be instances where athletic footwear is also not appropriate. The agriculture teacher should conduct a safety inventory to identify potential dangers and hazards before using laboratory instruction.

Chemical Storage   Chemicals are used throughout the curriculum in agricultural education. The agriculture teacher should be trained in the proper storage, handling, and use of all chemicals used in the program. A system should be established for filing material safety data sheets and for ensuring that out-of-date chemicals are removed and properly disposed of. Lee (2000, p. 77) gave the following suggestions for properly storing chemicals:

1. An appropriate storage area should be provided. Often known as the stockroom, this area should be locked and kept off-limits to students and other school personnel.
2. Only compatible chemicals should be stored together. Some chemicals are incompatible and will react violently if they come together when accidentally spilled.
3. Chemical containers should always be labeled appropriately. Chemicals that are unlabeled invite trouble.
4. Chemicals that are corrosive and flammable should be stored in approved fireproof cabinets.
5. All containers should be properly covered. Tops should be appropriate for the materials being stored.

Safety Equipment   It is important that students have access to and use proper safety equipment. Lee (2000, pp. 79–80) gave the following recommended safety equipment list:

1. Emergency eyewash equipment—A handheld wash bottle that can be disposed of after one use is likely best; several are needed in each laboratory area.

    A fountain, permanently mounted, may work well (minimum of one per laboratory, with additional fountains in larger laboratories). Note: An alternative is a unit that requires no plumbing and uses a special eyewash solution. Either should be capable of flushing both eyes at the same time.

    Caution: Keep eyewash facilities clean. Microorganisms that cause infection can build up in facilities that are not properly maintained.
2. Drench shower—freestanding or mounted (minimum of one per laboratory).
3. Fire blanket—fiberglass; 43 × 39 inches (one per laboratory).
4. First-aid kit—metal cabinet complete with standard first-aid supplies (one per laboratory).

5. Tethered buoy ring (needed for aquaculture involving ponds, large tanks, and streams)—rope 30 feet or longer attached to sturdy post (one per worksite or tank).

6. Sanitizing goggle cabinet with goggles—wall-mounted, with ultraviolet lamp; twenty-goggle minimum capacity (one or more per laboratory).

7. Goggles—clear; scratch resistant (one pair for each student).

8. Fire extinguisher—Halon or dry chemical charge (one or more per laboratory).

9. Particle masks—bulk quantity.

10. Safety storage cabinet—approved double-wall construction; 18-gauge welded steel; must conform to OSHA regulations (one per laboratory).

11. Sand bucket—metal; 5-gallon size, with sand (one per laboratory).

12. Smoke or heat detector—standard detector (one per laboratory).

13. Aprons—vinyl, rubberized, or leather, depending upon the requirements of the laboratory (one per student or twenty per laboratory, depending on the number of students).

14. Safety gloves—disposable latex gloves (quantity supply for laboratory); leather gloves as appropriate (one pair per student or twenty pairs per laboratory, depending on the number of students).

15. Trash containers—metal or plastic lined; 10-gallon size (one per workstation).

16. Broom with dustpan—for cleaning nonhazardous spills (one set per workstation).

17. Mitts—nonslip; used for handling hot materials (one pair for each workstation).

18. Safety charts—wall-mounted signs or charts appropriate for laboratory area (one set per laboratory).

19. Cleanup kits—approved kits for caustic, solvent, and acid spills (one of each for each laboratory).

20. Other safety equipment as needed for laboratory activities.

Vandalism and Violence   Classroom management techniques are discussed in Chapter 14. There are additional considerations regarding student management in the laboratory. In all cases, the agriculture teacher must follow local procedures regarding violence, vandalism, and student behavioral problems. Some students may be prone to threaten or cause harm to other students or the teacher. These students may be denied access to the laboratory. Vandalism can render some equipment inoperable and, depending on budgets, not repairable. Vandalism also disrupts the aesthetics of the laboratory and if left untreated can lead to further vandalism. Theft, besides being illegal, creates a need for expensive replacement of items stolen and disrupts learning, as students may not be able to work until items are replaced.

10–12. Displaying posters in a laboratory alerts people to safety areas or hazards.

The agriculture teacher can take some actions to alleviate potential vandalism and violence problems. First, tools and other materials should be kept in locked storage rooms or cabinets. See Chapter 21 for suggestions on end-of-class cleanup procedures. Second, the laboratory should be arranged so the teacher can see all students as they work. And finally, the teacher should never leave the laboratory or classroom unsupervised. If the teacher must be out for a time, another teacher or appropriate substitute should be in the room to supervise.

REVIEWING SUMMARY

Facilities are important for instruction in agricultural education. The facilities must match the type and size of program that the school and the community need. The agricultural education program plan should guide the facility planning, maintenance, and updating processes. Building agricultural education facilities involves input from the community, school administrators, the agriculture teacher(s), architects, and others.

Four types of facilities are typically found in an agricultural education program. One or more dedicated classrooms assist in the instructional program. Laboratory facilities are a must and need to match the instructional program. The agricultural education program requires storage facilities for the various equipment, tools, and supplies needed for the instructional program. Finally, an office and workspace facilitate the agriculture teacher doing his or her job.

Organization of agricultural education facilities aids the instructional process and gives them an attractive and professional appearance. Careful thought must be put into whether the facilities will be designed for a one-teacher or multiple-teacher agricultural education program. If possible, the design should allow for future additions or modifications as the program focus expands or changes.

The agriculture teacher is responsible for selecting and ordering equipment, tools, and supplies. Using local suppliers is good for positive public relations and community support but may not always be possible or allowed. The teacher is also responsible for the maintenance and inventorying of these items. Regular maintenance greatly extends the life and usefulness of equipment and tools.

Safety is critical when using agricultural education facilities. Teachers may be held liable for student injuries because of actions they did or did not take. Teachers must be familiar with safety rules and practices themselves and must instill these in their students. Most important, agriculture teachers must model safe practices and must not violate safety procedures.

QUESTIONS FOR REVIEW AND DISCUSSION

1. What are the questions to be asked when designing agricultural education facilities?
2. What are the four categories of agricultural education facilities? Why are these needed?
3. Why are tables and moveable chairs preferable for an agricultural education classroom?
4. What are the areas of agricultural education for which laboratories are needed?
5. What are special storage needs in agricultural education?
6. Why does an agriculture teacher need an outside telephone line and an Internet-connected computer?
7. What questions should be answered when purchasing equipment, tools, and supplies?
8. Why is maintenance of equipment and tools important in agricultural education?
9. Why is a beginning inventory upon initial employment important? Why is an annual inventory important?
10. Why is safety so important in agricultural education?

ACTIVITIES

1. Investigate the guidelines and requirements for agricultural education facilities in your state. Begin with the Web site of the state department. Contact the individual in charge of school facilities at the state level and request information, including copies of relevant state laws.

2. Debate the necessity for agriculture teachers to have equipment and facilities that may not be available to other teachers. Examples include a private office, an outside telephone line, an Internet-connected computer with printer in the office, additional storage space, a workroom or conference room, vehicles, and specialized equipment or rooms.

3. Using the *Journal of Agricultural Education, The Agricultural Education Magazine*, and the proceedings of the National Agricultural Education Research Conference, write a report on research about safety within agricultural education.

Current and past issues of these journals can be found in the university library. Selected past issues can be found online at the American Association for Agricultural Education and the National Association of Agricultural Educators Web sites.

4. Consider an ideal facility for the area of agricultural education you want to teach. Sketch the layout of the classroom and laboratory areas. Be sure to include storage and office areas. Be sure to include doors, windows, bench arrangements, and other details. Make your sketch on poster paper for display and discussion.

CHAPTER BIBLIOGRAPHY

Binkley, H. R., and R. W. Tulloch. (1981). *Teaching Vocational Agriculture/Agribusiness*. Danville, IL: The Interstate Printers & Publishers, Inc.

Frick, M., and S. Stump. (1991). *Handbook for Program Planning in Indiana Agricultural Science and Business Programs*. Unpublished manuscript, Purdue University.

Indiana Department of Education. (1999). *Teacher/Local Team Self-study of Standards and Quality Indicators for Agriscience and Business Program Improvement*. Indianapolis: Author.

Kahler, A. A., B. Morgan, G. E. Holmes, and C. E. Bundy. (1985). *Methods in Adult Education* (4th ed.). Danville, IL: The Interstate Printers & Publishers, Inc.

Lee, J. S. (2000). *Program Planning Guide for Agri-Science and Technology Education* (2nd ed.). Upper Saddle River, NJ: Pearson Prentice Hall Interstate.

National Association of Agricultural Educators (NAAE). (2003). NAAE Web Site. Accessed on August 22, 2003.

National Association of State Directors of Career Technical Education Consortium (NASDCTEc.) (2002). *Focusing Education on the Future: What Are Career Clusters?* Accessed on November 14, 2003.

National Electrical Installation Standards. (2012). *The National Electric Code and NECA*. Accessed on November 7, 2012.

National FFA Organization. (2005). *Local Program Resource Guide: 2005–2006* [CD-ROM]. Indianapolis: Author.

National FFA Organization. (2011). *Local Program Success* (online). Accessed on November 30, 2011.

National Fire Protection Association. (2012). NFPA Web Site. Accessed on November 7, 2012.

Newcomb, L. H., J. D. McCracken, J. R. Warmbrod, and M. S. Whittington. (2004). *Methods of Teaching Agriculture* (3rd ed.). Upper Saddle River, NJ: Prentice Hall.

Occupational Safety and Health Administration. (2012). Web Site. Accessed on November 7, 2012.

Phipps, L. J., and E. W. Osborne. (1988). *Handbook on Agricultural Education in Public Schools* (5th ed.). Danville, IL: The Interstate Printers & Publishers, Inc.

U.S. Environmental Protection Agency. (2012). *Indoor Air Quality* (online). Accessed on November 7, 2012.

11
Instructional Resources

Have you ever attempted to do a job without adequate tools or materials? Hard to do, right? Think of the situation you were in as related to that of a teacher without needed instructional resources.

Achieving goals in education requires resources just as much as making achievements in other areas of our lives and society. A student needs "tools" in order to learn; a teacher needs "tools" in order to promote learning. Teachers need resources to efficiently produce skilled and knowledgeable students just as veterinarians need instruments and equipment to promote animal health and floral designers need flowers and supplies to construct an arrangement. Without the needed "learning tools," the teaching-learning process will be inefficient and fail to reach its potential in promoting goal achievement.

Some schools provide an abundance of resources for teaching and learning; other schools may not. Teachers are in the important position of using resources to promote student learning in an attempt to gain the desired outcomes of education. Further, teachers have the responsibility of obtaining the needed resources to the fullest extent possible. This will often require extra initiative to make needs known to school administrators and promote the allocation of school resources so that students have the "tools" they need in order to learn.

objectives

This chapter covers the roles, kinds, selection, and filing of instructional resources. It has the following objectives:

1. Describe the roles of instructional resources in accountability
2. Identify kinds and sources of instructional resources
3. Discuss e-learning media in agricultural education
4. Apply selection criteria in obtaining instructional resources
5. Explain the management of instructional resources

terms

academic software	instrument
activity manual	interactive whiteboard
ancillary instructional resource	laptop
basal instructional resource	lesson plan library
computer-based module	netbook
desktop computer	paper-based material
display panel	reference
e-book reader	school management software
e-learning	smartphone
electronic-based material	student-use material
equipment	supply
facility	tablet computer
instructional materials	teacher's manual
instructional materials adoption	teacher-use material
instructional resource guide	textbook
instructional resources	tool

11–1. A student teacher is using a commercially prepared CD in teaching. (Courtesy, Education Images)

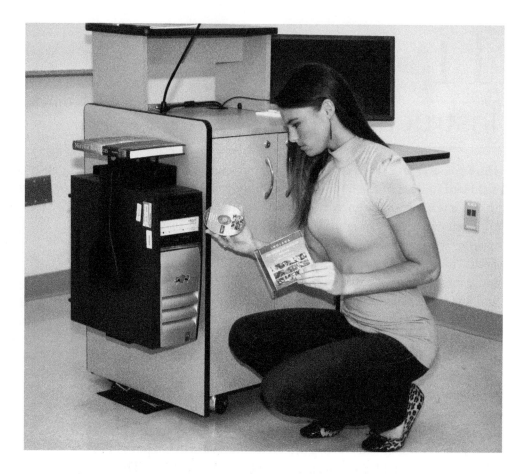

ROLES OF INSTRUCTIONAL RESOURCES IN ACCOUNTABILITY

A resource-rich learning environment in the classroom and laboratory promotes teaching and learning. An agricultural education teacher has the responsibility of requesting and obtaining appropriate instructional resources. A wise teacher always maintains a list of needed materials just in case funds become available. A teacher also has the responsibility of preparing precise requests for instructional materials (similar to bid specifications) and justifying their importance and usefulness.

Instructional resources are the materials used in the teaching process. They are among the "tools" that teachers use in providing instruction and that students use in learning. In a broad sense, instructional resources include many kinds of materials, ranging from those that are published on paper or as computer-based materials to plants, animals, lab instruments, supplies, and safety equipment.

Some materials and teaching approaches are much more effective than others. Good assessment of materials is needed before obtaining and using them in a teaching-learning environment. Just because certain materials are available does not mean that they will make a strong contribution to student achievement. Teachers should be prudent in making resource choices. Careful investigation of materials is needed before obtaining and using them with students. Determine the experiences of other teachers, the approaches used in the development of the material to assure authenticity, correlation to content standards, and contribution to high assessment scores of students.

Instructional Materials

Some instructional resources are known as instructional materials. *Instructional materials* are materials that have the content or skill information that is being taught.

11–2. Students should have an educational environment with resources that are useful in promoting efficient teaching and learning. (Courtesy, Education Images)

They include textbooks, activity/lab manuals, transparencies, electronic presentations, brochures, magazines, reference manuals in either print or electronic format, and many others but not laboratory facilities, equipment, tools, and supplies.

A teaching and learning environment rich in relevant, up-to-date instructional materials promotes increased achievement. As tools in the teaching and learning process, instructional materials are sometimes compared to the tools carpenters use in construction, such as hammers, levels, and squares. Without these tools, construction cannot progress; carpenters cannot achieve their goals. Likewise, without instructional materials, learning cannot progress; teachers and students cannot achieve their educational goals.

Students in an environment that is deficient in instructional materials will not likely perform well on accountability testing. Low scores by students may result in the teacher not being viewed as successful in promoting student achievement. The school administrators may also not appear successful as instructional leaders. The bottom line: A resource-rich learning environment promotes student educational achievement and provides motivation for future success.

Facilities, Equipment, Instruments, Tools, and Supplies

Many kinds of instruments, tools, equipment, and supplies are instructional resources that may be used in agricultural education. These vary with the nature of the instructional program. The curriculum outline and courses are established first. Appropriate instructional resources are then obtained to implement the curriculum. A particular course should not be offered just because a school has the facilities for doing so. Courses are offered to fill educational needs.

Facilities are essential in carrying out the instruction of a course of study. A *facility* is the school building (including classrooms and laboratories), grounds around the building, and any structures on a permanent foundation. Greenhouses and other useful structures are sometimes defined as facilities and other times as equipment. As a facility, a greenhouse, livestock barn, or other structure would need to be situated on a permanent foundation. Facilities should provide a modern and safe educational environment for students and teachers. Many states have facility plans for agricultural education facilities. In constructing facilities, local school boards engage the services

11–3. Schools with animal science instruction typically have facilities that include a barn or other structure and equipment for keeping animals.

(Courtesy, Education Images)

of an architect. Agriculture teachers and advisory groups should provide input to the architect in the design of facilities for agricultural education.

Equipment includes instructional resources that are larger, often stationary, and frequently operated with motors or engines. Equipment with motors or engines may be known as power equipment. In an agricultural mechanics lab, equipment may include welding machines, radial arm saws, and air compressors. Equipment in horticulture may include trimmers, mowers, soil mixers, and components of irrigation systems. With care, equipment is long lasting and can be used repeatedly. Equipment often poses hazards to students and teachers. Proper instruction in safety and safe usage is essential. In addition, instructional equipment includes items used in the classroom or laboratory to provide information and reinforce learning.

Instruments, tools, and equipment are used repeatedly in teaching and learning. An *instrument* is a device used for a particular purpose. An example is a caliper used to make precise measurements. Instruments are often small and require careful storage to maintain and protect them from damage. Some instruments are fairly expensive; others are inexpensive.

11–4. Instructional resources should support the intended outcomes of teaching and learning, such as those used in extracting DNA from strawberries.

(Courtesy, Education Images)

A *tool* is a handheld device that aids in doing a task. An example is a claw hammer. A claw hammer is designed to drive and pull nails. Without a hammer, nail driving is difficult. Proper hammers, for example, are needed so that students can be taught selection, use, and care of them. Tools used in agricultural mechanics classes vary from those used in horticulture, veterinary science, or forestry, though a few may be the same.

Schools that fail to allocate funds for instructional resources are planning to fail in student achievement. Quality instructional resources allow students to learn on their own and help develop self-motivation toward learning. Relevant instructional resources designed for student use empower students as learners. Yes, students can learn on their own without the continuous intervention of a teacher.

The instruments, tools, and equipment needed vary with the course and curriculum outline. A general agriscience class will have instructional resources that support achieving the educational objectives of the course. These may include microscopes, slides, soil test kits, and numerous other items.

Supplies are needed in order to use most instruments, tools, and equipment. A *supply* is a consumable material used to carry out instructional activities and/or work. Supplies vary with the nature of the instruction. Horticultural supplies may include potting soil, seeds, fertilizer, herbicide, insecticide, and other materials used to grow plants in the greenhouse or land lab. In agricultural mechanics, supply needs vary with the areas taught. For example, welding requires rods or wire, metal, and/or other supplies; construction requires lumber, plywood, nails, and other supplies; and machinery maintenance requires oil, filters, and other supplies.

A list of instruments, tools, equipment, and supplies for a particular course is often included with the curriculum outline. Otherwise, a teacher would get input from local advisory council members, other experienced teachers, and state supervisory personnel. Teachers gaining employment with existing programs will often go

11–5. Equipment and supplies should be appropriate to the area of instruction, such as animal science.
(Courtesy, Education Images)

to schools that have some resources remaining from previous years. A new teacher should inventory what is on hand and in workable condition and then request resources needed to implement the curriculum. In all cases, teachers should follow school procedures in making requests.

KINDS OF INSTRUCTIONAL MATERIALS

Instructional materials can be classified in a number of ways, such as by intended user and by medium of presentation. Teachers choose materials that provide the best fit in their schools and with their students. (Note: This section of the chapter refers to materials used in communicating subject matter and achieving subject-matter objectives. It does not include instruments, tools, and equipment used in laboratory skill development.)

Commonly used materials include textbooks and the ancillaries that accompany them. An ***ancillary instructional resource*** is a secondary or supplementary material that accompanies another material, most often a textbook. The material an ancillary accompanies is known as basal. A ***basal instructional resource*** is a material that provides the base or foundation for the organization of a subject. All else revolves around the basal resource. Textbooks are considered basal instructional resources.

User Classification

Instructional resources can be classified by who uses them. Some materials are designed for use by students as important resources in learning. Other materials are designed for use by teachers in efficiently directing the learning of students often focused on using student materials. Either may be published on paper or available electronically.

Student-Use Material   ***Student-use material*** is an instructional resource prepared for use by students. Usually, a great deal of care has been taken to make the material appealing to students. The writing level is within the range of readability of the students. Numerous photographs and line drawings are used to depict important concepts. Textbooks are the most common student-use materials, though activity manuals, lab exercise manuals, and self-paced CD materials are also used.

11–6. Textbooks are basal instructional resources that guide content delivery and promote student mastery. This shows a published program, including a student edition with accompanying teacher's manual, activity guide, and instructor's guide to the activity manual.
(Courtesy, Education Images)

- **Textbooks**—A *textbook* is a book designed for use by students that deals with a specific subject. Textbooks have usually been prepared following a systematic format, carefully reviewed to assure technical accuracy, and correlated to appropriate course objectives or standards. Textbooks are usually organized into parts or units. Each part or unit is composed of several chapters. The chapters are further divided with headings and subheadings. Motivational approaches to gain student attention and interest are used. End-of-chapter evaluations and hands-on activities are presented. A textbook is usually prepared for students at a particular grade level or range of grades, such as grade 10 or grades 9 through 11. Textbooks are typically printed in color and bound with hard covers. Typical subjects of textbooks used in agricultural education classes include agriscience, horticulture, biotechnology, landscaping, veterinary science, and plant and soil science.

Research (Lee, 2001a) correlating the standardized test scores of students with the teaching strategies used in their classes found that students in classes where textbooks are an integral part of the instruction score at least 12 percent higher on achievement tests than students in classes where other teaching strategies not involving textbooks are used. The study included only paper-printed textbooks and not electronic versions.

Paperback and hardbound textbooks when used with care are not destroyed after one use and can be used repeatedly with several classes. Students should be encouraged to take good care of books so that the books have long, useful lives. Paper books can be used in locations where Internet service and computer equipment are not available.

An alternative to textbooks printed on paper is online textbooks. Many paper textbooks are available online. An access code is purchased by the school and provided to students so that they can log on at school or at home (if Internet is available at home, and most homes today have such access). In most cases, the online materials are little more than scanned pages of printed books. Some students and teachers do not feel that the online copies of paper books are very user-friendly. A few online publishers

11–7. A student edition (textbook) may be used at school or at home. This shows a student placing a textbook into a book bag for return to school.
(Courtesy, Education Images)

have developed student information sheets so that the design is user-friendly and varies from that of a traditional textbook. An example is the Student Content E-Units of CAERT, Inc.

- **Activity manuals**—An *activity manual* is an ancillary resource that has activities organized around a basal resource, such as a textbook. The activities often focus on several goals and approaches in learning. Overall, the activities reinforce the achievement of objectives in the textbook, review terms, and provide hands-on experiments or other investigations. Students carry out the investigations and record their observations or experiences in space provided in the manual. Some of the investigations may involve using precise laboratory instruments. Other activities may involve collecting data from crops, animals, streams, forests, and other phenomena. Activity manuals often have instructor's guides to aid the teacher in directing student learning. Each student needs his or her own activity manual. Most activity manuals are consumable; they can be used only once because students answer questions and record information in them. Some activity manuals may be available as online materials.

- **Computer-based modules**—A *computer-based module* is instructional material that involves the use of units or modules guided by a computer program. The computer program directs students from one activity to another and through the activities. Textbooks, video clips, and hands-on activities may be part of the instruction. Some observations show that students with this type of instruction lose interest after a few weeks.

- **Record books**—Supervised experience and FFA participation accomplishments should be recorded. This may be with paper-printed record books or computer-based record systems. Records are essential for students to apply for advancements in FFA. States and/or local school districts may have record books or systems developed specifically for their use.

- **Online Reinforcement and Assessment**—Online approaches for reinforcing rote learning, providing student tutorials, and assessing student learning are available. Most widely used is the online assessment of student learning. Some are offered through state education agencies; others are from commercial assessment sources.

Teacher-Use Material   *Teacher-use material* is an instructional resource designed for use by a teacher. Its purpose is to aid the teacher in planning, delivering, and evaluating instruction. Some teacher-use materials stand alone, such as program planning materials. Other teacher-use materials are companions to student-use materials. A few examples of teacher-use materials are as follows:

- **Teachers' manuals**—A *teacher's manual* is a teacher-use material prepared to accompany a student-use material, such as a textbook. A teacher's manual or teacher's edition provides useful information for the teacher to use in delivering instruction using a student edition textbook. A teacher's manual will typically summarize content, offer teaching suggestions, include additional recommended resources, and provide answers to questions in the student edition. Teachers' manuals may be provided free of charge by publishers to teachers who adopt and purchase classroom sets of student edition textbooks. Teachers' manuals may be printed on paper or available in electronic form as a CD/DVD or for downloading from the provider's Internet site. (In some cases, editions of textbooks are published that incorporate the teacher's manual in the same binding as the student edition. These may be known as teachers' editions.)

- **Instructional resource guides**—An *instructional resource guide* is a collection of material that helps a teacher plan and deliver instruction. Instructional resource guides often contain detailed lesson plans, sample tests, transparencies or electronic-presentation images, lab sheets, and other material for use in delivering instruction. The guides may be prepared as loose-leaf notebooks, on CD-ROMs, or a combination of notebook and CD-ROM formats. Increasingly, such materials may be available on the Internet by using access codes. Such materials are usually sold, though they are occasionally offered free of charge if teachers adopt and use specific student materials.

- **Lesson plan libraries**—A *lesson plan library* is a collection of teaching plans focused on a fairly broad area, such as landscaping or wildlife management. The lesson plans are complete with a summary of content, suggested instructional strategies, sample tests, lab sheets, and transparencies or electronic-presentation images. Lesson plan libraries are often correlated to state standards or course blueprints. They may also be closely correlated to end-of-course or other forms of competency tests. These lesson plans may be available on a CD or by subscription to an online site. Some state teacher organizations and education agencies provide lesson plan libraries; others are available through commercial sources. Most commercial sources have rigorous review processes to assure that materials have been field-tested and are technically accurate and educationally sound.

- **Reference materials**—Reference materials are needed for use by the teacher in preparing lesson plans and answering questions. A *reference* is material that is more advanced than students would normally use. References may be in a variety of formats, including books, manuals, brochures, notebooks, CDs, and DVDs. They are sometimes publications from the Extension Service. References on science, agriculture, horticulture, veterinary medicine, and other areas are often needed, depending on the technical areas included in the instructional program and the nature of the agricultural industry in the local community. References on FFA are needed in all programs. Web sites can often be used as sources of information. An example is the publication of the *National Agriculture, Food and Natural Resources Career Cluster Content Standards*, which is available at the Team Ag Ed Web site.

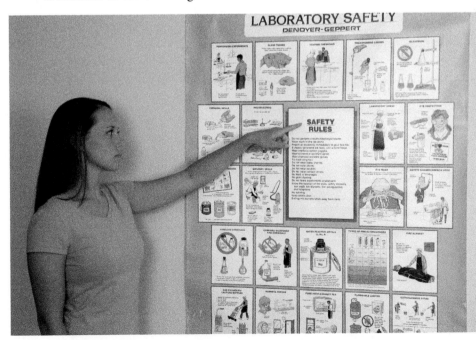

11–8. Posters often have important educational roles in agricultural education.
(Courtesy, Education Images)

Medium Classification

Medium classification refers to the form in which instructional resources are used by students and teachers. Some instructional resources are printed on paper; others may be in electronic form on CD or available through the Internet.

Paper-based material is instructional and learning material that is printed on paper. This medium of material has long been used and has yielded good educational results. The material may be printed in color or in black and white. Almost all are bound in some way—for example, as loose-leaf notebooks, hardback textbooks, or softback activity manuals.

Electronic-based material is instructional and learning material that is used in conjunction with a computer or computer network system. The material typically includes CDs or DVDs. Each student needs access to a computer. Students may work alone or work together in pairs. Some schools establish instructional technology labs, where groups of twenty or so students work at the same time.

Electronic materials available through the Internet typically involve the teacher having an access code to a Web site. The availability of an access code may require payment of an annual fee to the Web site provider. Computer hardware is needed for students to use these instructional resources. (The next section of this chapter, "Materials for E-Learning," provides more information about electronic-based materials.)

MATERIALS FOR E-LEARNING

E-learning is electronically based teaching and learning that usually involves some application of computer technology. It may be used by a teacher to supplement classroom instruction or as a stand-alone source of instruction. In most cases, e-learning approaches are integrated as supplements or enhancements in the traditional agricultural education classroom and laboratory. In agricultural education, they typically do not replace the instruction of the teacher. As Roblyer and Dowering (2010) indicated, "simply having students use technology does not raise achievement." They further indicated that the impact of technology depends on the ways it is used.

Essentially, e-learning involves a computer or smart device (such as an iPad) and CD or DVD materials or network-enabled transfer of knowledge and skills.

11–9. A student has entered school access information on a networked personal computer in the agricultural education classroom.
(Courtesy, Education Images)

E-learning approaches are also used in the evaluation or assessment of student performance. Interaction may involve a mainframe computer thousands of miles away from the school, home, or wherever it is being used. The fundamental principles of how people learn apply to e-learning just as they do to other forms of instruction.

E-learning, and all that it embraces, has received considerable attention in recent years. Schools have made major financial investments to obtain, install, and inservice teachers in the use of e-learning approaches. Whether or not it is as effective and efficient in promoting learning as traditional instruction by a teacher is an area that needs research. Little research has been done examining teaching approaches involving e-learning in secondary agricultural education. Research reported in *The Agricultural Education Magazine* (Lee, 2001a) suggested that students taught via computer modular approaches would not score as high on standardized tests as those taught with traditional on-task classroom methods. Earlier research by Marrison and Frick (1993) reported in the *Journal of Agricultural Education* found that postsecondary students who used computer multimedia modules had lower feelings toward the module than those taught using lecture approaches in a traditional class. They indicated that computer modules should be used to supplement traditional classroom instruction. Research by Dahlgran (2008) at the University of Arizona with college instruction in agricultural economics found that online homework resulted in greater allocation of study time and, hence, increased learning. More research is needed on the use of computer-based learning. As computers have become more widely available, e-learning materials have most likely improved in educational qualities.

Agriculture teachers often find computer-based and online materials and approaches excellent supplements to traditional methods. Visual materials are increasingly computer-based and have mostly replaced older materials such as transparencies, 35-mm slides, and videotapes. Online information can be accessed that is up to date and relevant to instructional needs.

With e-learning, the quality of the content and the engagement of learners in actively mastering the information are very important. The teacher must be quite involved in delivering instruction and using e-learning approaches to enrich, enhance, review, and assess student learning and not as the major source of instruction. Technology is not to be used in teaching just for the sake of having technology. It is never to be the major source of instruction. This chapter focuses on instructional resources and not on the approaches in their use.

Presentation Media

Classroom approaches may involve computer-based presentation materials. Two major ways are available for presentation enrichment of classroom learning using computer-based materials: interactive whiteboards and electronic display panels or monitors.

An *interactive whiteboard* is a presentation medium that uses a computer that is connected to a projector to display materials on a white surface that acts as a touch screen. (Note: An interactive whiteboard is sometimes referred to as a Smart Board, which is a registered trademark of Smart Technologies.) The components are connected wirelessly or with USB or serial cables. A laptop or PC at the teacher's workstation often serves as the source of information. The information may be from a CD/DVD, online sources, or entered by the teacher. A projector displays a computer's video output on the interactive whiteboard. It becomes interactive when special pens or other devices are used write or draw on the presentation.

Projectors for interactive boards are typically mounted to the ceiling. They are positioned to assure the best possible projection without distortion. They are typically connected to the computer with a cable, though some operations may involve wireless technology.

11–10. An agriculture teacher is using a wireless device to operate an interactive whiteboard, shown with a horse in the background.
(Courtesy, Education Images)

A *display panel* is an electronic presentation device connected to a computer, camera, or other equipment for the presentation of visual information. Large tube-type devices have been replaced with liquid crystal display flat panel displays or plasma displays. The images may be prepared by the teacher or obtained from a commercial source on a CD/DVD, from an online source, by way of satellite or broadcast television, or from other sources. In some cases, display panels may have audio capabilities. For classroom use, the area of display should be large enough for the most distant students to readily view what is being shown. In larger rooms, display panels may be positioned around the room from the ceiling or wall. Most are rather permanently attached to the wall or ceiling, though some are on carts that make them transportable. With individual or small group use, smaller versions such as personal computer or laptop screens may be adequate.

Arrangements may be established so that teachers have access to smaller screens at their desk or station. Portable devices with wireless connectivity are sometimes used. Information may be recorded or saved for future use. A teacher station may also include a camera to present visual information to the group. The camera may be on a small platform for ready access to documents with images, formulas, or other illustrations.

11–11. A ceiling-mounted camera for projection to an interactive whiteboard.
(Courtesy, Education Images)

11–12. Large display panels in the front of an agricultural teacher education classroom allow students to view and interact with visual material.
(Courtesy, Education Images)

Various remote devices are available to operate and prepare interactive materials. Some of these are used by the teacher; others are used by students. Wireless pens may be used on whiteboard surfaces or other devices. Providers of classroom display equipment will be able to provide the needed software and demonstrate its use. In some cases, the materials may be prepared and saved using an online source, with Knoodle® being an example.

The "heart" of an interactive system is the computer software. The software may be designed for teachers and/or students to prepare and present information. It may be used to prepare visuals and save presentations for future use. This software is typically stored in the computer located in the classroom near the equipment. The software selected must be compatible with the interactive surfaces and other devices that are used. One presentation software program that is often used in education is PowerPoint®, a product of Microsoft. This software allows teachers and students to prepare presentations using summaries ("bullets" or very brief statements) of the content and includes various illustrations such as line art and photographic images. Software programs may also make virtualization possible. These often use virtual materials prepared by

11–13. A teacher educator's station in a classroom equipped with electronic presentation media.
(Courtesy, Education Images)

commercial or government sources. The processes or items depicted are not real but are realistic. Mostly, it is modeling through the use of a computer, such as the virtual tail docking of a pig or construction of a flower arrangement.

Networking of all classrooms may be used to share useful information with all students and teachers. In a school district, all school sites may be networked so that multiple schools are involved. School districts often employ instructional technologists to establish and operate networks and assist teachers with individual classroom needs. In some cases, a school program in radio or television may allow students to prepare and present information by way of the network.

Environmental quality is increasingly being explored with the use of computer and interactive technologies. A major concern among some educators is classroom air quality. Inhaling poor-quality air has a negative effect on learning. Electrical devices may emit magnetic fields and substances such as volatile organic compounds. Furniture that promotes good posture and keeps the computer from direct body contact is preferred. "Toasted skin syndrome" may occur if a laptop is kept on the lap over time, according to AolHealth (2011). Wireless connections, modems, laptops, and other devices may lower male fertility, as reported in a study of the effects of electromagnetic radiation on sperm quality (MSNBC, 2011).

Learner-Focused Media

Several electronic devices are available for individual e-learning and/or ready access to information. Individuals may purchase these devices or schools may purchase them in bulk and check them out to individual students. Changes frequently occur with these devices as new applications are developed. Several examples are included here.

- **Desktop computers**—A *desktop computer* is a personal computer with limited portability and essentially for use in a single location. Schools typically place networked desktops in agricultural education classrooms, though some are stand-alone. They may be connected to a modem for Internet access and to a printer for producing copies on paper. Programs may be loaded into each individual computer. In some cases, a central server might be used for several keyboards and screens. Students would be provided access codes for networked desktop computers. Use by students would occur at school and during times made available for computer use. The uses might be for Internet research, instructional program

11–14. A student is placing a school-issued laptop into a protective case used in transporting from school to home and back.

(Courtesy, Education Images)

11–15. A student is using a laptop to access an online textbook.
(Courtesy, Education Images)

use, preparation of presentations, and entering record information. Schools often restrict access on a computer to certain Web sites and other information.

- **Laptops**—A *laptop* is a personal computer that is easily transported for mobile use and has the capabilities of a typical desktop computer. A rechargeable battery system is located inside of a laptop. Such computers can also be operated by plugging into a 110-volt receptacle, provided an adapter is used. Laptops may be issued to students for use at school or to take home for use. Schools would typically obtain a quantity of laptops so that each student has one for his or her use. An instructional technologist would typically be in charge of school-issued laptops. Students can generally view the same information and perform the same operations on a laptop as on a desktop. Internet access may be wireless or by cable depending on how the laptop is configured. School Web sites and FFA chapter Web sites may have educational information or present links for accessing educational sites. Laptops can be used in vehicles using battery power or 110-volt plug-in as found in some vehicles. Vehicles with "hot spots" may also allow mobile access to the Web. (For safety purposes, the operator of a vehicle on a highway should not attempt to use a laptop. An operator should stop and safely park the vehicle before using a laptop.) The *netbook* is a category of small and inexpensive laptop computer that has found some favor for educational purposes. Netbooks are often purchased by schools and issued to students.

- **Smartphones**—A *smartphone* is a mobile telephone with a computing platform that allows features such as Internet and e-mail capability far beyond telephone conversations. Some are combined with the functions of personal data assistants (PDAs). Most smartphones allow the downloading of apps (software for specific purposes) that make it possible to access agricultural or other information that is not otherwise available on a mobile basis. Agriculture apps allow individuals on farms—in pastures and fields—to access information for a specific use. For example, the USDA has the app known as SoilWeb, which is a GPS-based program that provides soil survey information. Another app is The Plant Doctor, which aids in the diagnosis and treatment of plant diseases.

11–16. An app on a smart-phone is being used to assess the condition of a shrub outside the classroom.
(Courtesy, Education Images)

A QR (quick response) code reader app may be installed on a device for scanning QR codes to gain near-instant connectivity to a site or source of information. Rural service may be limited by the absence of ready cell phone service required for smartphone connectivity.

- **Tablet computer**—Best known as the iPad by Apple, Inc., tablet computers have made major advances in use and popularity. A ***tablet computer*** is a device that has features of a small computer and of a smartphone. The operation platform is similar to that of a smartphone. Apps can be added to achieve specific goals. Some schools are issuing iPads to students for access to online information, including textbooks and other e-learning materials. The devices are small, lightweight, resilient, and yet have an adequate-sized screen. E-mail and Web work can be done with an iPad in a location with wireless Internet access. Mini iPads are also used with connectivity to Wi-Fi or cellular sources for sharing information.

- **E-book reader**—Also referred to as an e-reader, an ***e-book reader*** is a mobile electronic device designed for reading of digital e-books and periodicals. E-book readers are similar to tablet computers. Some uses with textbooks may occur, but the potential is yet to be realized. The Amazon Kindle is the most popular e-book reader. Others include the Nook by Barnes and Noble and the Sony Reader.

School Software

The two kinds of software, according to Lever-Duffy and McDonald (2011), most widely found in schools are administrative software and academic software. Administrative software may be referred to as school management software.

School management software is a computer-based program used in schools to achieve a wide range of functions related to delivering education, recording data, and

11–17. A tablet computer can be used to download books, much as an e-book reader.
(Courtesy, Education Images)

reporting information. Such software is typically networked throughout a school system, with access limited to authorized individuals. Access is granted to various sites in the system to teachers, students, parents, and others using access codes and passwords restricting what can be seen and done. The system may include administrative functions of the school, student enrollment and class scheduling, posting announcements and homework, and recording and reporting the performance of students. For example, teachers may report student attendance, grades, and other information via online software management. Some systems have built-in parental notifications if students are absent, fail to achieve a minimum performance level on a test, or do not turn in homework.

An example is RenWeb, which has Web-based school management software programs. In addition to school and education management, RenWeb also has reporting features for parents and students. Alerts can be quickly communicated to all school personnel, parents, and students through posting and calling systems. Access codes restrict use to authorized individuals. A feature gaining attention is that of an app for mobile devices such as iPhones, iPads, and similar devices. In addition to RenWeb, examples of other systems are Rediker Software and QuickSchools.

Though not widely used in e-learning, school management software is used in one way or another in almost all school districts. Links within school systems may direct students to e-learning sites. Some school systems are linked to regional and state education agencies. FFA record systems are typically separate from school management.

Academic software is software that is useful to teachers and learners in the teaching and learning processes. This software is used to enrich the education environment. Examples of uses of academic software according to Lever-Duffy and McDonald (2011) include desktop publishing, graphics, reference, tutorials and drill-and-practice, educational games, and simulations. Agricultural educators have some software applications with unique roles for agricultural education, such as supervised experience record keeping and curriculum planning.

In the use of online e-learning materials, all educators should be aware of the provisions of the Children's Online Privacy Protection Act of 1998 (COPPA). In short, this act is administered through the Federal Trade Commission (FTC) and prohibits the online gathering of information on children under the age of 13 unless parental consent has been provided. This would particularly apply to some middle school agricultural education programs. The National FFA Organization requires that a completed parental permission form be on file for students to access certain areas of its Web site, such as the FFA Agricultural Career Network. Many school systems have local policies and procedures that also apply.

SELECTING AND OBTAINING INSTRUCTIONAL RESOURCES

Choosing the instructional materials to use in teaching is an important decision. The content of materials should be closely correlated to the state or local curriculum requirements. These should carefully follow end-of-course or other testing programs used in local schools and may be mandated by the state education agency. It would be unwise to use materials that do not cover content in the tests that are administered. The materials must be appropriate for the levels of the learners and up to date in content. Teaching out-of-date information is a serious compromise of student time and other resources.

Instructional materials adoption is the process used by school districts and state education agencies in selecting materials and allocating resources. Local school boards or school administrators often establish procedures to be followed. Some states (particularly those that provide funds for purchasing textbooks) have statewide adoptions carried out under the direction of their state boards of education. A key part of the process at all levels is to obtain the input of lay people. The lay people are stakeholders in the schools and may be parents, business leaders, and others who are not educators. A selection committee or advisory group is often involved. Public hearings may be held.

Selection Criteria

Selecting instructional materials is an important decision. School funds are limited, and teachers want to get the best possible materials. The same criteria are typically

11–18. The content of an activity manual should be organized parallel to that of the textbook.

(Courtesy, Education Images)

applied in the selection of computer-based and e-learning materials as to those that are printed on paper. Here are some questions to ask in the selection process:

1. Is the material correlated with the objectives and end-of-course test for a course? (In other words, does the material contribute to the achievement of state-specified objectives and good performance by students on end-of-course tests?)
2. Is the content coverage in sufficient detail?
3. Is the content technically accurate?
4. Is the content current (up to date)?
5. Is the content sequenced properly?
6. Is the content written on an appropriate level?
7. Is the material appealing to students?
8. Is the material appropriately illustrated with photographic images and line art?
9. Is the material durably constructed/prepared?
10. Has the material been field-tested and reviewed as part of development?
11. Are ancillaries available to support use of the material?
12. Is the material priced within range for the funds available?
13. Is the material appropriate to the values of community citizens?

Specific criteria may be used in evaluating textbooks, activity manuals, and other materials. It is important to be aware of local school district and state procedures on materials acquisition.

Consumable materials are often purchased following a somewhat different procedure. Consumables often accompany the basal materials and are used up, so to speak, after one class. State funds may not be available in some states to purchase consumables.

In selecting an ancillary, ask the following questions:

1. How well does the material correlate with the basal textbook/material?
2. Does the material promote systematic learning?
3. Does the material reinforce important concepts in the basal textbook?
4. Are the activities appealing to learners?
5. Has the material been field-tested?
6. Is the material current?
7. Is a companion instructor's guide available? (This may include online support for the ancillary.)

Obtaining Materials

Always follow the procedures of a local school in obtaining materials. The school principal, director, or department chair can provide information on how the process works. Some schools may have textbook coordinators that stay current with the situation on adoptions and bidding processes.

School systems vary, but many solicit instructional materials in the spring for use the following year. The requests are obtained from all teachers and compiled into a master list. Administrators and others review the list and set priorities on what is to be obtained. Once purchasing decisions have been made, the recommendations may be taken to the local school board for approval before a purchase order is issued. Depending on the budgetary year, purchases are often made in early July with funds budgeted for the next school year.

In some states, state funding is available for instructional materials in local schools only if materials on state-adopted lists are selected. Such lists may include textbooks, CD-based materials, Web-based materials, and other kinds of

11–19. Sample form to use in evaluating textbooks.

**Agricultural Education
TEXTBOOK EVALUATION FORM**

Title of Book: _____ Edition: _____

Date Published: _____ Class Quantity Cost (per copy): _____

Name and Address of Publisher/Source: _____

_____

Instructions: Rate the following items as acceptable (YES) or unacceptable (NO) as they relate to the textbook. Place an X in the column that represents your evaluation.

The textbook	YES	NO
1. Is written at the appropriate level.	_____	_____
2. Has a durable binding.	_____	_____
3. Is technically accurate.	_____	_____
4. Is technically up to date.	_____	_____
5. Has modern photographs.	_____	_____
6. Has helpful and good-quality line drawings.	_____	_____
7. Has a companion teacher's manual.	_____	_____
8. Has a student-friendly layout.	_____	_____
9. Facilitates systematic instruction.	_____	_____
10. Has useful questions at the end of each chapter.	_____	_____
11. Suggests activities to apply the content.	_____	_____
12. Equitably treats all individuals.	_____	_____
13. Emphasizes safety and ethics.	_____	_____
Overall, the book would be useful in my program.	_____	_____

Comments:_____

_____

_____

Signature of Reviewer: _____ Date: _____

---

instructional materials. A state list may have two or more alternatives for a particular course or subject. State adoptions often occur on a cycle of five years or so. Local school districts appoint local adoption committees. Local districts that buy materials not on the state-adopted list must usually use local school funds for these purchases.

In any school, developing a strong justification for instructional materials is often needed. It is a teacher's responsibility to provide a rationale for obtaining new materials. Developing support from advisory committee members is usually helpful. As mentioned earlier, it is always wise to have a list of needed instructional materials available just in case you are informed at the last minute that some funds are available. It is the teacher's responsibility to be aggressive in obtaining quality, up-to-date instructional materials for agricultural education.

MANAGING INSTRUCTIONAL RESOURCES

Instructional materials need to be managed so that they are readily available and are protected from unnecessary wear and tear. The appearance of an agricultural education office, classroom, or laboratory can be greatly influenced by how materials are filed. Piles of worn books, stacks of out-of-date brochures, broken computer equipment, damaged CDs and DVDs, and dilapidated racks of magazines detract from appearance and lower the overall perception of agricultural education facilities by students and others. Keep facilities neat, clean, and well organized!

Books

Greatest learning efficiency occurs when there is a textbook for every student and every student has a textbook checked out. The textbook may be taken home or to a study area as well as used in the classroom. Having students use textbooks as an integral part of learning has a major upward impact on student achievement scores. In the case of online textbooks, students may be provided access codes for use at home.

When not in use or checked out to students, textbooks and other books used by students can be organized on bookshelves in the classroom or storage room. Stored books should be arranged by title, and all turned the same way. The place where books are stored should be dry and away from direct sunlight.

Reference books used by students may be kept on bookshelves in the classroom area. Students should be instructed to return each book to its proper location after use. Reference books used by the teacher may be kept in the agricultural education office or on the teacher's desk at the front of the classroom.

Brochures and Magazines

Brochures and magazines related to the instruction should be displayed in the classroom. Extension Service bulletins, as well as handbooks on agricultural chemicals, plant and/or animal diseases, and related topics, may also be included. These are often in racks or on shelves that allow students to easily see titles or subjects and return the items after use. Brochures and magazines for teacher reference can be kept in the agricultural education office.

11–20. A teacher is providing individual instruction to a student.
(Courtesy, Education Images)

Brochures and magazines should be periodically checked for damage and age. Tattered materials should be discarded, and new ones obtained. Out-of-date materials should also be discarded. Older materials may promote the use of practices that are no longer permissible, such as the common use of methyl bromide as a soil fumigant. Magazines older than one year should be discarded unless there is a very special reason to retain them.

School libraries or media centers can often support the agricultural education program by obtaining brochures and magazines. Libraries and media centers have facilities for organizing such materials and making them available to students. The agriculture teacher can work with the librarian or media center director to see that appropriate materials are obtained and made available.

Electronic Forms

CDs, DVDs, videotapes, and other electronic materials used in an agricultural education program should be carefully and securely stored to provide protection, prevent loss, and facilitate locating them when needed.

Organizer cabinets or shelves can be obtained for CDs and DVDs. Videotapes, though not widely used any more, can be stored on regular shelving. Turn all materials so that the titles are easily seen. Materials may be organized by subject in alphabetical sequence or by the class in which they are used.

To assure security, such materials are often kept in the agricultural education office or stored in an instructional materials room. Conditions of storage should protect the materials from damage. Such materials may be in locked storage cabinets or rooms to assure protection from pilferage and loss. Increasingly, information is being stored online or in, as commonly termed, cloud storage. Cloud storage is a network of online storage where data is stored in virtualized pools by third parties.

Records

An agricultural educator has several roles related to records. In some cases, records are kept in the central office of the school, and the teacher must provide the information that is to be recorded. In other cases, the teacher enters the information in a networked computer information system. Some information is recorded on a daily

11–21. A teacher and student are discussing supervised experience program management with the aid of a computer-based system.
(Courtesy, Education Images)

11–22. A USB flash drive (thumb drive) allows transportability of student projects, records, and other information. (Always keep a backup file on an external hard drive or other location!)
(Courtesy, Education Images)

basis, such as class attendance. Other information is recorded on a weekly, semester, or annual basis.

Records of FFA participation and supervised experience activities are often kept in the agricultural education office or classroom. Students may have access to their records and regularly record information to keep them up to date. Good records are essential for the teacher in preparing state reports on the program. They are important for individual students in working toward FFA advancements. Students may save SAE records onto a USB flash drive (thumb drive) for transportability and reduce information on the hard drive of a computer. (It is wise to keep backup files in case a USB flash drive is misplaced.)

Some records should be retained indefinitely in an agricultural education department. Records of FFA membership may need to be retained for at least ten years but may be kept longer depending on school procedures.

In all cases, records must be accurate and protected from tampering. Inaccurate records result in later problems and in decision making based on false information.

Agriculture teachers need to be aware of FERPA guidelines in dealing with student records. FERPA is an abbreviation of the Family Educational Rights and Privacy Act, or Buckley Amendment. The federal act protects the privacy of student education records. FERPA gives parents certain rights until the student reaches the age of 18 or attends a school beyond the high school level. The law provides that parents may inspect records, ask that errors be corrected, and sign a written release in order for any information to be disclosed. In some instances, uses of student records are exempt from FERPA regulations, such as when a student transfers to another school or the school prepares a directory. FERPA applies to all schools that receive funds through the U.S. Department of Education.

Laboratory Materials

Laboratory materials must be managed properly to assure safety and security and to prevent unauthorized use and damage. How materials are stored and organized varies with the kinds of tools, instruments, equipment, or supplies. (Note: Chapter 10 provides additional information related to tools and equipment.)

Tools and instruments are typically stored in a locking cabinet or tool room that is opened and made available to students as needed. Each item should have a

11–23. Appropriate personal protective equipment is essential in most laboratory learning activities. (This shows a student wearing goggles in preparing to do wet-lab work. These goggles seal around the eyes to prevent liquid from getting into the eyes should a splash occur. Safety glasses are not adequate in a wet lab.)

(Courtesy, Education Images)

specific place in a cabinet or on a wall panel or shelf. The name of the item or a drawing may be used to identify the location.

Equipment is managed to assure a long, useful life. Covers may be put on some types of equipment, such as microscopes, when they are not in use. Other equipment may be locked in cabinets, with voltage meters and DO (dissolved oxygen) meters being examples. Larger equipment, such as welding machines and table saws, may be stationary on a counter or lab facility floor. Management also includes being sure that the equipment is cleaned after use, protected from damage, and kept in a safe operating condition. All safety guards, goggles, and other personal protective equipment (PPE) must be in good condition and accessible before some laboratory learning activities.

Supplies are stored based on the nature of the material. Gasoline, for example, must be stored in approved containers and in locations away from open flames or sources of ignition. Chemicals may be kept locked in cabinets or closets to prevent unauthorized access. Other materials may need only to be protected from the weather, such as some potting media that should be protected from rain. Living forms, such as seeds, should be stored to assure viability.

REVIEWING SUMMARY

Many kinds of instructional resources (materials used in teaching) are used in agricultural education. Student achievement is promoted by having a resource-rich learning environment. It is the teacher's responsibility to have appropriate materials available for efficient teaching and learning. Further, these materials should support the achievement of learning standards established for the agricultural education program and promote student performance on assessments.

Instructional resources include both published and nonpublished materials. Each is obtained to promote the achievement of the goals of the instructional program. Some materials are designed for student use; other materials are for teacher use.

Instructional materials are published kinds of instructional resources. These include textbooks, activity manuals, teachers' manuals, references, and similar materials in paper or electronic formats. Materials that provide the base or foundation are known as basal materials. A textbook is an example. Materials that accompany and enrich basal materials are known as ancillary materials. An activity manual is an example.

Nonpublished instructional resources are used to add hands-on learning opportunities for students. A wide range of instruments, tools, equipment, and supplies are needed. Instruments, tools, and equipment are lasting and can be used over and over. Supplies tend to be consumable and include such things as seeds, scions, paint, lumber, nails, and welding rods. These are obtained to implement instruction as specified in the curriculum guide provided by the state or local school district.

E-learning is the use of computer technology to promote teaching and learning. In most all cases, it should supplement and enhance instruction provided by the teacher—not replace the teacher. Presentation media include the interactive whiteboard with associated computer and projector or a display panel or set of panels, depending on room size. Individual learner-focused media may be used with desktops, laptops, smartphones, and tablet computers.

Materials are selected after the courses and the curriculum have been identified. The materials should help implement the curriculum and help achieve the instructional objectives. All materials should have appropriate content and be up to date, technically accurate, sequenced properly, written at the appropriate level, and appealing to students. Illustrations, durable construction, evidence of field testing, and availability of ancillaries are also factors in selection.

Once materials are obtained, the teacher is responsible for managing them. This includes organizing and storing them, protecting them from damage or loss, and issuing them to students. It also includes obtaining, organizing, maintaining, and promoting use by students of the appropriate personal protective equipment (PPE).

QUESTIONS FOR REVIEW AND DISCUSSION

1. What are instructional resources?
2. Distinguish between consumable and nonconsumable instructional materials.
3. Discuss the responsibility of a teacher in obtaining instructional resources.
4. What is instructional material? Give three examples.
5. Distinguish between instruments, tools, and equipment used in agricultural education.
6. What is a supply? How do supplies vary with the instructional program?
7. What is student-use material? Name and briefly explain four examples.
8. What is teacher-use material? Name and briefly explain four examples.
9. Distinguish between paper-based and electronic-based instructional material.
10. What is e-learning? How is it used in agricultural education classes?
11. What presentation media may be used in e-learning? Name and describe two main presentation mediums.
12. Learner-focused e-learning media may be used in agricultural education. Name and briefly describe four media.
13. What questions should be answered in selecting instructional materials?
14. What practices may be used in obtaining instructional materials?
15. Discuss the educational benefits of issuing textbooks to students.
16. What practices should be followed in managing instructional materials?

ACTIVITIES

1. Investigate the instructional materials selection and purchasing procedures in a local school. Interview an administrator, the chair of the materials adoption committee, or a teacher. Determine the practices in adopting materials, the use of citizen input, the schedule followed, and other details relating to agricultural education. Prepare a written report on your findings.
2. Assess the sources of instructional materials for agricultural education. Visit the Web sites of at least three sources and determine the kinds of materials available, the costs, and other details. Check out the websites of some of the following organizations:
   - Agripedia—**www.ca.uky.edu/agripedia**
   - Center for Agricultural and Environmental Research and Training (CAERT)—**www.caert.net**
   - Curriculum, Content and Assessments for CTE—**www.mycaert.com**
   - CEV Educational Multimedia—**www.cevmultimedia.com**

- ITCS Instructional Materials—**www.aces.uiuc.edu/IM**
- National FFA Organization—www.ffa.org
- Pearson Prentice Hall (career and technical–agriscience)— **www.pearsonschool.com**
- Instructional Materials Service—**www-ims.tamu.edu**
- Delmar Cengage Learning—**www.cengagesites.com/academic/?site=4373**
- Curriculum for Agricultural Science Education—**www.case4learning.org**

3. Select a secondary agriculture textbook (basal material) of your choosing. Any title or subject will be fine. Assess the book. Apply criteria stated in this chapter. Write a one- or two-paragraph assessment of the merits of the book for use in a high school class on a subject in line with that of the book.

4. Assume you have just taken an agriculture teaching position at a school that has not previously had an agricultural program. Your administrator has indicated that funds are available to purchase materials. What would you request? Choose a typical class (such as introduction to agriscience, livestock, or horticulture) and develop a specific list for a class of twenty students. Prepare details such as a list of bid specifications. Include estimated total cost, not to exceed $3,000.

5. Investigate the possible impact of the Children's Online Privacy Protection Act of 1998 on students in agricultural education. Prepare a brief report on your findings. The act can be accessed at Federal Trade Commission Web site.

6. Explore the use of QR codes to access information in agricultural education. Identify possible QR codes and assess the quality of the information that is obtained as related to teaching and learning.

7. Investigate the following programs/apps and prepare a short summary statement on each in terms of potential use in agricultural education: Google Drive®, Dropbox®, Box, SkyDrive®, CloudOn®, NetTexts®, Jot®, Edmodo®, Schoology®, RenWeb®, Prezi®, Educreations®, ShowMe®, Puffin®, Photon®, and SkyFire®.

8. Explore the practice of requiring students to bring their own technology devices into a school or classroom that has Wi-Fi. This is sometimes referred to as BYOT—bring your own technology. Why is this gaining in popularity? What issues are involved? Prepare a brief written report on your findings.

CHAPTER BIBLIOGRAPHY

AolHealth. (2011). Warm laptops can cause 'toasted skin syndrome.' Accessed on December 14, 2011.

Dahlgran, R. A. (2008). Online homework for agricultural economics instruction. *Journal of Agricultural and Applied Economics, 40*(1), 105–116.

Emmer, E. T., C. Evertson, and M. E. Worsham. (2000). *Classroom Management for Secondary Teachers* (5th ed.). Needham Heights, MA: Allyn & Bacon.

Kellough, R. D., and N. G. Kellough. (1999). *Secondary School Teaching: A Guide to Methods and Resources*. Upper Saddle River, NJ: Prentice Hall.

Lee, J. S. (2001a). Gaining high school achievement in agriscience. *The Agricultural Education Magazine, 74*(3), 12–13.

Lee, J. S. (2001b). *Teaching AgriScience*. Upper Saddle River, NJ: Pearson Prentice Hall Interstate.

Lever-Duffy, J., and J. B. McDonald. (2011). *Teaching and Learning with Technology* (4th ed.). Boston: Pearson Education.

Marrison, D. L., and M. J. Frick. (1993). Computer multimedia instruction versus traditional instruction in post-secondary agricultural education. *Journal of Agricultural Education, 34*(4), 31–38.

MSNBC. (2011). Wifi-enabled laptops may be nuking sperm. Accessed on December 14, 2011.

Newcomb, L. H., J. D. McCracken, J. R. Warmbrod, and M. S. Whittington. (2004). *Methods of Teaching Agriculture* (3rd ed.). Upper Saddle River, NJ: Prentice Hall.

Potter, R. L. (1999). *Technical Reading in the Middle School* (Fast Back 456). Bloomington, IN: Phi Delta Kappa Educational Foundation.

Roblyer, M. D., and A. H. Dowering. (2010). *Integrating Educational Technology into Teaching* (5th ed.). Boston: Allyn & Bacon.

Sadker, M. P., and D. M. Sadker. (2005). *Teachers, Schools, and Society* (7th ed.). New York: McGraw-Hill, Inc.

part three
Instruction in Agricultural Education

12

The Psychology of Learning

Franklin Kowalski wishes he could "get inside the heads" of his students and see what they are thinking. He wonders why he can teach the same the lesson he did last year, the same way, but where last year's group fully understood the material and wanted to go deeper into the subject, this year's class struggles with the basics.

Mr. Kowalski remembers the learning theory he studied in college. Fortunately, he can recall the important areas that will help him improve on student learning. In recalling the learning theory, he asked himself three questions:

- How do students learn?
- Do students have different ways of learning?
- Why are some students "naturally" motivated to learn?

He is looking back in his college notes and books to brush up his answers. He is beginning with his *Foundations of Agricultural Education* book.

terms

affective	learning style
behavioral learning theory	motivation
brain-based learning theory	operant conditioning
cognitive	Premack principle
cognitive learning theory	psychomotor
constructivism	schema
hierarchy of human needs	self-efficacy
learning	theory of multiple intelligences

12–1. Understanding how people learn helps teachers be more effective.
(Courtesy, Education Images)

WHAT IS LEARNING?

Learning is a permanent change in behavior as the result of an experience. If the change does not persist, then learning has not occurred. Learning results in people viewing situations differently in the future. They adapt their behavior based on what they have learned and the demands of their new environment.

Behavior is more than a physical, outward activity. It is broadly defined and includes the ***cognitive*** (knowledge or "stored" information), ***affective*** (attitude), and ***psychomotor*** (manipulative skill) domains. Learning must also be the result of experience. So, although a student changes as he or she ages, such as growing in height and weight, this is not learning.

Learning is not limited to school-based experiences. Students learn from their environment, parents, peers, and numerous other sources. Agricultural education, through supervised experience and FFA, is designed to expand school-based learning beyond the classroom and laboratory. Through this structure, agricultural education places learning within real-world social settings.

MAJOR LEARNING THEORIES

In the twentieth century, two dominant types of theories were held of how people learn: behavioral learning theories and cognitive learning theories. These theories continue to impact education at the beginning of the twenty-first century.

A ***behavioral learning theory*** is a theory of learning that focuses on observable changes in outward behavior and on the impact of external stimuli to effect change. From this type of theory, we get behavioral learning objectives, as discussed in Chapter 13.

A ***cognitive learning theory*** is a theory of learning that focuses on the internal mental processes, how they change, and how they affect external behavior changes.

Behavioral Learning Theories

An influential behavioral theory explaining how students learn is operant conditioning, based on the work of Edward Thorndike and B. F. Skinner. This theory expands on two earlier behavioral principles: contiguity and classical conditioning.

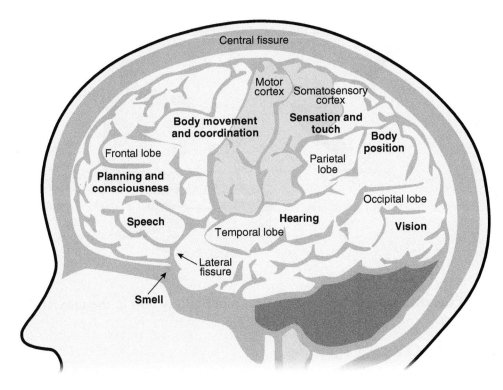

Contiguity states that if two things, a stimulus and a response, are paired together often enough, the learner will make an association between them. This occurs in the classroom with drill-and-practice activities. A student who pairs a picture of a steer "red with white face and no horns" with the beef breed "polled Hereford" is following the principle of contiguity.

Classical conditioning is best known by the work of Pavlov and his experiments with dogs salivating. Dogs naturally salivate when food is placed in front of them. However, Pavlov found that dogs could associate other things, such as a sound, with getting food. Eventually, the dogs would salivate upon hearing the sound although no food was present. Classical conditioning occurs regularly in the classroom. A student who becomes nervous or anxious when a test is given is an example. Teachers who have open, inviting classrooms are using classical conditioning by making students feel safe and accepted and associating that with learning. In classical conditioning, the behavior is involuntary and typically emotional or physiological. The stimulus is first, followed by the behavior.

In *operant conditioning*, the student has control of the behavior. The stimulus is a consequence of the behavior. In the classroom, a teacher who praises students for providing correct answers to questions is using operant conditioning. The consequence (praise) follows the behavior (correct response), leading to greater classroom participation. Teachers use reinforcement to increase or strengthen behaviors and punishment to decrease or weaken behaviors.

An important principle regarding reinforcement is called the ***Premack principle***. This states that a preferred activity can be used as a reinforcer for a less preferred activity (Santrock, 2008). The Premack principle works by placing the less preferred activity before the preferred activity. Agricultural education examples abound. Mr. Jones's animal science class dislikes calculating feed rations and mixtures for the aquaculture tanks but enjoys testing and analyzing the water and recording the results. Mr. Jones is using the Premack principle when he says to the class, "Let's do a good job of accurately calculating feed mixtures during the first

12–3. Classical conditioning explains how external stimuli can trigger involuntary responses.

part of the class period, then we'll conduct water-quality tests in the aquaculture laboratory."

Cognitive Learning Theories

Cognitive learning theories look at how people process information, organize information, and construct knowledge. In the cognitive view, students are active learners who seek out information and are constantly reorganizing new and old information. Associated with cognitive learning theories is the information processing model, which focuses on how people gather, code, and store information in their memory and subsequently retrieve it. The information processing model compares human information processing to the computer and how it works. Teachers use cognitive learning theories when they direct student attention, enhance student perception, and provide students with rehearsal tools. Teachers who use frequent examples, clear objectives and directions, and varied interest approaches are assisting students in cognitive learning. Organizational tools such as notes and learning webs are other examples of this theory in practice.

Bandura's social cognitive theory (1986) added social factors to learning theories. Bandura's theory recognizes that behavior, person/cognitive, and environment factors all play a reciprocal part in learning (Santrock, 2008). Teaching through Career Development Events (CDEs) can use social cognitive theory. The agriculture teacher establishes an environment of modeling, guided practice, and peer teaching. Students develop cognitive strategies allowing them to problem solve the CDE, which improves performance. Mock events develop behavior that is culminated at the CDE competition. The person/cognitive factor most recently emphasized is self-efficacy. *Self-efficacy* is the belief that one can accomplish something and have a positive outcome (Bandura, 1997). Self-efficacy influences behavior, so increasing student self-efficacy should lead to positive behavior and more effective learning.

Constructivism is a set of learning theories that emphasize how students actively make sense of the information they receive (Woolfolk, 1998). One view of constructivism is concerned with how students use internal schemas to represent accurately the outside world. A *schema* is an abstract guide used to organize an experience or concept. A second view of constructivism focuses less on the accurate representation of the outside world and more on how old knowledge and new knowledge are transformed to be useful to the individual. A third view of constructivism states that knowledge reflects the outside world but is filtered and influenced by culture, language, teachers, and other factors.

Agriculture teachers use constructivism when they follow a "learning by doing" approach. They are encouraging their students to make sense of the outside world through actively engaging in their environment. An example is the agriculture teacher who teaches parliamentary procedure by organizing the class as a mini-FFA chapter with activities and with decisions to be made. Some students may construct the parliamentary procedure knowledge around a schema of "This is how my youth group could operate." Other students may construct a schema of "This is what it means to be an FFA member or officer."

Brain-Based Learning

The *brain-based learning theory*, developed by Caine and Caine (1991, 1994, 1997, n.d.), resulted from a synthesis of theories and literature regarding the brain from multiple disciplines, including biology and psychology. From this synthesis, Caine and Caine developed twelve principles of brain/mind learning. These are listed in Figure 12–4, and a brief description of each follows.

Principle 1 states that both the brain and the body are involved in learning. Teachers using this principle will engage students in activities that require them to use their senses and bodies. Principle 2 states that learning has social aspects and is influenced by social interactions. FFA, with its student, chapter, and community activities, provides multiple opportunities to combine learning with social interactions and relationships. Principles 3, 4, and 5 are concerned with how learners acquire and store information. Learners need to make sense out of their environment and experiences (Principle 3). They do this by organizing information into unique mental schemas (Principle 4). Emotions and mind-sets are a part of learning and are critical in establishing mental patterns or linkages (Principle 5).

Principle 6 states that our brains perceive information both in parts and in its wholeness at the same time. In addition to facts and information, agriculture teachers should organize instruction so real-world examples and stories are also included. Principle 7 states that although the brain perceives what it is focused upon, it also perceives information that is outside of this focus. Teachers use this principle when they place posters and pictures on the classroom walls. Principle 8 states that although learning involves conscious activity, much learning occurs afterward in unconscious

---

The Caines' 12 Brain/Mind Learning Principles

1. Learning engages the physiology.
2. The brain/mind is social.
3. The search for meaning is innate.
4. The search for meaning occurs through patterning.
5. Emotions are critical to patterning.
6. The mind/brain processes parts and wholes simultaneously.
7. Learning involves both focused attention and peripheral perception.
8. Learning always involves conscious and unconscious processes.
9. We have at least two ways of organizing memory: A spatial memory system and a set of systems for rote learning.
10. Learning is developmental.
11. Complex learning is enhanced by challenge and inhibited by threat.
12. Each brain is uniquely organized.

12–4. Student learning is enhanced when teachers utilize brain/mind learning principles.
(Source: Cain & Cain, n.d. Used with permission.)

processes. Therefore, learning is facilitated when time is provided for reflection. Principle 9 states that we organize in two different categories of memories. One, rote memorization, is most useful for isolated information. The other, dynamic, is engaged best through experiences and infotainment.

Principle 10 states that the brain develops in a predetermined way during childhood yet never loses its capacity for learning. Educators who espouse a lifelong learning approach are following this principle. Principle 11 states that the best learning environment is one that appropriately challenges learners. Caine and Caine used the term "downshifting" to describe what happens to students who are threatened in the learning environment. Students who downshift feel helpless and fatigued and then give up. Principle 12 states that every student is unique. In structuring the learning environment, the teacher must recognize different learning styles, multiple intelligences, and student diversity in its broadest definition.

ROLE OF LEARNING THEORIES IN AGRICULTURAL EDUCATION

Agricultural educators use learning theories to organize and structure subject matter, to motivate students, and to determine what teaching methods to use. The ultimate goal is to provide the optimal learning environment for students of agricultural

---

Principles of Teaching and Learning

1. When the subject matter to be learned possesses meaning, organization, and structure that are clear to students, learning proceeds more rapidly and is retained longer.

2. Readiness is a prerequisite for learning. Subject matter and learning experiences must be provided that begin where the learner is.

3. Students must be motivated to learn. Learning activities should be provided that take into account the wants, needs, interests, and aspirations of students.

4. Students are motivated through their involvement in setting goals and planning learning activities.

5. Success is a strong motivating force.

6. Students are motivated when they attempt tasks that fall in a range of challenge such that success is perceived to be possible but not certain.

7. When students have knowledge of their learning progress, performance will be superior to what it would have been without such knowledge.

8. Behaviors that are reinforced (rewarded) are more likely to be learned.

9. To be most effective, reward (reinforcement) must follow as immediately as possible the desired behavior and be clearly connected with that behavior by the student.

10. Directed learning is more effective than undirected learning.

11. To maximize learning, students should "inquire into" rather than "be instructed in" the subject matter. Problem-oriented approaches to teaching improve learning.

12. Students learn what they practice.

13. Supervised practice that is most effective occurs in a functional educational experience.

14. Learning is most likely to be used (transferred) if it occurs in a situation as much like that in which it is to be used as possible and immediately preceding the time when it is needed.

15. Transfer of learning is more likely to take place when what is to be transferred is a generalization, a general rule, or a formula.

16. Students can learn to transfer what they have learned; teachers must teach students how to transfer learning to laboratory and real-life situations.

12–5. Student learning is enhanced when teachers follow the principles of teaching and learning.
(Source: Newcomb et al., 2004.)

education. Learning theories help explain why supervised experience and FFA are critical components of the total agricultural education program.

As discussed in Chapters 3 and 4, agricultural education over the years has fluctuated between a vocational focus and a science/business focus. Another way of characterizing this is between instructing students in agriculture for future careers and instructing students about agriculture for agricultural literacy. It is important to consider the focus of the agricultural instruction in our view of student learning.

To have the optimal impact on educational practice and student learning, learning theories must be translated into language that is relevant and applicable to teachers. Newcomb et al. (2004) have described sixteen principles of teaching and learning to guide teachers in the instructional process. These are summarized in Figure 12–5.

STUDENT LEARNING STYLES

Howard Gardner (1983), in his *theory of multiple intelligences*, proposed that there are seven dimensions of intelligence. These are bodily-kinesthetic (skillful use of the body and handling of objects), interpersonal (ability to sense emotions, needs, and motivations in others), intrapersonal (ability to use self-awareness to guide decisions and behavior), linguistic (skillful use of words and language), logical-mathematical (skillful use of numbers, logic, reasoning, and patterning), musical (ability to sense sound, tone, pitch, and rhythm), and spatial (ability to perceive the visual world and recreate based on one's own perceptions). Gardner believed that everyone possesses intelligence in all seven dimensions; however, most people excel in only a few. Students are more successful when engaged in learning using the intelligence they excel at than when there is a mismatch. Because of this, teachers must provide learning opportunities that engage students' multiple intelligences.

Cognitive styles describe how people perceive and organize information from their environment (Woolfolk, 1998). The most widely described cognitive styles within agricultural education are field dependence and field independence. Field-dependent people tend to see a pattern as a whole rather than separating out individual elements. They tend to learn best in social situations involving interaction. Field-dependent learners prefer the teacher to provide structure and organization to the instruction. Field-independent people can see the individual elements of the pattern. They tend to be more analytical and tend to prefer working as individuals. Field-independent learners prefer to provide their own structure and organization to the learning. A teacher attuned to these differences in students will provide structure to those students who need it and flexibility to others.

Learning styles are often included in how we view teaching. A *learning style* is a description of how a person prefers to learn and under what environment he or she prefers to learn. Some students are intrinsically motivated and approach education as a "learn for the sake of learning" situation. These are the students who easily take notes, ask questions on their own initiative, and delve deeper into subjects than required by assignments. These students appreciate teachers who provide time for class discussions and who give assignments with flexible requirements. Other students are more extrinsically motivated and approach learning from a "what is required" perspective. These students are motivated by grades, rewards, and structure. Extrinsically motivated students appreciate teachers who use a businesslike approach and who give assignments with specific requirements.

Another way to characterize learning styles is as auditory, visual, or kinesthetic. Auditory learners prefer to be in an educational setting where there is talking, speaking, and auditory clues regarding the content. Visual learners like to read, see notes, and observe. Kinesthetic learners like to touch, manipulate, and do in order to learn. Regardless of primary learning style, students learn best when instruction is varied and includes all three learning styles.

STUDENT MOTIVATION

Motivation is the energy and direction given to behavior. Some students appear to be highly motivated; others appear to lack motivation. Of course, this interpretation of motivation is in terms of learning a particular subject or skill.

There are two sources of motivation for students in the classroom: those things internal to them (intrinsic motivation) and those things external to them (extrinsic motivation). Although a teacher has the most control over extrinsic motivational factors, understanding intrinsic motivational factors is also important.

Intrinsic Motivation

Maslow (1970) developed a *hierarchy of human needs*, which is depicted in Figure 12–6. The most basic needs (physiological) are those of shelter, food, sleep, water, and survival. Transferring this to the school setting, the classroom should be in good repair, have adequate furniture, and provide light, heating, and cooling. Students who come to school hungry or sleep-deprived are less motivated to learn until these needs are met.

Next in the hierarchy are safety needs. Personal safety concerns related to gangs, bullies, or crime may detract from student learning.

Belonging is the third step in the hierarchy. Students need to feel that they belong in the classroom, are accepted by their peers, and have something of worth to contribute.

The next step is self-esteem. Students need opportunities to build their self-confidence, obtain approval, and be recognized.

The top three steps in the hierarchy are called growth needs. When these needs are met, the person is motivated to seek even more. For example, when knowing and understanding needs are met, the student has an even greater motivation to know and understand more in that area.

12–6. Maslow's hierarchy of needs.
(Adapted from Maslow, 1970)

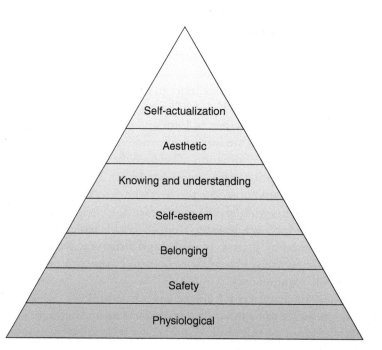

PRINCIPLES OF INTEREST
Primary

1. Much of the interest a person exhibits comes from natural impulses within the person. These include:

   - Love of nature
   - Altruism
   - Curiosity
   - Self-advancement
   - Creativeness
   - Competition
   - Sociability
   - Ownership
   - Desire for approval
   - Activity

2. Something is interesting if it affects us, others about us, or humanity at large.

3. Interest increases with an increase in related knowledge of any subject, provided such knowledge is well understood.

4. Interest increases with the acquisition of any given ability or skill.

5. Interest flows, or spreads, from any interesting thing into any uninteresting thing whenever the two are clearly connected in thought.

Secondary

1. Thinking is essentially interesting; memorization, uninteresting.

2. Interest is contagious in the sense that when one or more persons show interest in something, others will tend to "catch" that interest.

3. Interest is strengthened by a sense of progress.

4. Interest is created and sustained by a state of suspense.

5. An idea, when fully accepted, becomes a new interest center, from which interest will spread to any other thing that is seen to be connected with it.

6. The novel and unexpected are interesting.

7. Humor creates interest.

12–7. Student learning is enhanced when teachers utilize the principles of interest.
(Source: Crunkilton & Krebs, 1982.)

Extrinsic Motivation

Crunkilton and Krebs (1982) described practices teachers can use to affect student motivation. These are outlined in Figure 12–7 as primary, or universal, principles and as secondary principles. These principles can be used to gain student attention and motivation at the beginning of the class period. However, teachers should use these principles throughout their instruction.

Agricultural education is well suited for teachers to use all the principles of interest but especially the natural impulses. Anything living, plants and animals, or other real objects from nature touch on the love of nature within students. Hands-on learning taps into students' creativeness, activity, and self-advancement. When structured appropriately, hands-on learning can also involve ownership, competition, and approval. Community service activities are one example of altruism. Group projects allow students to be sociable but also involve activity, creativity, and ownership.

12–8. A variety of realia (real things or likenesses of real things) stimulates interest and helps motivate student learning.
(Courtesy, Education Images)

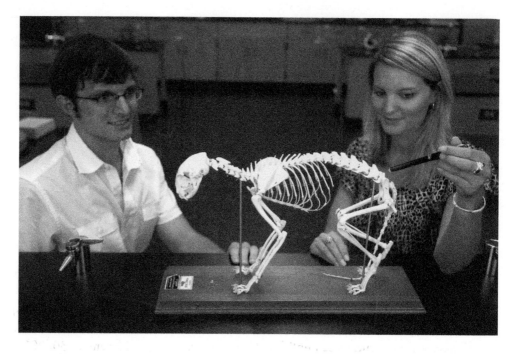

REVIEWING SUMMARY

To be effective, agriculture teachers need to know how students learn. The goal of instruction is to engage all students actively.

Educational practice in agricultural education has been influenced by both behavioral and cognitive learning theories. Brain-based learning theory gives an explanation of why some agricultural education tenets work and an impetus to change those teaching practices that do not work.

Students are intelligent in multiple ways. Agriculture teachers need to structure instruction so students of differing intelligences can all succeed. The best teachers are cognizant of student learning styles and tailor instruction to meet student needs.

Motivation is a key factor in the learning environment. Without motivation, there will be no learning. The goal is to achieve intrinsic motivation.

QUESTIONS FOR REVIEW AND DISCUSSION

1. What is learning?
2. Where does learning occur?
3. Describe how an agricultural education class is conducted by a teacher who is a behavioral learning theory adherent.
4. Describe how an agricultural education class is conducted by a teacher who is a brain-based learning theory adherent.
5. What is the Premack principle?
6. What is downshifting?
7. Explain Gardner's theory of multiple intelligences.
8. How do learning styles affect student learning?
9. Explain the impact of Maslow's hierarchy of needs on agricultural education classrooms.
10. How are the primary and secondary principles of interest used in agricultural education classrooms?
11. What is motivation? Differentiate between the two types.
12. How does classical conditioning occur in the classroom? Explain.

ACTIVITIES

1. Investigate brain-based learning theory. Chose an agricultural topic and develop a lesson to teach the topic using brain-based learning theory.
2. Investigate Gardner's theory of multiple intelligences. Write a report on the seven dimensions using a different FFA Career Development Event to highlight each dimension.

3. Using the *Journal of Agricultural Education, The Agricultural Education Magazine*, and the proceedings of the National Agricultural Education Research Conference, write a report on field-dependent and field-independent cognitive styles of research in agricultural education. Current and past issues of these journals can be found in the university library. Selected past issues can be found online at the American Association for Agricultural Education and the National Association of Agricultural Educators Web site.

CHAPTER BIBLIOGRAPHY

Bandura, A. (1986). *Social Foundations of Thought and Action.* Englewood Cliffs, NJ: Prentice Hall.

Bandura, A. (1997). *Self-Efficacy: The Exercise of Control.* New York: W. H. Freeman.

Bereiter, C. (2002). *Education and Mind in the Knowledge Age.* Mahwah, NJ: Lawrence Erlbaum Associates.

Binkley, H. R., and R. W. Tulloch. (1981). *Teaching Vocational Agriculture/Agribusiness.* Danville, IL: The Interstate Printers & Publishers, Inc.

Bloom, B. S. (Ed.). (1956). *Taxonomy of Educational Objectives.* New York: Longman, Green, & Co.

Caine, R. N., and G. Caine. (1991). *Making Connections: Teaching and the Human Brain.* Alexandria, VA: Association for Supervision and Curriculum Development.

Caine, R. N., and G. Caine. (1994). *Making Connections: Teaching and the Human Brain* (rev. ed.). Menlo Park, CA: Addison-Wesley Publishing Company.

Caine, R. N., and G. Caine. (1997). *Education on the Edge of Possibility.* Alexandria, VA: Association for Supervision and Curriculum Development.

Caine, R. N., and G. Caine. (n.d.). *Overview of the Systems Principles of Natural Learning.* Accessed on November 8, 2012.

Crunkilton, J. R., and A. H. Krebs. (1982). *Teaching Agriculture Through Problem Solving* (3rd ed.). Danville, IL: The Interstate Printers & Publishers, Inc.

Eggan, P., and D. Kauchak. (1997). *Educational Psychology: Windows on Classrooms.* Upper Saddle River, NJ: Prentice Hall.

Gardner, H. (1983). *Frames of Mind: The Theory of Multiple Intelligences.* New York: Basic Books.

Kahler, A. A., B. Morgan, G. E. Holmes, and C. E. Bundy. (1985). *Methods in Adult Education* (4th ed.). Danville, IL: The Interstate Printers & Publishers, Inc.

Maslow, A. (1970). *Motivation and Personality* (2nd ed.). New York: Harper & Row.

McCormick, F. G., Jr. (1994). *The Power of Positive Teaching.* Malabar, FL: Krieger Publishing Co.

National Association of Agricultural Educators (NAAE). (2011). *NAAE Web Site.* Accessed on November 30, 2011.

National FFA Organization. (2003). *Local Program Resource Guide: 2003–2004* [CD-ROM]. Indianapolis: Author.

National FFA Organization. (2011a). *Local Program Success* (online). Access on November 30, 2011.

National FFA Organization. (2011b). *Official FFA Manual.* Indianapolis: Author.

Newcomb, L. H., J. D. McCracken, J. R. Warmbrod, and M. S. Whittington. (2004). *Methods of Teaching Agriculture* (3rd ed.). Upper Saddle River, NJ: Prentice Hall.

Phipps, L. J., and E. W. Osborne. (1988). *Handbook on Agricultural Education in Public Schools* (5th ed.). Danville, IL: The Interstate Printers & Publishers, Inc.

Santrock, J. W. (2008). *Educational Psychology* (3rd ed.). New York: McGraw-Hill.

Taylor, G. R. (2002). *Using Human Learning Strategies in the Classroom.* Lanham, MD: The Scarecrow Press.

True, A. C. (1929). *A History of Agricultural Education in the United States, 1785–1925.* U.S. Department of Agriculture, Miscellaneous Publication No. 36. Washington, DC: GPO.

Woolfolk, A. E. (1998). *Educational Psychology* (7th ed.). Needham Heights, MA: Allyn & Bacon.

13

The Teaching Process

objectives

This chapter explains the importance of understanding the teaching process. It has the following objectives:

1. Explain teaching methods
2. Identify considerations in selecting a method
2. Describe Bloom's taxonomy and relate its use to teaching
3. Discuss methods of group instruction
4. Discuss methods of individual instruction
5. Describe the process of developing and/or adapting appropriate lesson plans in teaching
6. Discuss the use of instructional technologies in teaching

Kim Daniels is an excited beginning agriculture teacher. It's the week before school begins, and her facilities are clean and organized. Her classroom has tables and chairs for thirty students and is equipped with an interactive whiteboard and Internet-connected computer. Countertops with base storage cabinets and strategically placed sinks are along two of the walls. Against another wall are five Internet-connected computer stations.

A walk through her office and into the laboratory reveals two aquaculture tanks with a hydroponics station and grow lights. Spaced throughout the remainder of the laboratory are workbenches and tool/equipment storage cabinets for various agricultural mechanization experiments and projects.

Kim is continually thinking about one overriding question: "How will I teach my students?" Now, that is a good question. It is time she got serious and figured out what she is going to do.

terms

Bloom's taxonomy
demonstration
discussion
experiment
field trip
group teaching method
independent study
individual teaching method
learning module
learning objective
lecture
lesson plan
project
resource people
role-playing
supervised study
teaching method

13–1. The teacher is the leader and facilitator of classroom learning activities.
(Courtesy, Education Images)

CHOOSING THE TEACHING METHOD

Several teaching methods or strategies are available to agriculture teachers. A ***teaching method*** is the overall means that a teacher uses to best facilitate student learning. Most methods comprise a number of techniques or details for delivering the instruction.

Major Methods

Teaching methods can be placed into two major classes: group and individual.

Group teaching methods are commonly used in agricultural education classrooms. A ***group teaching method*** is a teaching approach that instructs students together as a group. Group methods overcome some of the concerns of teaching students individually, such as time and resource requirements. In terms of accountability, group instructional methods ensure that all students are taught the same content.

13–2. Small group instruction is used in this Georgia horticulture class. (Note that the teacher is on the left and the student teacher is on the right.)
(Courtesy, Education Images)

Individualizing instruction is a goal of many educators. An ***individual teaching method*** is an approach that recognizes that each individual learns at a different pace and possibly at a different level. Whereas individualized instruction can be more time consuming than group methods, the reward is in greater potential for student interest and deeper learning.

Teacher and Facility Factors

Considerations in choosing the overall teaching method include the abilities and styles of the teacher. Teachers should be always striving to build their repertoires of teaching methods. Beginning teachers tend to utilize those methods they are most comfortable with and feel are the strongest. However, these teachers are cautioned not to overuse a particular method.

Although facilities, equipment, and supplies are a factor in determining the method of instruction, these are less of a limiting factor than some might believe. An agriculture teacher who lacks microscopes to conduct an experiment may be able to borrow some from the science teacher. The problem of outdated equipment might be solved through field trips, computer simulations, or supervised experience at businesses with up-to-date equipment.

The abilities to improvise, borrow, and create are extremely valuable ones for an agriculture teacher to possess. Chapter 7 discussed utilizing advisory and citizen groups. These are helpful in getting needed resources for effective teaching. Chapter 10 discussed in further detail the role facilities play in teaching, and Chapter 21 discusses laboratories. Chapter 24 explains the importance of community resources for an effective agricultural education program.

Finally, considerations such as class size, time available to devote to a topic, and class structure influence the decision on what method of instruction to choose. As detailed below, some methods are best suited for individual or small-group instruction. Other methods are designed to cover a large amount of information as efficiently as possible. Class structural factors include time of day, time of year, proximity to holidays or vacations, and type of schedule system. Class periods under a block system, which can range from 85 to 100 minutes in length, require great variety in both instructional methods and learning tasks.

13–3. Teachers consider a number of factors in selecting the methods they will use. (Courtesy, Education Images)

Choosing the appropriate method is not merely using a formula but involves numerous considerations. What principles of learning should be applied to the lesson? At what developmental level are the students? What are their learning styles and dimensions of multiple intelligences? Are there physical, mental, or emotional considerations? What resources are available to use in implementing a teaching method? What should students know and be able to do at the end of the lesson? At what level of learning are students expected to demonstrate knowledge, skills, or dispositions?

BLOOM'S TAXONOMY

The question of level of learning expected of students is important. In 1956, Benjamin S. Bloom classified educational objectives into what is commonly referred to as **Bloom's taxonomy**. He first stated that there were three domains of learning. The cognitive domain is those things involving mental activity. The affective domain is those things involving attitudes, emotions, and values. The psychomotor domain is those things involving body movements, mechanical skills, and manual skills. Figures 13–4, 13–5, and 13–6 give examples of verbs associated with the domains.

Categories of Cognitive Skills

Within Bloom's taxonomy for the cognitive domain are six categories. Although Bloom did not present these as a hierarchy, the term "higher-order thinking skills" is sometimes applied to the last three of the six. The categories (from lowest to highest) are Knowledge, Comprehension, Application, Analysis, Synthesis, and Evaluation. Krathwohl (2002) presented a revised taxonomy. The three lower levels were

Cognitive					
Knowledge **Acquire**	**Analysis** **Analyze**	**Comprehension** **Associate**	**Synthesis** **Arrange**	**Application** **Apply**	**Evaluation** **Appraise**
count	construct	classify	categorize	calculate	assess
define	detect	compare	combine	change	compare
draw	diagram	compute	construct	classify	critique
identify	differentiate	contrast	create	complete	determine
indicate	explain	convert	design	demonstrate	evaluate
label	infer	describe	develop	discover	grade
list	outline	differentiate	explain	employ	judge
match	separate	discuss	formulate	examine	justify
name	subdivide	distinguish	generalize	illustrate	measure
outline	summarize	estimate	generate	manipulate	rank rate
point		explain	integrate	operate	recommend
quote		extrapolate	organize	practice	select
read		predict	plan	prepare	support
recall		rewrite	prepare	produce	test
recite		translate	prescribe	relate solve	
recognize			produce	use	
record			propose	utilize	
repeat			rearrange		
state			reconstruct		
tabulate			specify		
trace			summarize		
write					

13–4. Verbs associated with the cognitive domain of learning. Adapted from Bloom (1956).

Affective				
Receiving	**Organization**	**Responding**	**Valuing**	**Characterization**
accept	abstract	answer	argue	act avoid
accumulate	adhere alter	approve	assist	change
ask	arrange	commend	debate	complete
choose	balance	comply	deny	display
combine	combine	conform	help	rated high by peers, superiors, or subordinates
control	compare	discuss	increase measure of proficiency in	require
differentiate	define	follow	increase numbers in	resist
follow	discuss	help play	join	resolve
listen (for)	formulate	practice	protest	revise
reply	integrate	read	read	serve
select	organize	volunteer	relinquish	solve
separate	prepare		select	verify
set apart	theorize		specify	
share			support	

13–5. *Verbs associated with the affective domain of learning.*
Source: Krathwohl, D. R., Bloom, B. S., & Masia, B. B. (1973).

Psychomotor			
apply	dismantle	make-up	select
assemble	draw	manipulate	sense
build	fabricate	manufacture	service
calibrate	fasten	measure	sharpen
change	fell	mix	simulate
clear	follow	operate	smell
compose	form	organize	touch
connect	gap	perform	trace
construct	hear	plan	troubleshoot
correct	imitate	position	try
cut	install	put together	use
demonstrate	lay out	recognize	visualize
design	locate	remove	
desire to respond	maintain	restore	
discover	make	see	

13–6. *Verbs associated with the psychomotor domain of learning.*
Source: Simpson E. J. (1972).

renamed as Remember, Understand, and Apply. The three higher-level categories were renamed as Analyze, Evaluate, and Create, with the top two category places switched. The original categories are further explained next.

The Knowledge category is focused on facts, specifics, and observed phenomena and on the learner's ability to remember those as they were learned. The Comprehension category focuses on whether learners can explain information. Learners are expected to express information in a different form than that learned, explain information, and recognize relationships. The Application category is the ability to transfer information. Learners are expected to apply principles to new situations, solve problems, or demonstrate correct usage of a method or procedure.

The Analysis category focuses on whether learners can take complex information apart. Key concepts for this category are whether the learners can break a whole into its parts, discover relationships and principles, and determine similarities and differences. The Synthesis category focuses on whether the learner can take parts of information and create something new, look at information or a situation creatively, or devise a plan of action given disparate pieces of information. The Evaluation category focuses on whether learners can make judgments about information. Learners are expected to make judgments when given evidence or to evaluate the acceptability of something based on criteria.

Writing Objectives

Teachers use Bloom's taxonomy to write learning objectives (cognitive, affective, and psychomotor) to guide instruction, inform students, insure coverage of subject matter, and guide evaluation. A ***learning objective*** is a carefully written statement of the intended outcomes of a learning activity. Several variations are used, including behavioral objectives. Teaching proceeds best if the learning objectives have been carefully specified.

Learning objectives are always written in terms of what the student will be able to do at the conclusion of the instruction. Writing objectives using Bloom's taxonomy is as simple as ABCD.

Consider this example: "Students will be able to quote the FFA Creed from memory with 80 percent word accuracy."

- "A" stands for audience and is almost always students.
- "B" stands for behavior, or what is expected of the students. In this example, the students are to quote the FFA Creed. (This is sometimes known as the "terminal behavior.")
- "C" stands for the condition under which the behavior is to be done. In this example, students are to quote from memory.
- "D" stands for the degree or level to which students are to perform to succeed. In this example, students must be 80 percent accurate, or can misspeak only 20 percent of the creed.

METHODS OF GROUP INSTRUCTION

Different methods of instruction assist students in accomplishing the learning objectives. Methods of group instruction are discussed first. Six common group teaching methods are widely used in agricultural education. These are lecture, discussion, demonstration, field trip, role-play, and resource people.

Lecture

Very few agriculture teachers lecture in the formal definition of the methodology. An accomplished lecturer should be an expert in the subject, be skilled in oratory, and have experience and presence that earn respect from the audience.

However, the use of lecture in its many forms is a frequently used instructional method. As commonly defined, ***lecture*** is an oral presentation by the teacher or other individual. It may include a wide range of techniques. Lecture involves these five qualities:

1. More teacher talk than student talk.
2. Greater degree of flow of communication from teacher to students than from students to teacher or from student to student.

13–7. Presentations by students can be enhanced with the use of instructional technology. (Courtesy, Education Images)

3. Linear flow of information, with the structure determined by the teacher.
4. Use of illustrative materials to supplement what is being spoken. These include PowerPoint presentations, chalkboard/whiteboard notes and illustrations, overhead transparencies, slides, real objects, audio and video presentations, and pictures.
5. Use in combination with discussion to create lecture-discussion. This is typified by periods of lecture with note-taking followed by class discussion, analysis, or application of the preceding content.

When does a teacher choose lecture as a preferred instructional method? Lecture is best suited to large groups as is often seen in college classrooms. Lecture is also best when the content is such that all students need to have the same information on the same level. Within Bloom's taxonomy, knowledge and comprehension objectives are well suited for the lecture methodology. When time is a concern, lecture allows for disseminating a large amount of information in the least amount of time.

A good lecture requires organization. Unprepared teachers who believe they can just "throw together" a lecture at the last minute are fooling themselves and stealing instructional time from their students. Class needs to begin with a well-thought-out introduction. What are the students expected to know and be able to do after the instruction; in other words, what are the objectives? Why should students want to learn the content? What do students already know in this area, or what have they studied in previous lessons related to this area? The introduction should transition into a structured presentation of new content. If the topic is new, the lecture should first present the whole picture, then break it into its components, then present the whole again.

High school students should be provided an organizational outline to guide their note-taking from the lecture. This outline could be done on the chalkboard/whiteboard, as a PowerPoint presentation, or in the form of a note-taking guide. When lecturing to high school students, the teacher needs to check for understanding through questions, exercises, and frequent reviews.

Figure 13–8 shows the characteristics of a good lecturer. Teachers should evaluate themselves on each of these characteristics and work to improve those in which they have a deficiency.

1. Thoroughly prepares for the lecture.
2. Sees that the room is properly arranged for the lecture.
3. Has knowledge of the audience that helps in relating to its interests.
4. Is enthusiastic, which promotes student interest.
5. Speaks clearly and fluently; varies rate and intensity of speech.
6. Avoids distracting mannerisms.
7. Emphasizes major points and uses appropriate realia (real things) to promote learning.
8. Properly introduces and summarizes the subject.

13–8. Selected characteristics of a good lecturer.

Discussion

Discussions are commonly used in agricultural education. A **discussion** is a teaching process that involves student sharing of information. Typically, discussion might involve the agriculture teacher posing questions to the students, which they answer directly, offer their opinions about, or debate. Discussion may be combined with lecture, as described in the previous section, or be the sole instructional method used.

In terms of structure, discussion tends to be less formal than lecture. This does not mean that discussion can be effectively done without planning. On the contrary, a well-conducted discussion requires extensive planning by the teacher. The teacher needs to determine the objectives of the discussion ahead of time and structure the discussion to accomplish the objectives within the time frame selected. Nothing is more frustrating than an endless discussion with no conclusion or resolution of the topic discussed. A good discussion involves student-to-student interaction in addition to student-to-teacher and teacher-to-student.

Figure 13-9 presents selected characteristics of a good discussion. All participants—the students as well as the teacher—need to be ready for the discussion. The teacher should have a lesson plan for the discussion, just as he or she would for any other lesson using another instructional method.

1. An appropriate situation exists for discussion.
2. Students possess the background knowledge to give informed answers or opinions.
3. The classroom climate is such that students feel safe in voicing opinions and answering questions; they do not fear ridicule or embarrassment.
4. The teacher asks appropriate-level questions, asks probing questions when students give one-word answers, and provides answer prompts when necessary.
5. Incorrect information is corrected while the dignity of the student who gave the wrong information is maintained.
6. All students are actively engaged in the discussion.
7. Students have ample opportunities to contribute to the discussion, with no one student monopolizing the time.
8. The teacher actively listens, provides feedback, and clarifies when necessary.
9. Key points are highlighted during the discussion and summarized at the end.
10. The discussion is brought to a conclusion.

13–9. Selected characteristics of a good discussion.

A properly organized discussion will flow from point to point logically yet allow for student insights and questions. Students need to possess the necessary background information to make informed comments. This may require students to have read materials for homework or during previous class instruction. During the discussion, the teacher serves as a facilitator, prompter, and conversation referee. Depending on the class size, the teacher may need some system for making sure that all students participate in the discussion. Before the allotted time expires, the teacher needs to bring the discussion to closure, highlight key points, and assist the students in formulating conclusions.

The types of questions asked in a discussion are critical to the types of answers students will give and the level of their learning. There are five basic types of questions: closed, open, probing, higher-order, and divergent. Closed questions are used to regulate answers or obtain specific answers. For example, "Lecture is best used in what situations?" Open questions are useful for starting a discussion and seek a variety of answers. Such questions many times allow students to describe their own experiences or express opinions, so there are no wrong answers. Probing questions are used when students give one-word answers, more information is needed, or there is a desire to bring other students into the discussion. Sometimes probing questions mirror what a student has said. For example, "Are you saying you think lecture is overused?" Higher-order questions require students to think rather than recite memorizations. Higher-order questions are designed to make students discover rules and principles as they express their ideas. These questions use verbs from the upper three levels of Bloom's taxonomy, such as analyze, generalize, and critique. Divergent questions have no "right" answers. These require students to use their imaginations and think beyond set borders or parameters. Divergent questions lead to creativity and new ideas and are sometimes referred to as open-ended-answer questions.

Demonstration

The agricultural education curriculum has many topics that are well suited for demonstrations. There are skills in each agricultural area—mechanics, horticulture and landscaping, plant and soil science, natural resources, agricultural business, animal science, and food science—in which demonstrations are appropriate. A ***demonstration*** is a teaching process that involves showing students how to do something before they do it for themselves or that involves showing them what would happen as the result of a particular action. Practice by students of the skill demonstrated is essential in order to assure their skill development.

As in other group teaching methods, planning is essential to the success of a demonstration. All materials need to be gathered ahead of time, the teacher needs to have practiced the demonstration, and care must be taken that all students can see the demonstration. The teacher may want to have parts of the demonstration completed in advance, especially if time is a consideration. For example, demonstrating proper painting techniques may require that the primer dry twenty-four hours before putting on the coat of paint. Instead of taking multiple days to do the demonstration, the teacher may demonstrate applying primer, then pull out a previously primed wall section that has dried and use it to demonstrate applying the coat of paint.

Field Trip

Field trips are a useful instructional method for making the "real world" part of the classroom. A ***field trip*** is a learning experience that involves traveling away from the agriculture classroom. An example is a field trip to an agricultural experiment station to observe research projects that are underway.

13–10. *Students learn animal care skills by hands-on animal care work.*
(Courtesy, Education Images)

Field trips can be used for one of three purposes. At the beginning of an instructional unit, field trips are beneficial as an interest approach and to help the students to see the whole of a topic. Field trips during the middle of a unit can serve as fact-finding missions for problem solving or provide opportunities to test hypotheses. At the end of an instructional unit, field trips are useful for demonstrating theory into practice and bringing the parts back into a whole picture.

Field trips that involve travel must be planned and approved well in advance. School systems sometimes restrict when field trips can be taken and how many can be taken within the school year. School administrators may insist that trip planning and approval forms be submitted at the beginning of the school year for any field trips taken that year. This requires the teacher to be organized and to prioritize which trips are crucial to student learning. Another consideration is transportation. Details such as cost, a bus driver, parental permission, and notification of other teachers must all be planned.

Field trips do not have to involve taking buses to faraway sites. Mini-trips can be taken right on the school grounds. For schools on block scheduling, the longer class periods allow mini-trips to and from sites close to the school within the agriculture class period. Virtual field trips may not involve the sensory experiences that real field trips provide, but they can be an acceptable substitute. Videos, the Internet, and student presentations can all provide some sense of "being there."

Role-Play

Role-playing is the act of an individual assuming the real or imagined role of another individual. It is a useful instructional methodology for the agricultural education classroom.

The FFA component of agricultural education provides many opportunities for student role-playing. When studying parliamentary procedure, students are better able to comprehend officer roles and FFA ceremonies by role-playing officers and members and conducting FFA business. Role-playing is also a useful tool in the classroom when discussing sensitive topics. A debate on the place animal rights have in the animal production industry could quickly become heated and unruly. If, instead, students are assigned to role-play various characters, such as animal activist, rancher, feedlot owner, and consumer, then the situation becomes less personal and more objective.

13–11. Teachers are inspecting transportation equipment before going on a field trip. Reserve transportation well ahead of time and be sure that a properly licensed driver is available.
(Courtesy, Education Images)

Resource People

Agricultural education can be a difficult curriculum to teach. Very few agriculture teachers, if any, feel comfortable in their knowledge of every agricultural subject area. **Resource people**, who are experts in their area, can provide the knowledge and insights that the agriculture teacher may not have. Resource people can also lend credibility to information or instruction that students doubt or about which they have heard conflicting opinions. In addition, by utilizing resource persons in the classroom, the agriculture teacher is introducing his or her students to key community contacts and potential employers.

Resource people may be used in several ways. An individual resource person may be invited to present to the class in a lecture, discussion, or demonstration format. Another common procedure is to invite three to five resource persons to conduct a panel discussion. In either instance, the agriculture teacher should contact each resource person several weeks in advance of the presentation date. This should be followed by a letter describing the class, the objectives of the class period, and suggested topics of discussion. A day or two before the presentation, the agriculture teacher should contact each resource person to confirm date, time, and location and answer any last-minute questions.

METHODS OF INDIVIDUAL INSTRUCTION

Individual teaching methods are widely used in agricultural education. They include experiments, supervised study and guided practice, independent study, learning modules, and projects.

Experiments

Because agriculture is an applied science, it is especially important that experiments be a part of instruction. All areas of agricultural education lend themselves to experiments. An **experiment** is a trial using a definite and planned procedure. The outcome of the trial is predicted as part of the scientific method. Experiments can be carried out using very formal procedures or less-formal ones.

What is the value of experiments? Experiments are intertwined with the scientific method, which is the basis for the problem-solving teaching approach.

13–12. A Montana teacher and his students are discussing the results of an experiment.
(Courtesy, Education Images)

Experiments teach observation skills, measurement, analysis, and reasoning. Experiments utilize a "learning by doing" approach that actively engages the students physically as well as mentally.

Teacher planning and preparation are keys to the success of using experiments as a teaching method. An experiment needs to be planned thoroughly. The teacher must have done the experiment beforehand in order to anticipate student questions and frustration areas. The teacher must decide whether supplies and equipment will be needed for each student or whether students will work together in groups. In addition, the students need to be prepared for the experiment. What are the procedures to follow? What are students to observe? How are students to record and analyze data? Students also need to know that experiments don't always yield perfect data. Sometimes experiments that yield outlier data are the most interesting. It is not enough to simply do the experiment; the students need to understand why the experiment yielded the results.

Figure 13–13 shows the typical steps in conducting an experiment. Figure 13–14 is a sample lab sheet for student use when doing an experiment. Experiments in agriculture may last for multiple days or even weeks, so several data collection sheets may be necessary.

---

1. Interest is developed. This appeal may flow from students' SAE, an FFA Career Development Event, or personal interests. The teacher may also create interest through techniques discussed in Chapter 12.

2. The problem is defined. What do the students want to discover?

3. Information is gathered. Students find out what is already known about this problem. This is also known as a literature review.

4. Hypotheses are formed. What are possible explanations for the observed phenomena?

5. The experiment is planned. Supplies and equipment are gathered. Procedures are determined. A timeline for data collection is developed.

6. The experiment is conducted. Data are collected through observations and measurements.

7. Data are analyzed, and conclusions are drawn.

8. A report is written.

13–13. Steps in conducting an experiment.

Name: _____   Date: _____

Class: _____   Unit: _____

1. Why do you want to do this experiment?

2. What is the problem?

3. Give citations for at least three references regarding this problem.

4. What is your hypothesis?

5. What supplies and equipment are needed to conduct the experiment?

6. What are the procedures you will follow?

7. Collect, record, and analyze data. Data can be recorded on a chart or in a table on a separate sheet of paper.

8. What conclusions can be drawn from the results of the experiment?

13–14. Sample student lab sheet for use with an experiment.

Supervised Study and Guided Practice

Supervised study is a method in which students are directly responsible for their own learning while under the direction of the teacher. When used effectively, supervised study enhances student interest and develops students' problem-solving abilities. To guide the students, most supervised study involves the use of worksheets or of questions to be answered.

Although supervised study is typically thought of as students reading textbooks, students' use of references can take many forms. Videos, pictures, Extension publications, commercial brochures, journals and magazines, Internet resources, and resource people are all examples of valid references.

Care should be taken when the Internet is used as a resource. Students should understand that on the Internet

- Not all Web sites are appropriate.
- Not all information is accurate.
- Not all information is unbiased.

The agriculture teacher should work with the school media specialist in determining appropriate resources and teaching students information-searching procedures.

The supervised part of supervised study is critical. The agriculture teacher is responsible for directing the supervised study. The teacher should be in a location where he or she can see all the students and should not be using supervisory time for tasks such as making telephone calls or grading assignments. Occasionally the teacher should circulate around the room and monitor student progress. This does not mean interrupting students with chitchat or needlessly giving answers. Instead, the teacher

13–15. Students in supervised study using up-to-date textbooks and receiving individual instruction from their teacher.
(Courtesy, Education Images)

can provide encouragement, assist with misunderstandings, and assess individual comprehension of the topic.

Although supervised study is categorized as individualized instruction, there may be times when students can work together—for example, on group projects or presentations and during peer-assisted instruction. Grouping students is also appropriate when reference materials are limited.

Guided practice is a special form of supervised study. Whereas supervised study is typically associated with cognitive and affective learning objectives, guided practice is used for psychomotor objectives. In guided practice, the agriculture teacher demonstrates a skill or activity and then provides opportunity for students to practice the skill while under the watchful eye of the teacher. The teacher can quickly correct bad habits or safety problems individually or bring the students together for group instruction if appropriate.

Independent Study

There are occasions when it is appropriate to teach an agriculture course or topic as an independent study. An *independent study* is a learning activity that is not part of an organized class learning experience. Maybe a student wants to take an advanced level of an agriculture course and the number enrolled is not enough to justify a class period for the course. In such a case, the student would take on more responsibility for his or her instruction.

To be successful, independent study must have clear objectives, guidelines, and timelines in place at the beginning. Although the student assumes greater responsibility for finding resource materials and studying those materials, the agriculture teacher is still responsible for evaluation and overall direction.

Increasingly, independent study opportunities are available through distance education. Instruction may be delivered via packets of printed material, videos, the Internet, Internet Protocol (IP) video, or combinations. Students may be eligible to receive high school or college credit or both.

Learning Modules

A *learning module* is a self-contained activity that a student can complete independently. Learning modules are most often used for cognitive and psychomotor

Name: _____

Date: _____

Class: _____

Project: _____

1. What tasks will you do today on your project?

2. What equipment, tools, supplies, and materials will you use today?

3. What did you accomplish today?

4. Do you need additional supplies or materials, and if "yes," what are the specifications?

5. What problems or concerns did you encounter today?

13–16. An example of a project accountability sheet.

objectives. Learning modules may be purchased or may be developed by the agriculture teacher. Modules placed around the room can be combined into stations that students rotate through as they complete the activities. An advantage of learning modules is that students can proceed at their own pace and instruction can be modified to meet individual learning needs of students. Learning modules are also useful in meeting FFA membership requirements for non-enrolled students.

Projects

Agricultural laboratories are ideal for individual student projects. A *project* is a significant, practical activity of educational value with one or more definite goals. An advantage of projects is that they can maximize student interest because the student has flexibility in choosing a project to complete. The project can also be geared to the ability of the student. Thus, instruction is available at the appropriate level for all students.

As with other methods discussed, projects require planning and active participation by the agriculture teacher. The teacher must instruct students on wise use of time; responsibility for safety, cleanliness, and productivity; and the importance of doing a quality job. The agriculture teacher must answer several questions before using individual projects as an instructional method. What is the learning purpose behind a project? What are the minimum requirements for an acceptable project? What is the maximum time allowed for project completion? What are the procedures for using tools/equipment, cleaning up the laboratory, and obtaining materials/supplies? What safety and behavior rules need to be emphasized?

Figure 13–16 presents an example of a project accountability sheet that allows the student to record accomplishments during a class period and makes daily grading by the teacher easier. The student answers the first two questions at the beginning of the class period and the last three questions at the end of the class period.

DEVELOPING A LESSON PLAN

A *lesson plan* is the road map an agriculture teacher uses to ensure effective, efficient, and empowering instruction. It is much like the plans used in building a house or the script used in making a movie or television program. Doctors

13–17. Background reading and reviewing of textbooks are often needed before beginning to write a lesson plan.

© Jasmin Merdan/Fotolia LLC

performing surgery devise plans of action, sometimes even placing dotted lines where incisions should be made. Football coaches develop plays, playbooks, and game plans. Teachers also need direction and vision! It should be made clear, however, that teachers are not movie actors and lesson plans are not scripts. A well-developed lesson plan allows the teacher to be flexible and creative and to take advantage of the "teachable moment."

There are many styles and formats of lesson plans, yet they all have some common components. The essential components of a lesson plan are presented here.

- **Preparatory block**—The first essential component is the preparatory block. This contains organizational information useful for filing and indexing, documentation, and quick referencing. The block also includes unit and lesson titles, educational goals and objectives, references and materials needed, key terms, safety matters requiring attention, and a listing of agricultural and academic standards met by the lesson.
- **Introductory block**—The next component is the introductory block. This part of a plan contains the interest approach, review of previous material, pretests, and other activities that occur at the beginning of a lesson.
- **Content and teaching–learning activities block**—The content and teaching–learning activities block is next. The two sections of this block may be set up one on top of the other or side by side. Depending on the lesson plan style, the sections of this block may be very detailed or serve as outlines.
  - At a minimum, the content section contains an outline of the subject matter, with key information highlighted. There is no need to repeat word for word information that is already found in a textbook or on a worksheet.
  - The teaching–learning activities section describes the strategies used to teach the content. These include instructions to the teacher on methods, questions to be asked to students, student activities, and procedural instructions, such as when or how to do a task. This section should include checks for understanding and scaffolding techniques.
- **Summary/conclusion block**—The summary/conclusion block provides details on how the lesson will be summarized and what conclusions can be drawn from the lesson. This block may also contain details on how this lesson links back to previous lessons and forward to the next lesson. It is important

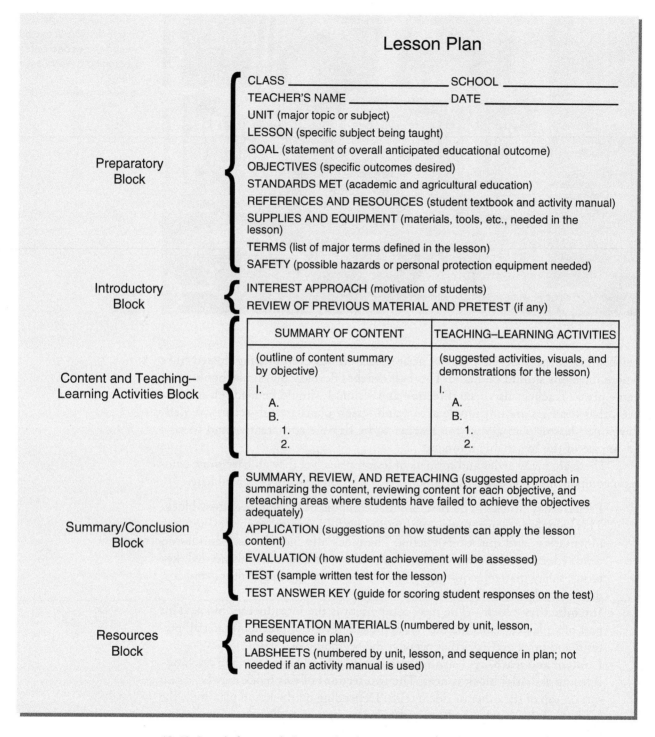

Lesson Plan

Preparatory Block
- CLASS _____ SCHOOL _____
- TEACHER'S NAME _____ DATE _____
- UNIT (major topic or subject)
- LESSON (specific subject being taught)
- GOAL (statement of overall anticipated educational outcome)
- OBJECTIVES (specific outcomes desired)
- STANDARDS MET (academic and agricultural education)
- REFERENCES AND RESOURCES (student textbook and activity manual)
- SUPPLIES AND EQUIPMENT (materials, tools, etc., needed in the lesson)
- TERMS (list of major terms defined in the lesson)
- SAFETY (possible hazards or personal protection equipment needed)

Introductory Block
- INTEREST APPROACH (motivation of students)
- REVIEW OF PREVIOUS MATERIAL AND PRETEST (if any)

Content and Teaching–Learning Activities Block

SUMMARY OF CONTENT	TEACHING–LEARNING ACTIVITIES

(outline of content summary by objective)
I.
  A.
  B.
    1.
    2.
	(suggested activities, visuals, and demonstrations for the lesson)
	I.
	  A.
	  B.
	    1.
	    2.

Summary/Conclusion Block
- SUMMARY, REVIEW, AND RETEACHING (suggested approach in summarizing the content, reviewing content for each objective, and reteaching areas where students have failed to achieve the objectives adequately)
- APPLICATION (suggestions on how students can apply the lesson content)
- EVALUATION (how student achievement will be assessed)
- TEST (sample written test for the lesson)
- TEST ANSWER KEY (guide for scoring student responses on the test)

Resources Block
- PRESENTATION MATERIALS (numbered by unit, lesson, and sequence in plan)
- LABSHEETS (numbered by unit, lesson, and sequence in plan; not needed if an activity manual is used)

13–18. *Sample format of a lesson plan showing relationships of plan parts to the major blocks.*

to include suggestions on how students can apply the lesson content. The evaluation section contains a description on how the students will be evaluated, evaluation instrument(s), and answer key(s).

- **Resources block**—The resources block is typically at the end of the lesson plan. This block contains worksheets, handouts, printouts of PowerPoint presentations, and note-taking guides.

In general, a lesson plan will cover one to five days of instructional time (one day equals a fifty-minute period). A lesson plan that takes less than one day to cover means the topic is probably too narrow for a complete lesson. A lesson plan that takes more than five days to cover may mean the topic is a unit of instruction that needs to be divided into two or more lessons. Also, the teacher may develop daily plans to guide the introduction, teaching–learning, checks for understanding, and evaluation for each class session. These are broad guidelines and do not apply to all situations.

Commercially and Professionally Developed Lesson Plans

In a traditional schedule, the agriculture teacher instructs students for 180 days and has five to six class periods a day. If the average lesson plan lasts for three days, then this typical agriculture teacher will need to develop 60 lesson plans for each class, or up to 360 separate lesson plans. Of course, time frames and lesson plan needs vary somewhat with semester-long classes and alternative scheduling. Why should an agriculture teacher develop his or her own lesson plans if these can be purchased? How should an agriculture teacher use commercially developed lesson plans?

Many states, either through curriculum centers or through private companies, provide lesson plans for state-approved curriculums to their agriculture teachers. These lesson plans are typically professionally done and include illustrations, Power-Point presentations, premade tests with answer keys, and suggested teaching–learning activities. These lesson plans can be a great help to both beginning and experienced teachers. Beginning teachers are saved the time of creating lesson plans from scratch. Experienced teachers may gain new ideas for teaching certain content.

These purchased lesson plans do not relieve the teacher of the responsibility of planning instruction. Commercial lesson plans may be generic in nature, requiring the agriculture teacher to make the content locally applicable. The plans may be written at the level of a typical student, requiring the agriculture teacher to modify them to differentiate instruction to reach all students in the classroom. In addition, each agriculture teacher has a style of teaching that he or she develops over time.

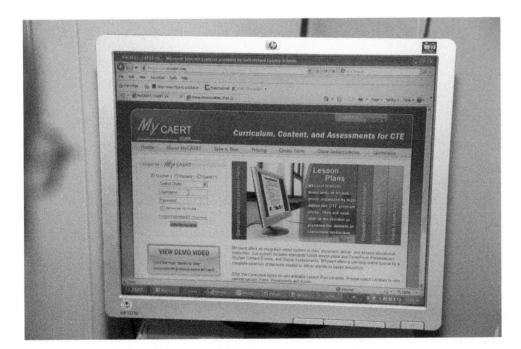

13–19. Commercially available lesson plans, such as those from CAERT, Inc., include information that relieves a teacher of a great deal of preparation time.
(Courtesy, Education Images)

The commercial lesson plans are not written for a particular teacher's style, so modifications will need to be made in this area. Finally, commercial lesson plans, just as teacher-developed lesson plans, should be updated yearly with new content and teaching modifications.

USING INSTRUCTIONAL TECHNOLOGIES AND THE INTERNET

We live in a technological, information-rich society. Technologies seem to rise and become obsolete within a decade. The rate of information production continues to increase so rapidly that the time it takes for the volume of information to double is measured in years and months. Education is adopting instructional technologies at an ever-increasing rate. However, there are principles that apply, whether one is using a chalkboard/whiteboard, an overhead transparency, a PowerPoint presentation, or some other audiovisual aid.

The principles of interest discussed in Chapter 12 apply to using instructional technologies. Instructional technologies can be used to bring nature, humor, creativity, human interest, and the unusual into the classroom. When used effectively, instructional technologies can spread interest from interesting things to uninteresting ones. A caution is that the technology used should not be the focus but instead should enhance interest in the topic being studied.

Visuals

Well-designed visuals are important regardless of the instructional technology used. Visuals need to be visible and comprehendible to all students regardless of their location in the classroom. Guidelines for designing visuals or for evaluating premade visuals are given in Table 13–1.

Instructional technologies need not be new to be valuable. The chalkboard/whiteboard is a useful tool in almost every agricultural education classroom. Because of its large size, ease of use, and ease of correction, this instructional tool is extremely functional.

13–20. Preparing a lesson presentation.
(Courtesy, Education Images)

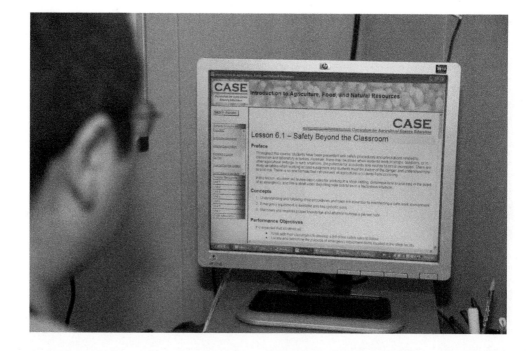

Table 13-1
GUIDELINES FOR USING VISUALS IN THE CLASSROOM

Element	Description
Letter size (on paper)	Use 18- to 20-point font size (72 points = 1 inch) when developing computer slides or transparency masters.
Letter type	Use a sans serif font (e.g., Arial) for television, computer, or video.
Color	Use background and foreground (text) colors that have good contrast, are pleasing to the eye, and are not saturated (two intense colors used together).
Letter case	Use upper- and lowercase letters. All uppercase is difficult to read.
Word layout	Place words horizontally, left to right. Avoid stacking letters (words placed vertically).
Volume of text	Use key words rather than whole sentences. Limit visuals to 5–7 words per line and 5–11 lines. Allow for white space.
Letter size (projected)	When projected on the screen, letters should be 1 inch in height for every 20–25 feet the farthest student is from the screen. Example: The back of the classroom is 40–50 feet from the front of the classroom. Projected letter height should be 2 inches.
Screen size	For every unit of screen width, the closest student should be twice that distance away, and the farthest student six times. Example: The screen is 3 feet wide. The closest student should be 6 feet from the screen and the farthest student 18 feet.

Some instructional technologies do become obsolete. Filmstrips and 16-mm films gave way to first videotape and laserdiscs, then DVDs. Reel-to-reel audiotapes were replaced by audio cassettes, which were then replaced by audio compact discs. Some of the instructional technologies discussed in this section will probably become obsolete within the near future.

13–21. Student supervision is an important part of classroom computer use.
(Courtesy, Education Images)

Computers in the Classroom

The computer, in all of its forms and sizes, continues to be a valuable instructional tool in the classroom. Its uses cut across all areas of instruction. PowerPoint presentations are used to deliver notes, present procedures, and provide pictures to supplement lectures. Word processing, spreadsheet, and database programs are used in all aspects of teaching. Specialized computer programs are available to assist with agronomic, agribusiness, and management applications. The use of GPS and GIS has allowed agriculture teachers to teach about site-specific farming and other applications.

The computer has made the greatest impact by bringing the Internet into the agriculture classroom. It has been estimated that there are millions of Web sites and a billion Web pages. More than 50 percent of U.S. companies sell products online. In U.S. public schools, there is one computer for every four students, with almost 100 percent of schools having Internet access. Agricultural education classrooms may have Internet-connected computers, or classes may have to go to the media center to access the Internet. The Internet provides students access to electronic mail; public domain software and shareware; discussion groups, such as communities of practice and bulletin boards; information and news on a variety of topics; and research tools, such as online journals, databases, and search engines.

Schools are increasingly using laptop or tablet computers as instructional equipment. There may be a classroom set that students share, or each student may receive one to use throughout the school year similar to a textbook. These computers may have access to electronic textbooks and curricular materials as well as apps to assist with classroom assignments.

As students use school computers and access the Internet, they must use the Internet safely and with proper netiquette. Figure 13–22 gives guidelines for using the Internet.

Safety Guidelines:

1. Not all materials are appropriate for minors. Although filters, blocks, and ratings are useful, nothing takes the place of an observant, visible teacher.

2. Students should not give out personal information unless under supervision of the teacher.

3. Not all people in social media sites, chat rooms, and discussion groups are who they claim to be. Students should not give out information about themselves, send pictures of themselves, or arrange to meet someone. Only social media sites, chat rooms, and discussion groups of an educational nature should be accessed using school computers.

Use Guidelines:

1. Copyright laws must be followed. Areas of concern include music, movies, images, and books.

2. Plagiarism from the Internet must not be tolerated. A student's copying from a Web page and pasting the material into his or her document is plagiarism. Proper citing is critical.

3. Hacking, creating a virus, or damaging computers is not allowed.

4. Unauthorized access to others' accounts, data, files, or communications is not allowed.

5. Students should be polite on the Internet. Sending obscene, sexually explicit, or hate messages is not allowed. Cyber-bullying is not tolerated. It is a good idea to limit any type of personal e-mail messages sent or read using school computers.

13–22. Internet safety and use guidelines.
Source: Tippecanoe School Corporation, 2003.

REVIEWING SUMMARY

Teachers who organize their instruction using lesson plans, who carefully select teaching methods, and who appropriately utilize instructional technology have a positive impact on student learning. Teachers should strive to build continually upon their instructional repertoire and experiment with new techniques and technologies.

Bloom's taxonomy has guided teachers for fifty years in writing appropriate learning objectives. Although more recent research has shown that students learn through their whole experience, Bloom's three domains of learning are still important for planning, assessment, and accountability purposes.

Teaching methods can be divided into those designed for group instruction and those designed for individualized instruction. Effective teachers learn which methods work best in what situations, with what students, and for what purposes.

Lesson plans are critical for teachers to be effective and efficient. Although numerous styles of lesson plans exist, they all have certain similar components. Commercially or professionally developed lesson plans hold advantages for both beginning and experienced teachers; however, the local teacher must still personalize the plans.

Instructional technologies enhance the classroom environment. Through the use of these technologies, students are exposed to content and resources they would not have access to otherwise. Teachers need to take care that the subject matter, not the instructional technologies, is the focus of student attention. Computers, although extremely useful in the classroom, are merely another tool the teacher uses to enhance the student learning experience.

QUESTIONS FOR REVIEW AND DISCUSSION

1. How is Bloom's taxonomy used in the agricultural education classroom?
2. How are learning objectives written using the ABCD approach?
3. What are the six common methods of group instruction?
4. What are the five common methods of individual instruction?
5. What is a lesson plan?
6. Why is it important for all agriculture teachers to use lesson plans?
7. What are the common components of all lesson plan styles?
8. What are the advantages and disadvantages of using commercially or professionally prepared lesson plans?
9. How do the principles of interest discussed in Chapter 12 relate to instructional technologies?
10. What are major uses of the computer in the agricultural education classroom?

ACTIVITIES

1. Investigate the affective and psychomotor domains of learning. Why have these not been developed as fully or emphasized as much in education as the cognitive domain? How is this different or the same for agricultural education?
2. Investigate agricultural education resources on the Internet. Develop and organize a links Web page or a WebQuest.
3. Using the *Journal of Agricultural Education*, *The Agricultural Education Magazine*, and the proceedings of the National Agricultural Education Research Conference, write a report on computers in classroom research in agricultural education. Current and past issues of these journals can be found in the university library. Selected past issues can be found online at the American Association for Agricultural Education and the National Association of Agricultural Educators Web sites.

CHAPTER BIBLIOGRAPHY

Agricultural Communicators in Education. (1983). *Communications Handbook* (4th ed.). Danville, IL: The Interstate Printers & Publishers, Inc.

Binkley, H. R., and R. W. Tulloch. (1981). *Teaching Vocational Agriculture/Agribusiness.* Danville, IL: The Interstate Printers & Publishers, Inc.

Bloom, B. S. (Ed.). (1956). *Taxonomy of Educational Objectives.* New York: Longman, Green, & Co.

Heinich, R., M. Molenda, J. D. Russell, and S. E. Smaldino. (1999). *Instructional Media and Technologies for Learning* (6th ed.). Upper Saddle River, NJ: Merrill/Prentice Hall.

Information Please. (2007). *Education Facts at a Glance.* Accessed on April 16, 2012, at http://www.infoplease.com/spot/schoolfacts1.html.

Kahler, A. A., B. Morgan, G. E. Holmes, and C. E. Bundy. (1985). *Methods in Adult Education* (4th ed.). Danville, IL: The Interstate Printers & Publishers, Inc.

Krathwohl, D. R., B. S. Bloom, and B. B. Masia. (1973). *Taxonomy of Educational Objectives, the Classification of Educational Goals. Handbook II: Affective Domain.* New York: David McKay Co., Inc.

Magid, L. J. (2003). *Child Safety on the Information Highway.* Arlington, VA: National Center for Missing and Exploited Children. Accessed on April 16, 2012.

McCormick, F. G., Jr. (1994). *The Power of Positive Teaching.* Malabar, FL: Krieger Publishing Co.

National Association of Agricultural Educators (NAAE). (2003). NAAE Web Site. Accessed on August 22, 2003.

National FFA Organization. (2003). *Local Program Resource Guide: 2003–2004* [CD-ROM]. Indianapolis: Author.

National FFA Organization. (2011a). *Local Program Success* (online). Access on November 30, 2011.

National FFA Organization. (2011b). *Official FFA Manual.* Indianapolis: Author.

Newcomb, L. H., J. D. McCracken, J. R. Warmbrod, and M. S. Whittington. (2004). *Methods of Teaching Agriculture* (3rd ed.). Upper Saddle River, NJ: Prentice Hall.

Phipps, L. J., and E. W. Osborne. (1988). *Handbook on Agricultural Education in Public Schools* (5th ed.). Danville, IL: The Interstate Printers & Publishers, Inc.

Simpson E. J. (1972). *The Classification of Educational Objectives in the Psychomotor Domain.* Washington, DC: Gryphon House.

SOFWeb. (2002). *Using the Internet: Research on the Net. Victoria, Australia.* Accessed on August 28, 2003.

3M. (1993). *Brilliant Meetings: The Art of Effective Visual Presentations.* Austin, TX: Author.

Tippecanoe School Corporation. (2003). *Tippecanoe School Corporation Student Internet Use Agreement. West Lafayette, IN.* Accessed on August 28, 2003.

True, A. C. (1929). *A History of Agricultural Education in the United States, 1785–1925.* U.S. Department of Agriculture, Miscellaneous Publication No. 36. Washington, DC: GPO.

14
Classroom Management

"John, where is your homework?" Mr. Watson asked.

"I forgot it," was the reply from John in the back of the room.

This was not the answer Mr. Watson wanted, so he tried a little harder to voice his displeasure. "Do I need to give you a kick in the seat of your pants to motivate you?" asked Mr. Watson, sitting up behind his desk to get a good look at John in the back row.

"I'd like to see you try," replied John, under his breath but just loud enough to cause his classmates close by to exchange glances and smile.

"What did you say?" said Mr. Watson, rising from his chair.

"Nothing," replied John, still avoiding eye contact with his teacher.

Unfortunately, similar scenarios occur many times a day in classrooms across the nation. The learning environment for all students has been disrupted. At least two people have problems: the student and the teacher. Neither has performed properly.

When a student cannot cope in school and when the teacher uses inappropriate discipline techniques, misbehavior becomes an even larger problem. Teachers must be professional in dealing with student misbehavior so that instruction continues and material presented in the lesson is learned.

objectives

This chapter explains practices associated with establishing and maintaining a quality teaching–learning environment. It has the following objectives:

1. Discuss unique attributes of agricultural education as related to establishing a quality learning environment

2. Describe the characteristics of student behavior in a learning environment

3. Identify effective approaches in student discipline

4. Describe how to handle misbehavior when it occurs

5. Identify student behavior trends in agricultural education

terms

discipline
engagement
learning environment

misbehavior
proximity control
punishment

CLASSROOM MANAGEMENT: AN OVERVIEW

The American people like their local schools, and they have trust and confidence in their children's teachers. In fact, three of four Americans believe that teachers should have the authority and flexibility to design and teach the curriculum they think is best. The American public does not see student misbehavior as a major problem in schools, and generally believe that schools are fulfilling their intended purpose (Bushaw & Lopez, 2011). With this in mind, agricultural education teachers should approach classroom management knowing that they have the public's support necessary to create and maintain a positive classroom environment.

Classroom management starts with the environment the teacher creates in the classroom. Students require clear direction toward learning goals. The mechanism for achieving those goals is *engagement*. Engagement occurs when students make a psychological investment in learning. The lessons must be interesting and relevant to the interests of the students. Oftentimes, the instructor needs to help students see the connection between the lesson content and the personal interests and experiences of students. When students are engaged in meaningful learning experiences, there is less time and opportunity for misbehavior. Although often associated with misbehavior, *discipline* can best be thought of as teaching structure, routine, and behaviors to students to facilitate their learning.

Students tend to respond well to routines and clearly established norms for behavior in the classroom. Teachers should help the students develop these norms on the first day of classes and reinforce them throughout the academic year (Marzano, 2011).

Misbehavior and Punishment

Misbehavior is the condition that results from a person's inappropriate attempt to solve a problem or adapt to a situation (Maslow, 1970). This inappropriate coping behavior is purposeful and motivated. It is often a learned response to situations in the student's environment. In Maslow's words, "It is an attempt to make up for internal deficiencies by external satisfiers" (Maslow, 1970, p. 132). Misbehavior is any behavior that is inappropriate for the moment. For instance, certain classroom activities may require students to stand up and move about the classroom, and that might be an acceptable behavior at the moment. However, at other times, it may be more conducive to the learning environment if the students remain seated and attentive to a demonstration or lecture. Standing up and moving about the classroom then becomes misbehavior in that moment.

Students cope with internal stress in a number of ways. For instance, a student with a need for group affiliation might cope with this need by expressing foolish behavior in the classroom in order to gain acceptance from his or her classmates. Another student who works a part-time job well into the late evening hours may cope with a physiological need for rest by being frequently absent from school (Dahl, 1999). When the coping mechanism violates social norms within the classroom, it becomes misbehavior (Croom & Moore, 2003).

Recognizing the seriousness of behavior in the classroom is an essential part of teaching. However, teachers rarely communicate among themselves to any depth about the subject of student misbehavior, even though the stress generated by misbehavior is of greater concern than problems with other working conditions (Abel & Sewell, 1999). Because most teachers spend the majority of their workday almost exclusively with pupils, they tend to formulate their own definition of misbehavior and handle misbehavior accordingly (Borg & Riding, 1991).

14–1. What features do you see in this learning environment? Are students demonstrating on-task behavior? (Courtesy, Shutterstock © auremar)

Thus, there is a wide range of behavioral modification techniques used to redirect or repress misbehavior.

Engaging the Students as Partners in Classroom Management   Some teachers and school administrators value the input of students in developing fair and effective policies for behavior. If the school does not have an approach that involves students in the development of fair and acceptable norms for behavior, the teacher can provide students the opportunity for input in the development of classroom rules and procedures. By having the students "own" the classroom behavioral norms, the teacher is displaying respect for the students and providing them with an opportunity to direct their learning experiences.

There are good reasons for having students take charge of their classroom management strategies. A student-focused classroom management strategy places the responsibility for both learning and behavior squarely in the hands of the student. Students are less inclined to break their own standards for behavior. When they do behave in a contrary manner to their own classroom rules, the teacher and student can have a useful and meaningful discussion about their behavior. *Punishment* is the administration of a penalty or undesirable action designed to discourage unacceptable behavior. Punishment can involve having to wait until the end of the lunch line, staying after school, picking up trash on the school grounds, or doing other activities that are deemed undesirable.

Three Variables of a Discipline Problem

DeBruyn (1983) found that three variables (the student, the teacher, and the class) are present in almost every discipline problem. These variables determine the frequency, intensity, and duration of misbehavior. Further, these three variables influence the type of misbehavior that occurs. These variables, along with a fourth environmental variable, the teaching and learning facility, are summarized in Table 14–1.

Table 14–1
ENVIRONMENTAL VARIABLES INFLUENCING STUDENT BEHAVIOR

The Student	Socioeconomic status and family structure
	Personality
	General physical and mental health
	Dietary habits and sleep patterns
	Level of interest in the class or subject matter
	Learning style
	Effect of peer pressure
The Teacher	Teaching ability
	Knowledge and expertise in student behavior management
	General physical and mental health
	Dietary habits and sleep patterns
	Personality
	Quantity and quality of student supervision
The Rest of the Class	Time of day
	School activities (prom day, last day of school, etc.)
	Size of class
	Peer pressure
	Subject matter or teaching method employed
The Teaching and Learning Facility	Cleanliness
	Noise pollution and air quality
	Safety
	Size of classroom
	Student desk style, arrangement, and quantity
	Teaching method being employed

Source: Adapted from DeBruyn (1983).

The Student   On any given day and in any given situation, a student's behavior could change from socially acceptable for the classroom into misbehavior. As human beings, students are often influenced or motivated by certain external factors in their environment. A student who did not get enough sleep the night before is going to be sleepy in class. The student who failed to eat a nutritious breakfast might be lethargic in class, or if the student loaded up on caffeine-laced soft drinks and candy bars, he or she might be hyperactive at some point. Student behavior is likely to be different on the day of the school prom and on an early-release day than it is on an ordinary school day. The experienced teacher can frequently determine the type of behavior students will exhibit by observing them as they come into the classroom. A major goal should be to help students practice self-discipline and exercise self-control.

The Teacher   Though misbehavior has been identified as an individual choice, it may be influenced by factors in a person's environment. Obviously, the teacher sets the tone and pace of the class and exerts a significant influence over students. Therefore, the teacher's actions may influence student behavior either positively or negatively. A good example of how a teacher can negatively influence student behavior can be seen in this example. A student puts his head down on his desk and falls asleep during a classroom lecture. The teacher notices the behavior but fails to wake the student or somehow intervene in the behavior. In this case, the teacher is contributing to the misbehavior.

Classroom management is an important part of having a good learning environment. The presence of the teacher in the classroom or the supervision of students during lab activities is a significant factor in student behavior. Teachers who are constantly supervising students and directing their behavior generally have fewer behavioral problems.

The Class of Students   The third factor in student misbehavior is found in the actions of the class as a whole. Students have a need to be accepted by their classmates. Moreover, they have a strong need for belonging in the sense of acceptance by the members of the group. Therefore, students will often act in a manner that brings admiration and attention to them by other members of the group, even though their actions are classified as misbehavior.

Furthermore, this need for belonging and safety is so strong in students that often their response to corrective action by the teacher is an escalation of the misbehavior. For instance, the student who is disciplined in front of his or her peers might become defiant and argumentative with the teacher. That very same student might be altogether different and more socially acceptable in a private student–teacher conference. Effective teachers recognize that they greatly influence student behavior.

The variable that can be controlled with the greatest ease is the teacher's behavior. A teacher has the responsibility of not only attempting to diagnose the problem but also of taking steps to adjust instruction and interaction with students to help students make the right choices about socially acceptable behavior. Recognizing and coping with student misbehavior is an essential part of instruction. It is important that the beginning teacher develop tools for effectively dealing with behavior.

STUDENT BEHAVIOR AND THE AGRICULTURAL EDUCATION CLASSROOM

A number of studies found that student attitude was a significant factor in misbehavior. The ambivalent attitude of students, students being poorly prepared for class, and a general negative student attitude were the most frequently cited behavioral problems (Burnett & Moore, 1988; Camp & Garrison, 1984). Students were also perceived to be inattentive and irresponsible, but these behaviors were not deemed to be a substantial disruption in class. The most prevalent misbehavior in agricultural education was found to be student actions resulting from a dislike of school. Either the students do not like school or are ambivalent toward it (Croom & Moore, 2003).

Between kindergarten and secondary school, some students developed a dislike for schooling. Their attitude carries over into the thoughts, feelings, and reactions they have related to the efforts of instructors to teach them. In the high school agricultural education program, the task is to provide conditions under which student interest is maintained at a high level. The effective agriculture teacher constantly seeks effective methods to keep students engaged in learning at the highest possible level.

The agricultural education classroom generally has multiple learning environments, including, but not limited to, the standard classroom with desks and bookcases, a greenhouse and land laboratory, a mechanics lab, and a biotechnology lab. In addition, it may have storage areas, an instructor's office, and other facilities. The key to managing these facilities effectively is found in a sound classroom management plan based upon the best practices of discipline.

14–2. Agricultural education uses a wide range of learning environments. (Here a student is practicing safely in the application of a pesticide to a poinsettia crop in a South Carolina school greenhouse.) (Courtesy, Education Images)

Student performance is often derived from the level of expectation held by the teacher. A key in managing agricultural education is to establish and maintain a quality learning environment. The *learning environment* is all the conditions that influence teaching and learning. Some environments promote learning; others may impair learning. It is the teacher's responsibility to be in charge of the situation and establish a quality learning environment. A teacher may use a classroom management plan based on the best practices of a quality learning environment.

Agricultural education may have power tools, chemicals, lab equipment, and many other potentially hazardous devices and materials. Students who are misbehaving could seriously injure themselves and others around them. Teachers should develop organizational skills and lesson plans that minimize the risk of injury to students.

Furthermore, students enroll in agriculture classes with varying degrees of skill in the agricultural sciences. Some students know the proper use of tools, equipment, and chemicals, while others do not have very much skill in these areas. Some students have in-depth knowledge of some agricultural subjects but very little knowledge of others. For instance, the student who is an expert welder and machinist may not have much skill in handling livestock.

Lab work involves a significant amount of psychomotor ability, and students must be taught how to work safely and follow procedures. The hands-on nature of agricultural education demands much attention to the development of new psychomotor skills in addition to cognitive and critical-thinking skills.

Agriculture classes are elective courses. Students may enroll in these courses because they are personally interested in the subject matter. However, most experienced teachers recognize that not every student who enrolls does so because of a deep personal interest in an agricultural career. Some students take the courses because they have some interest in agriculture and wish to explore further. These students can either become more interested in the subject matter or lose interest as the courses progress. Some students take agriculture courses because they want to avoid other required subjects or because they need additional electives to graduate. Not every student who comes through the doorway arrives by his or her own accord. Sometimes, students are placed in agriculture classes without regard to their potential. Whatever the method by which students arrive in the classroom, the possibility of behavioral problems is real.

It would be an error to assume that discipline and classroom management are what teachers do in order to get the lesson accomplished and the content delivered. Student discipline is more than an ancillary duty; it is part of the learning process itself. The wise teacher learns to incorporate important lessons about behavior into the lessons about agricultural content. Furthermore, the teacher who understands the psychology of learning and the processes of human growth and development at the adolescent and preadult stages can effectively help students make the right choices regarding their behavior.

CREATING AN ENGAGED LEARNING ENVIRONMENT

The engaged learning environment includes high-quality and relevant lessons prepared and delivered by a highly trained and capable agriculture teacher. This learning environment begins at the classroom door and includes all of the events and activities from the moment the student enters the classroom until the student leaves for their next class. It includes all of the FFA activities planned for the day, as well as activities conducted in the student's supervised agricultural experience (SAE). The partners in this engaged learning environment are the teacher, the students, and everyone who provides input and resources that support the learning activities. It is difficult, but not impossible, to plan and conduct highly interesting and relevant learning experiences that meet the specific learning needs of the students in a class. Successful teachers develop procedures and routines that guide the learning process and provide structure to the learning experience.

And yet, it is not enough that the teacher establishes rules for the students to follow without considering the larger issues at hand. Students learn through a process involving inquiry and socialization. They have to learn how to think for themselves in a social environment. The complexities of human behavior demand a larger response to student misbehavior than a series of choices and punishments, because society will either benefit from or be diminished by the behavior of students once they become adults. As Kant reminds us in his treatise *On Education*, "If you punish a child for being naughty, and reward him for being good, he will do right merely for the sake of the reward; and when he goes out into the world and finds that goodness is not always rewarded, nor wickedness always punished, he will grow into a man who only thinks about how he may get on in the world, and does right or wrong according as he finds either of advantage to himself" (Kant, 1960, p. 84).

It is important to see misbehavior as an individual choice, but one with consequences for the whole class or learning community. One student's misbehavior often has a negative influence on the rest of the class. In the agricultural education lab and classroom, misbehavior could lead to serious physical injury. For instance, students playing in the mechanics shop could fall and injure themselves or cause others to be injured. In another instance, the student who puts her head down on her desk goes to sleep during a group assignment may not be disturbing other students, but that student is certainly not carrying her share of the work for the group.

DEVELOPING A STUDENT-FOCUSED MANAGEMENT PLAN

The teacher is in charge of the students, and is ultimately responsible for teaching them the subjects of agricultural education. This does not preclude the teacher from developing a student-focused model of classroom management. In the student-focused model, students take responsibility for setting up the rules and procedures that they perceive will help them learn best. This is a departure from the traditional teacher-focused model of classroom management. In the teacher-focused model, the teacher sets the rules, metes out rewards and punishments, and decides most instructional activities without input from students.

The student-focused model asks students to take leadership in the learning process. Students develop classroom rules that meet school policy, and that meet their inherent needs for learning, belonging, autonomy, and personal freedom.

A COMPARISON OF CLASSROOM MANAGEMENT STYLES

Teacher-Focused	Student-Focused
Teacher sets the classroom rules.	Students collectively establish norms.
Discipline is derived from the teacher.	Behavioral change rests with the student.
All actions have consequences for the student.	All actions have consequences for the student and for the class.
Rewards are extrinsic.	Rewards are intrinsic.
Teacher is the sole authority in the class.	Students share the responsibility for learning.

Derived from Davis, Summers, & Miller, 2012.

Have a Well-Prepared Lesson

Good preparation by the teacher helps prevent many misbehavior problems. With good planning, the learners are kept busy. Busy learners do not have the time to contemplate misbehavior.

A well-prepared lesson plan promotes student learning. The single greatest method for preventing misbehavior is to have a well-prepared, well-executed lesson delivered in an enthusiastic manner. Realistic, relevant lessons, presented in an interesting way, will engage the students, who will be inclined not to misbehave. When planning a lesson, it is important for the teacher to think like a student but act like a teacher. Answer the question, "How can I plan this lesson so that it is interesting to the students and meets their needs?" The same things that motivate adults also motivate students. Attractive visuals, interesting demonstrations, and real-life applications of learned skills are all things that motivate adults. These same methods can be used to engage student interest.

Through careful planning, the agriculture teacher can usually find a way to engage students in the lesson. By using their creative talents, teachers can find unique ways to keep students involved in a lesson, even when resources are limited. For example, a teacher is planning a lesson on taking a soil sample. The least preferable method is for the teacher to teach the use of the soil probe to students one at a time. Invariably, while the teacher is showing one student how to collect a soil sample effectively using the soil tube, the remainder of the class is talking, joking, or generally goofing off. Students observing this procedure generally learn it quickly and then must amuse themselves while waiting their turn to use the soil tube.

A more effective method for teaching this procedure would be this: The teacher gives each student a shovel for the purpose of collecting a soil sample. Most homeowners do not own soil tubes for taking soil samples, but they would have shovels that could be effectively used to complete the task. The teacher demonstrates the procedure to all students and then assigns them to practice the skill under supervision in a suitable location. Thus, students learn the technique for taking soil samples in a timely manner.

HOW WOULD YOU RESPOND?

Ms. Mary Watson sits at her desk grading the last of the animal science unit tests. The last test she picks up to grade belongs to Jeff. She immediately notices that most of the questions on the test are unanswered. She puts down the paper, picks up the grade book and finds Jeff's name in second period. Except for a few directed lab exercises where he worked with a partner, Jeff has few grades to show for his efforts. She gets up from her desk and walks to the classroom next door to speak to her teaching partner Tom Wells.

"Tom, did you have Jeff Wiggins in class last semester?"

Mr. Wells looked up from his paperwork and replied, "I sure did. Do you have him this year? I'll bet I know why you are asking about him."

"Yep, I have him in the animal science class this semester. How well did he do in your class?" asked Ms. Watson.

"Not good at all. It was the introduction to agricultural science class, and he barely made a D. He's not a great student," said Mr. Wells.

"What's the deal with Jeff, then?" asked Ms. Watson. "How can I get him engaged in this class? Every day he falls farther and farther behind. He doesn't bring a notebook to class, asks no questions, and often tries to sleep while I am teaching. I have to stay on him to keep his head up off the desk."

"How do the other students act around him?" asked Mr. Wells.

"See, that's the interesting thing. Even during labs when the other students are really engaged in what we are doing, Jeff just sits there like a bump on a log. He doesn't communicate well with the other students," explained Ms. Watson.

"What are you going to do?" asked Mr. Wells.

"I don't know, but if I don't do something soon, he is going to fail my class," replied Ms. Watson.

This misbehavior is a common experience for teachers. How would you address the problem that Jeff is having in this class?

Questions for discussion:

- What misbehavior is being expressed in this scenario?
- What do you think is the origin or cause of Jeff's behavior?
- How should the instructor respond to the misbehavior?

By taking the time to develop instructional activities that engage all students in a lesson, the instructor is able to prevent misbehavior while using instructional time more judiciously. The best defense against student misbehavior is a well-planned and well-executed lesson.

Use Interest Approaches Teachers seeking to engage students from the very first minute of the class period should use some type of focusing activity at the beginning of a lesson. The activity should direct the attention of the students to the lesson at hand. Before that, it is recommended that some type of "bell ringer" activity be planned for each class. This activity usually consists of a question or statement written on the chalkboard/whiteboard at the front of the classroom. Students are instructed to enter the classroom, get their notebooks and materials ready, and respond to the question or statement on the board. The bell ringer is an excellent method for reviewing at the beginning of a class, and it folds nicely into a more elaborate interest approach as the teacher begins to teach. Students should have something positive and constructive to do the moment they enter the classroom.

- What are the six classes of nutrients in livestock feed?

- Of the four pines we identified yesterday in class, which ones are naturally adapted to drier sites, and which ones are naturally found growing in wet soils?

- How many uses of the peanut can you list? How about cotton?

14–3. Examples of "bell ringer" questioning.

INTEREST APPROACH EXAMPLES

With each lesson, use interest approaches that gain the attention of students and begin developing genuine interest in the subject.

Three examples of interest approaches are included here.

Interest Approach #1

Lesson: "Grafting with Skill"
Class: This is a Horticulture 2 class with twenty students.

Interest Approach:

1. Bring in a bag of white seedless grapes and a couple of cut-up Red Delicious apples.

2. Share these with the students and ask them to identify what they are eating. Ask if they have ever eaten these items before.

3. Draw on their previous horticulture experience by asking them to find the seeds. (Note: They may or may not be able to find seeds in the Red Delicious apples.)

4. Ask them how many have apple trees at their houses or at their grandparents' houses. Explain that the seeds in the apples will not produce more Red Delicious apples.

   a. Ask: "Since there are no seeds, how do we get more seedless grapes?"

   b. Ask: "If we don't get more Red Delicious apples from the seeds, how do we get more?"

   c. Ask: "What do we get if we plant the Red Delicious apple seeds?"

5. Begin the discussion of grafting.

Interest Approach #2

Lesson: "Understanding Plant Terminology"
Class: This is a Horticulture 1 class with twenty-five students.

Interest Approach:

1. Give each student a flower and a large piece of paper. Tell the student to sketch the flower and write down the name of each part he or she recognizes.

2. Ask each student to identify his or her flower. Give the student the opportunity to identify someone else's flower with which the student might be familiar. Have a few students share their identified drawings.

3. Display a list of plant terms. This should not be an overwhelming list.

   a. Ask: "Who can identify a term from the screen?"

   b. Ask: "What is a calyx, what is a corolla, and how do you tell the difference?"

   c. Ask: "How many of these terms apply to the flower in front of you?"

4. Begin the discussion and definition of the plant terms.

Interest Approach #3

Lesson: "Identifying Plant Pests"
Class: This is a Horticulture 1 class with twenty-five students.

Interest Approach:

Bring in a display of plant pests that includes leaves with fungus and disease, whiteflies and aphids (these are probably readily available in the greenhouse), and weeds, like morning glory or stinging nettle. You may also include some pictures of mammals, like white-tailed deer. Altogether, there should be about ten specimens. Have students observe the specimens and describe what they see. Lead from this discussion into a presentation of the objectives for the lesson. Begin covering the content for the first objective.

Used with permission by Misty Lambert (Lambert, 2003).

Organize the Classroom to Promote Learning and Proper Behavior

The physical environment in which students learn is an important factor in preventing or encouraging student misbehavior. The arrangement of student desks can either alleviate or contribute to student misbehavior. Where the teacher is the primary focus, the students should all be able to face the teacher with ease. If the students are expected to take notes, then they should have adequate desk space for writing in their notebooks. If the primary method for the lesson requires student discussion, then the desks should be arranged to allow student groups to discuss the topic comfortably.

It is a good idea to modify instruction to allow for a variety of learning methods, such as lecture, small-group discussion, demonstration, debate, and role-play. To help ensure the success of these methods, the instructor might need to modify the seating arrangement. Often, minor misbehavior problems can be cleared up immediately by rearranging the students' desks.

14-4. Problems are minimal when students are engaged in the learning experiences.
(Courtesy, Education Images)

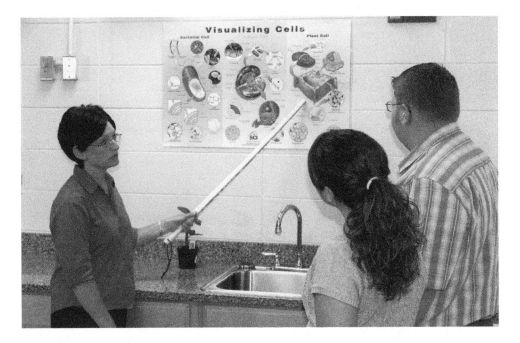

14-5. Engaging students through visual approaches promotes on-task behavior and minimizes misbehavior.
(Courtesy, Education Images)

14-6. Classroom arrangement for a presentation (lecture).

14–7. Classroom desk arrangement for small-group discussion and activity.

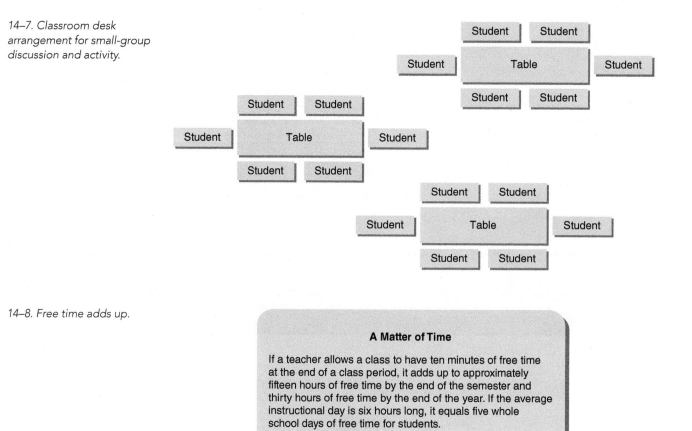

14–8. Free time adds up.

A Matter of Time

If a teacher allows a class to have ten minutes of free time at the end of a class period, it adds up to approximately fifteen hours of free time by the end of the semester and thirty hours of free time by the end of the year. If the average instructional day is six hours long, it equals five whole school days of free time for students.

What would the taxpayers think of the teacher who gives ten minutes of free time to students each day?

Teach the Entire Period

Making full use of class time is a companion to good instructional planning. Student teachers have often remarked that they were surprised at how quickly they went through their prepared lesson and that they were at a loss as to what to do when they finished the lesson before the class period ended.

Do not allow dead time in a lesson. The students should be working and thinking all the time. It takes less effort and causes less aggravation to prepare and deliver a lesson that engages the students for the entire class period than it does to constantly chase down and discipline students who have been given too much free time. Students dislike being bored. If the lesson being presented to them is irrelevant to their interests or needs, or if it is presented in a lukewarm manner, they may become bored. Boredom leads to disengagement with the lesson, and learning takes a backseat to other things that students might do to amuse themselves. By teaching the entire class period, the teacher can control student behavior in the best possible manner.

EFFECTIVE APPROACHES IN STUDENT DISCIPLINE

The teacher is important in having a quality learning environment. Misbehavior can be "teacher caused." Knowing appropriate ways to influence student behavior in the classroom effectively and positively will help prevent future student behavior problems.

Prevent Misbehavior

One of the best methods of managing student misbehavior is to prevent it from ever happening. The effective agriculture teacher recognizes the conditions that may cause misbehavior and modifies the instructional environment so that misbehavior cannot flourish. For instance, the teacher can often improve student behavior simply by reorganizing the seating arrangement in a classroom. To prevent students from playing with shop tools, the teacher can establish procedures by which the tools are locked in the tool room when not in use. If students appear sleepy and unfocused during a lesson on a warm day, the instructor can add activities to the lesson that cause students to move about the classroom. In some cases, simply changing the air temperature in the classroom can be conducive to good student behavior. A classroom that is too warm or too cool can discourage student participation.

Effective teachers know their students well enough to identify uncharacteristic behavior. It should be easy for a teacher to spot the behavior of students who come to class angry, worried, or excited about some prior experience. By making adjustments to the lesson, the teacher can minimize the effect of inappropriate student behaviors in the class. For instance, the student who enters the classroom angry over some problem could be pulled aside for a private conference with the teacher. By listening to the student's problem and offering constructive solutions in a nonthreatening manner, the teacher might be able to defuse that behavior before it affects the whole class. The key is to remove as many distractions from the learning environment as possible.

Develop Classroom Routines

On the first day of class, the teacher should explain how the classroom rules and procedures will be established during the course. Then, once the students and teacher have collectively established the ground rules for the class, the teacher should adhere to those policies strictly so that they become routine for the students.

14–9. Having students develop a dendrology notebook helps keep learning on-task.
(Courtesy, Education Images)

Teachers should also post their classroom rules in all instructional areas for students to see. Here is an example of classroom rules established with student input, and in the style of the U.S. Constitution:

We the students of Agricultural Mechanics II, in order to have a perfect learning experience, and promote the general welfare of all students, do ordain and establish this bill of rights and responsibilities. We will

- Be seated and have materials ready for class before the tardy bell rings.
- Respect the responsibility of the instructors to teach.
- Respect the rights of our fellow students to learn.
- Be kind and courteous to visitors and to each other.
- Make up work when we are absent from class.
- Tell our instructor if we have an injury in the shop.
- Tell our instructor if we notice broken shop equipment.
- Wear our safety glasses in the shop, and when operating any tools.
- Keep work areas, desks, and tables clean.

Questionnaire to guide students in developing classroom routines. Students can complete this worksheet individually, then meet as a whole class to develop classroom routines.

General questions about the classroom community:
- How do you want me to treat you?
- How do you want to treat one another?
- How do you think I want to be treated?
- How should we treat one another when there's a conflict?

General questions about how the class operates:
- What kinds of rules would help get the class started on time?
- What rules would help us get all of our lesson goals accomplished for the year?

Develop a Laboratory Safety Plan

Because an agriculture teacher usually manages some type of laboratory, it is important that a management plan be developed. Although managing labs and facilities is discussed in another chapter in this text, there are three essentials for effective shop or laboratory management as it relates to student behavior.

1. Position yourself in the laboratory so that you can observe all students working. Be where, with a minimal degree of effort, you can visually locate each student in the class. If the students are aware that you are monitoring their actions, then they might be less inclined to misbehave. Be highly visible in a shop or lab setting so that students can get answers to questions and assistance on projects.
2. Securely store unused tools and materials out of the reach of students. Store agricultural chemicals, fuels, paints, glues, and other toxic substances under lock and key to prevent accidental exposure. By following the same procedure for infrequently used tools, equipment, and supplies, you will ensure that these will be available for use the next time they are needed.
3. Keep all tools and equipment in safe working order, with all guards and shields in place. Broken equipment or equipment without the necessary safety guards is a serious safety hazard. This equipment should be tagged and locked out of service and should then be repaired at the earliest possible opportunity.

WHEN MISBEHAVIOR HAPPENS

What should be the teacher's response to misbehavior when it happens? Invariably, teachers will have to deal with some type of student misbehavior during the school day. All teachers are responsible for all of the students all of the time.

During the Lesson

Misbehavior during a lesson is particularly disruptive of both teaching and learning. One mistake a new teacher often makes is attempting to teach over the top of misbehavior. If students are talking, the teacher will attempt to talk louder in order to drown out the students. This procedure is both ineffective and exhausting for the teacher. Again, the use of interest approaches and high-quality lessons will negate a large portion of the discipline problems in a class.

Handle class disruption and misbehavior quickly and decisively. Do not let it continue, but try to deal with the student privately. A public reprimand will often result in a public spectacle. Keep teaching, but use gestures and eye contact to communicate messages to the misbehaving student.

If the student fails to heed the warnings of the teacher, the teacher should have a brief conference with the student about the misbehavior. The key is to be decisive and task-oriented.

Proximity Control   One effective method for encouraging good student behavior is for the teacher to move around the classroom, observing students as they work. Whenever a student is misbehaving, the misbehavior can be curtailed by the proximity or presence of the teacher. *Proximity control* is the ability to shape student behavior by being present in the classroom or laboratory area. The teacher need not say anything to the student, and often the simple presence of the teacher will cause the student's behavior to improve.

An additional benefit of moving about the class is that the variation of the teacher's location relieves student boredom. If the instructor stands for long periods in the same spot, habituation sets in, and the students begin to lose focus and attention. By moving around the classroom, the teacher breaks the monotony of the environment and causes students to remain attentive.

Setting an Example   Teachers must be good, professional role models for students. They must refrain from acting in a manner that encourages misbehavior. Teachers who fail to follow classroom rules are setting a poor example for the students to follow. Middle school and high school students have a keen sense of fairness, and they resent actions by the teacher that show a disregard for the rules. For example, the teacher who sends students to the school office for carrying knives on campus will be looked upon poorly by students if he or she continues to carry a pocketknife. Teachers who are following their own rules are good role models of self-discipline for the students.

First Class Meeting

A teacher's expectations should be made known at the first meeting of a class. By doing so, students understand the kind of behavior they must exhibit. It is far easier to have a good learning environment from the beginning than it is to try to impose it later.

Here are some things to do at the first class meeting:

1. Inform the students of your philosophy of teaching and learning. Take the time necessary to explain how you plan to teach, what you plan to teach, and why you will be teaching it. Students will understand very clearly your commitment to teaching.
2. Explain why general rules of conduct are necessary and useful for students in the class. Take the time to develop a list of classroom rules with student input.
3. On the first day of class, teach something from the content area. Do more than just distribute the syllabus and establish classroom rules. Set the tone and pace of the course the very first day by committing to teach agricultural subject matter.

Avoid Disruptions

A purpose of misbehavior with some students is to cause the teacher to stop teaching. It is in the teacher's best interest to thwart those efforts to stop instruction. Continue to teach if at all possible. Use proximity control to manage student behavior while continuing the forward motion of the lesson. Handle discipline issues quickly and quietly if possible so that the lesson keeps moving. Do not give the misbehavior more attention than necessary.

Under no circumstances should the teacher shout or lose his or her temper in class. By losing his or her temper, the teacher has used up all the tools of classroom management he or she was taught in school and has resorted to bullying to get the students to behave. The sad fact is that students consider the ranting and raving teacher to be great amusement and will often speak of the episode with great mirth in other classes during the day. If a student chooses to argue with you over a discipline issue in the classroom, defer the student until after class or meet briefly with the student in private outside of class.

If a teacher notices misbehavior, he or she should confront the student about the behavior and handle the matter appropriately. Under no circumstances should a teacher ignore misbehavior simply because it is not happening in his or her assigned work area. Student behavior is a shared responsibility between parents, students, and all teachers. Effective schools create the mind-set that there is no appropriate place on campus for misbehavior.

Engaging the Administration

Experienced teachers know how to effectively engage school administrators as partners in classroom management. Although the teacher has the responsibility for setting the standard of behavior in the classroom, the school administration has the responsibility for providing the foundation for behavioral expectations for the entire school. The teacher cannot effectively manage classroom behavior if his or

14–10. On-task behavior is an important part of the learning environment. What behavior is depicted by this learner—tired and bored. © lenelslan/Shutterstock

WHAT WOULD YOU DO IN THIS SITUATION?

Joanne entered the classroom just as the tardy bell was ringing, engaged in a loud argument with another student. Upon seeing Mr. Williams, the agriculture teacher, she immediately stopped arguing and loudly proclaimed, "Hi, Mr. Williams. My favorite teacher in the whole world!"

Joanne was a junior in Mr. Williams's horticulture class. Her loud voice was often matched by her fluorescent clothing and the bright red dye in her hair. Joanne had been a challenging student to teach, because she often picked fights with her teachers and with other students. She was frequently late for class, rarely turned in assignments on time, and usually asked inappropriate questions or used crude or inappropriate language. Her relationships with other students were strained at best.

"Hello, Joanne. Please take a seat so we can get started," replied Mr. Williams.

"I would, but Tom is in my seat again! Boy, you had better move before I ..." said Joanne, getting ready to take a swing at the student.

"Hold on, Joanne, just take any open seat. The late bell has rung and class has started."

"Okay, I'll take your seat, then," replied Joanne, as she sat down at the teacher's desk.

"Nope. Not that one, a regular student desk," replied an exasperated Mr. Williams.

"You said to take any seat," said Joanne. Several students laughed at Joanne's response.

"You know which seats you are supposed to sit in, so let's get moving," said Mr. Williams, as he moved to the front of the class to begin the lesson.

"Today we are examining the growth and development of ..." said Mr. Williams.

While moving to a student desk in the back of the room, Joanne loudly interrupted her teacher again, "Are we going out to the greenhouses today?"

"Joanne, you know the rules. Raise your hand if you have a question," replied Mr. Williams.

Joanne, still not in her seat at the back of the room, scowled and crossed her arms. "It was just a simple question," she responded, throwing down her school backpack and falling into a student desk at the rear of the class.

This scenario plays out every day in schools. Your response in this situation has the potential to mitigate student misbehavior or cause it to escalate into more serious misbehavior.

Questions for discussion:

- What misbehavior is being expressed in this scenario?
- What do you think is the origin or cause of Joanne's behavior?
- Which responses by Mr. Williams have been successful in helping Joanne manage her behavior?
- How might the instructor respond more effectively to the misbehavior?

---

her methods are in conflict with the school administration's policy. The opposite is also true. A school-wide approach to discipline and behavior is only effective if the teachers adhere to the same approach. Teachers should develop an open line of communication with administrators so that a team approach can be developed with regard to student discipline and behavior.

Make certain that the administrators are familiar with and approve of the classroom behavior plan. Ask administrators to help you find ways to engage problem students productively in the classroom. It is better to ask administrators to assist you in preventing misbehavior than to send students to the office for the school disciplinarian to handle. Good administrators appreciate a proactive approach to discipline and will applaud the efforts of the teacher in keeping students in class and on task.

Working with Parents

Teachers should contact parents when student misbehavior reaches the point where some intervention is necessary. Teachers should seek to balance their communication with parents by reporting student success when it occurs. Otherwise, the tendency to contact parents only when the students are behaving badly creates an antagonistic relationship between parents and teachers. Ask parents to help you with discipline issues, and keep them informed of their children's academic performance. Try to solve problems before they become serious issues.

REVIEWING SUMMARY

The learning environment is the nature and quality of the conditions in which teaching and learning occur. Some environments promote learning; others may impair learning. It is the teacher's responsibility to establish a quality learning environment.

Misbehavior may occur in a number of forms. Sometimes it does not disrupt the learning environment beyond that of the one individual who may be involved in something that detracts from learning. Misbehavior is defined as the condition that results from a person's inappropriate attempt to solve a problem or adapt to a situation.

Teachers use discipline to establish and maintain a quality learning environment. Discipline involves establishing structure and routine that facilitates classroom learning. Students learn behaviors, which through practice lead to self-discipline.

Three variables in a behavior problem are the student, the teacher, and the class of students. These three create an environment in which teachers and students carry out the learning process.

Prevention of misbehavior is by far the preferred approach. Teachers can use practices that promote desired behavior. Having well-prepared lessons that engage students is among the first steps. The arrangement of the furniture in the classroom can promote features of the physical environment. A teacher should teach the entire period and not have unused time at the end. Strong interest approaches help engage students and motivate them to learn what is being taught. Having classroom routines and safety plans also promotes a good learning environment.

Teachers must appropriately respond to misbehavior situations. It is best to avoid disrupting a lesson and the learning that is occurring. Proximity control is a useful strategy. Expectations should be presented during the first meeting of the class. Once learners know expectations, they are more likely to fall in line. Always follow school policies. Get to know parents and involve them in your program.

QUESTIONS FOR REVIEW AND DISCUSSION

1. What is meant by "learning environment"?
2. What are the features of learning environments in agricultural education?
3. What is misbehavior? Why is it said to be a "coping mechanism"?
4. What is discipline? How are punishment and reward parts of discipline?
5. What are the three variables in a discipline problem? Briefly describe each.
6. Why is prevention the best approach to student misbehavior?
7. How do well-prepared lessons minimize student misbehavior?
8. What should be considered in organizing a classroom to minimize misbehavior?
9. Why are interest approaches useful in minimizing misbehavior?
10. What classroom rules may be useful in establishing a good learning environment?
11. What should be a teacher's response when misbehavior happens?
12. What is proximity control? How is it useful?
13. Why is it useful to make expectations known during the first meeting of a class?
14. Do students misbehave in order to interrupt instruction? Explain.
15. How can working with parents be useful in minimizing misbehavior?
16. What trends related to student misbehavior are occurring in society?

ACTIVITIES

1. This activity focuses on quality learning environments as related to student achievement.
   - *Goal*: Identify principles and teaching strategies that yield good classroom and laboratory management results.
   - *Procedure*: Observe a teacher at work in a local agriculture program. Do not participate in class discussions or activities. Observe how the teacher manages the classroom or laboratory during the class period, and complete the following assignments:

   Draw a diagram of the classroom or laboratory setting in which you observe the teacher. Write a paragraph describing positive and negative attributes of the manner in which the classroom or laboratory is organized. Determine the mode for which the furniture is arranged: lecture mode, small-group mode, demonstration mode.

- *Observation*: Managing student work
Describe how the instructor grades student work. What are the procedures performed by the instructor to account for absent students? How does the instructor arrange for student work to be made up after an absence?
- *Observation*: Managing instruction
Describe how the instructor keeps students focused on instructional activities and how the instructor communicates what is expected of students. Also, include an analysis of the efficiency of the lesson (how time and other resources are managed).
- *Observation*: Managing special groups
Describe how the instructor manages instruction for students of varying ability levels. Evaluate the effectiveness of these methods.
- *Synthesis*: A final question
Who worked harder today, the students or the instructor? Why?

2. Conduct a research project on the use of corporal punishment. What is it? What are the pros and cons of using it? How do local policies of schools address corporal punishment?

CHAPTER BIBLIOGRAPHY

Abel, M. H., and J. Sewell. (1999). Stress and burnout in rural and urban secondary school teachers. *Journal of Educational Research, 92*, 287–94.

Borg, M. G., and R. J. Riding. (1991). Stress in teaching: A study of occupational stress and its determinants, job satisfaction and career commitment among primary schoolteachers. *Educational Psychology, 11*(1), 59–77.

Burnett, M. F., and G. E. Moore. (1988). *Student misbehavior in vocational agriculture and other vocational programs: A comparison.* Paper presented at the Proceedings of the National Agricultural Education Research Meeting.

Bushaw, W. J., and S. J. Lopez. (2011). Betting on teachers: The 43rd annual Phi Delta Kappa/ Gallup Poll of the Public's Attitudes Toward the Public Schools. *Phi Delta Kappan, 93*(1), 19.

Camp, W. G., and J. M. Garrison. (1984). The seriousness of student misbehavior in vocational agriculture. *Journal of the American Association of Teacher Educators in Agriculture, 25*, 42–47.

Croom, D. B., and G. E. Moore. (2003). Student misbehavior in agricultural education. *Journal of Agricultural Education, 44*(2), 14–26.

Dahl, R. E. (1999). The consequences of insufficient sleep for adolescents. *Phi Delta Kappan, 80*, 354–60.

Davis, H. A., J. J. Summers, and L. M. Miller. (2012). *An Interpersonal Approach to Classroom Management: Strategies for Improving Student Engagement.* Thousand Oaks, CA: Corwin/Division 15 (Educational Psychology) of the APA.

DeBruyn, R. L. (1983). *Before You Can Discipline: Vital Professional Foundations for Classroom Management.* Manhattan, KS: Master Teacher, Inc.

Kant, I. (1960). *On Education.* Ann Arbor, MI: University of Michigan Press.

Lambert, M. D. (2003). *Three Interest Approaches.* Raleigh: North Carolina State University.

Marzano, R. J. (2011). Classroom management: Whose job is it? *Educational Leadership, 69*(2), 85–86.

Maslow, A. (1970). *Motivation and Personality* (2nd ed.). New York: Harper & Row.

15

Agricultural Literacy

objectives

This chapter introduces agricultural literacy. The chapter covers the meaning and importance of agricultural literacy and describes several strategies for agricultural educators to utilize for developing agricultural literacy. It has the following objectives:

1. Explain the meaning and importance of agricultural literacy
2. Describe strategies for developing agricultural literacy

"My job is to prepare high school students for agricultural careers, and I just don't have time to worry about teaching third grade students," Barbara Green replied to a request for help with agricultural literacy materials by the third grade teacher at an elementary school in her community. "You can google agricultural literacy to find literacy materials," she added.

Does Barbara understand the true role of today's agricultural education teacher? Unfortunately, she possesses a very narrow view of the scope of the agricultural education profession. She is focused on one important aspect of the profession without giving consideration to the source of her future students and how they make decisions about their future. Interest in agriculture is important to Barbara's program, since agricultural education is an elective course in the high school curriculum. She is missing an excellent opportunity to generate interest in agriculture among potential future students for her program.

Just where do we get our food? The answer isn't easy and is far more involved than "at the local supermarket." Unfortunately, an increasing number of people have little idea about what is involved in producing wholesome, nutritious food and making it available. This is why we need to provide education for agricultural literacy.

With only about 1 percent of the nation's workforce currently involved in producing food and fiber, a large part of our society has limited exposure to agricultural production. Today, many people are unaware of the fact that agricultural producers are the primary source of the nation's food and fiber products. Agricultural educators can play a key role in educating about the importance of the plants, animals, and natural resources systems.

terms

agricultural awareness
agricultural literacy
agriculture
Agriculture in the Classroom

American Farm Bureau Foundation
 for Agriculture
infuse
literacy

15–1. Children are enjoying an agricultural literacy presentation from a high school student and FFA member. (Courtesy, National FFA Organization)

THE MEANING AND IMPORTANCE OF AGRICULTURAL LITERACY

Agriculture is the science, business, and technology of plant and animal production. It includes the systems for producing food and fiber and managing the environment and natural resources used by society. Today, we often refer to all of agriculture as the agricultural industry. This is because it includes agricultural supplies and services, marketing and processing, horticulture, forestry, wildlife, aquaculture, agricultural power, and other areas. The functions carried out that do not deal with production agriculture are called agribusiness. *Agricultural awareness* is the realization, perception, or knowledge of the agricultural industry, which is sometimes referred to as the plants, animals, and natural resources systems.

15–2. Using an agricultural literacy lesson plan library to prepare for teaching elementary school children. (Courtesy, Education Images)

Agricultural education is a systematic program of instruction available to students wanting to learn about the plants, animals, and natural resources systems. Approximately 6 percent of the nation's school population is enrolled in agricultural education programs. This leaves about 94 percent of public school students with no formal in-school instruction regarding the nation's plants, animals, and natural resources systems.

Literacy

Literacy is a term used to describe the ability to read and write. It can also refer to having more than average knowledge. A person who is literate is someone who is educated. An educated individual should have at least a basic understanding of the plants, animals, and natural resources systems and the role they play in the world. However, many "well-educated" individuals in the United States have no knowledge of the agricultural industry, which has impact on their daily lives and is essential for their health and well-being.

In 1988, the National Academy of Sciences released a study of agricultural education that made a distinction between education *in* agriculture and education *about* agriculture. This study defined "education in agriculture" as the vocational or career preparation component of agricultural education. It also surfaced the idea of *agricultural literacy* and defined it as "education about agriculture." In this chapter, agricultural literacy describes the attainment of knowledge and understanding of the plants, animals, and natural resources systems.

Importance

The National Academy of Sciences study stated that the topic of agriculture was so important that it should be taught to more than the small percentage of students enrolled in agricultural education programs. It concluded that all individuals should have a basic understanding of the nation's agricultural systems and that everyone should have some knowledge of where and how food and fiber are produced, processed, and distributed. The study also indicated that all individuals should be able to make informed choices about their own diet and health and possess the knowledge needed to understand the importance of caring for the environment.

The National Academy of Sciences study generated considerable interest among agricultural educators for agricultural literacy and served as a catalyst for numerous additional studies on this topic. It also influenced the current National Strategic Plan for Agricultural Education developed through the National Council for Agricultural Education. This plan includes a two-part mission statement that calls for preparing students for career success and creating lifetime awareness of the plants, animals, and natural resources systems. Agricultural education's stated vision is that all people will value and understand the important role agriculture plays in individual and world well-being. One goal of this strategic plan explicitly states that all students will attain agricultural literacy. The target audience for this goal is the 94 percent of the nation's school population not enrolled in agricultural education programs.

Few people realize that approximately 20 percent of the U.S. workforce is involved in the agricultural systems and that, worldwide, more people work in agriculture than in any other occupation. Broadly defined, the plants, animals, and natural resources systems include the science and business of plant and animal production, along with management of wildlife, forests, rangelands, rivers, oceans, and renewable natural resources. The systems also include horticultural crops, such as lawns, trees, gardens, landscaping, and turf for sports fields and golf courses.

The agricultural systems impact many aspects of each individual's everyday life and should not be taken lightly. The world's population is continuing to increase, which means that the demand for agricultural products used for food, shelter, and

15–3. Agricultural literacy can be a part of early childhood education. (Courtesy, U.S. Department of Agriculture)

clothing will also increase. It is important that all citizens have an awareness of the importance of agriculture in the world. This is a tremendous undertaking and requires the commitment and cooperation of many agencies, organizations, and groups, including agricultural education teachers and their students. Agricultural education teachers and students should be primary providers and facilitators of opportunities for increasing agricultural literacy. According to McDermott and Knobloch (2005), national leaders in agricultural education have identified literacy and awareness of agricultural education as major issues of importance to the future of agricultural education.

STRATEGIES FOR DEVELOPING AGRICULTURAL LITERACY

The vision presented in the National Academy of Sciences study was to have all students receive some instruction about agriculture beginning in kindergarten and continuing through grade 12. The current national vision for agricultural education is for all people to value and understand the role of agriculture and its impact on society. Achieving these visions will require many more classroom teachers to infuse agriculture into their curriculums. To *infuse* is to introduce one thing into another so as to affect it. In relation to agricultural literacy, it means to include agricultural lessons, examples, and topics in courses for other subjects. Leising, Igo, Heald, Hubert, and Yamamoto (1998) presented a framework for infusing agricultural literacy into core academic subjects across grade levels and described what young people should know about agriculture when they graduate from high school. A set of sample lesson units and a literacy test were included in the guide they developed.

Role of Secondary Agricultural Educators

As recognized professionals, agricultural education teachers have a large role to play in creating an agriculturally literate society. The major deterrent to agricultural education teachers advancing agricultural literacy to students other than those enrolled in their programs is finding sufficient time to deal with this issue. They generally view assisting other students as a good thing but outside the primary focus of the agricultural education program. However, many agricultural education teachers do realize there is value in spending time on agricultural literacy. For example, they understand that educating younger students about agriculture and agricultural education is a prime recruitment opportunity for attracting more students into the local agricultural education program.

Agricultural education teachers who view their role to be facilitators of the learning process are often more receptive to providing agricultural literacy than are teachers who focus on the more traditional role as primary deliverers of knowledge and information. Teachers who are facilitators recognize that there are many resources and strategies available for infusing agriculture into the public school curriculum.

Role of Agencies, Organizations, and Businesses

A number of agencies, organizations, and businesses have developed materials for use in teaching about many topics related to agriculture. However, these groups are not directly connected to the schools and often fail to get their materials into the hands of teachers most likely to use them. Most teachers, on the other hand, do not understand where to find good agricultural instructional materials and do not have the time to search for materials outside their subject areas. Agricultural education teachers can become links between teachers desiring to teach agriculture and groups desiring to have their materials utilized by classroom teachers.

USDA and State Agriculture Offices   The U.S. Department of Agriculture (USDA) and the state agriculture offices are good sources of information. Web sites, personnel,

15–4. Participation in animal shows promotes agricultural literacy. (Courtesy, U.S. Department of Agriculture)

materials, and other resources related to agricultural literacy are available. For example, the USDA Web site includes information for children, youth, and other groups.

Agriculture in the Classroom   *Agriculture in the Classroom* (AITC) is among the best-known examples of an agricultural literacy initiative. AITC exists in every state and has had a federal presence within the USDA. By merely introducing an AITC resource person to the faculty in an elementary school, the agriculture teacher can help open an entire school to agricultural literacy instructional materials, human resources, and teacher training opportunities.

Many resources are available through AITC. A good place to begin is the Agriculture in the Classroom Web site. This site features a wide range of materials and also has the "Kid's Zone" (for younger students) and the "Teen Scene" (for older students). Teachers can identify key contacts in their states who have resources available and who serve as resource persons for instruction. The resources often focus on the early childhood years as well as the middle school grades.

The American Farm Bureau Federation and its state affiliates have actively pursued AITC. Materials about agriculture have been provided to elementary school teachers. Workshops, field trips, and other means have been used to train teachers in AITC. Resource persons are available through local farm bureaus and other organizations.

American Farm Bureau Foundation for Agriculture   The *American Farm Bureau Foundation for Agriculture* is a source for materials, programs, and links to promote agricultural literacy. Its Web site has agricultural statistics, a teacher's toolbox, educational kits, books, DVDs, and videos.

Displays and Museums   Government agencies, organizations, and businesses often promote or establish exhibits. Most states have agricultural museums that help promote an understanding of agricultural history. Land-grant colleges and universities have experiment stations that are excellent for field trips and other activities. High-profile exhibits are featured at theme parks, such as Epcot at Walt Disney World in Lake Buena Vista, Florida. For example, the American Farm Bureau Federation maintains an exhibit at the Innoventions area of Epcot.

15–5. Taking children to farms and other agricultural enterprises is often used in developing agricultural literacy. (Courtesy, Agricultural Research Service/USDA)

National FFA Organization   Many agricultural education teachers have taught junior high/middle school courses. One of the more visible results of agricultural education's recent attempts to develop agricultural literacy programs has been the increasing number of junior high/middle school agricultural education programs. The National FFA Organization has made changes that have promoted middle school participation.

A change in the National FFA Constitution that officially allowed younger students to become members contributed to this increase. The junior high/middle school curriculums are generally exploratory and designed to increase awareness and knowledge of agriculture, whereas high school programs are more focused on career preparation. The junior high/middle school programs provide the most effective agricultural literacy education in the nation; however, they reach only a very small percentage of the total school population.

Teacher Education   Teacher training programs for agricultural education focus primarily on working with high school students. Thus, agriculture teachers sometimes fail to realize that children in the lower grades are the ones most receptive to accepting and applying new concepts. This means that agricultural awareness should be an important part of public education at the very beginning of a child's formal schooling.

Currently, the agricultural literacy efforts for the lower grades are less structured than those for the junior high/middle school levels. Agriculture teachers desiring to make an impact on increasing all students' awareness of agriculture should devote time to assisting with agricultural literacy efforts for students in the lower grades.

Most elementary and high school teachers have no background in agriculture and are, therefore, not inclined to try an area that is of little interest to them. A teacher's background and experience are important in determining the topics he or she is willing to teach. Teachers need an awareness of agriculture before they can be successful teaching it.

Universities that prepare agriculture teachers can assist in agricultural literacy efforts by helping teacher candidates in other subject-matter areas increase their understanding of agriculture and how it can be infused into academic courses. State education agencies and teacher preparation programs can help prepare for and support the teaching content of junior high/middle school agricultural education programs and provide leadership for articulating these programs with those offered at the high school level.

Food for America   Agricultural education teachers in many schools provide agricultural awareness programs for elementary students through the FFA "Food for America" program. This program offers opportunities for leadership development to agricultural education students while increasing the agricultural awareness of elementary students and teachers. Joint activities with the FFA Alumni, Young Farmers, and parent booster clubs are very effective strategies in which the teacher is the facilitator and not the primary deliverer of instruction. These joint efforts may include tours of school farms, petting zoos, school assembly programs, and/or demonstrations presented to individual classes. The activities are often conducted during National FFA Week. The Food for America curriculum can be found on the National FFA Web site under the "FFA Learn Resources" link.

Activity Orientation   Many students are very interested in animals and plants. They will respond positively to instruction that is hands-on and gets them involved. With the tremendous pressures on teachers and students to increase performance on state and national assessments, instructional materials that help teachers achieve the objective of higher student academic performance would be welcomed. Providing agricultural lessons that help students gain academic knowledge and skills measured by standardized assessment exams is critical for enticing other teachers to utilize agricultural instructional materials. Agriculture teachers can demonstrate how agriculture can be used as a motivational tool to increase student interest and enthusiasm for learning. Powell, Agnew, and Trexler (2008) recommend multiple approaches to delivering agricultural literacy.

Share Resources and Facilities   Agriculture teachers can team teach with other teachers to present agricultural information, and/or they can help other teachers find materials that focus on the subject of agriculture and on an increase in academic achievement in core subject-matter areas. The Internet enables agriculture teachers and students to more readily find high-quality instructional resources for increasing agricultural literacy. Agriculture teachers can also provide educational programs directed at students in other courses and programs at various grade levels by using today's technology. This technology allows lessons to be transmitted via distance delivery to multiple sources and/or captured in digital format to be presented in multiple classrooms throughout a school.

Another important strategy that is often overlooked is for the agriculture teacher to provide other teachers with access to facilities, resource people, equipment, and the local agricultural community. Other teachers do not have the connections with the agricultural community that the agricultural education teacher possesses. By connecting other teachers to key agricultural resources, the agricultural education teacher can facilitate the teaching of agriculture by numerous other teachers.

REVIEWING SUMMARY

All people need to value and understand the importance of agricultural systems and what they contribute to world and individual well-being. Agricultural literacy is a major goal of agricultural education; however, widespread cooperation and commitment are required to achieve nationwide knowledge and understanding of agriculture. Of necessity, agricultural literacy involves many agencies, organizations, and groups interested in creating awareness and understanding of the agricultural systems in this country.

Agriculture teachers and their students play an important role in educating all students about the plants, animals, and natural resources systems. In some instances, they are the primary deliverers of instruction about agriculture, and in many other endeavors, they serve as facilitators of the process.

QUESTIONS FOR REVIEW AND DISCUSSION

1. What is agricultural literacy? Why is it important to society?
2. What are some strategies agricultural educators use to increase agricultural literacy?
3. What should be the role of the agricultural education teacher in providing agricultural literacy?
4. What is Agriculture in the Classroom?
5. What are the ways an agriculture teacher can collaborate with other teachers to support agricultural literacy?

ACTIVITIES

1. Develop an agricultural literacy lesson appropriate for infusion into a middle school course. Identify the course for which the lesson is designed, the estimated time required to teach the lesson, and the best time of year to teach it.
2. Use the Internet to investigate agricultural literacy instructional materials. Develop a list of sources for the materials and identify the grade level(s) for which the materials are recommended. A good site for starting your search is at the Agriculture in the Classroom website.

3. Investigate the agricultural literacy structure, providers, and training opportunities within your state and prepare an informational handout you could share with your future teaching colleagues.

CHAPTER BIBLIOGRAPHY

American Farm Foundation for Agriculture. (2004). *Educating About Agriculture.* Accessed on April 22, 2004.

Binkley, H. R., and R. W. Tulloch. (1981). *Teaching Vocational Agriculture/Agribusiness.* Danville, IL: The Interstate Printers & Publishers, Inc.

Frick, M. J., A. A. Kahler, and W. W. Miller. (1991). A definition and the concepts of agricultural literacy. *Journal of Agricultural Education, 32*(2), 49–57. doi: 10.5032/jae.1991.02049.

Leising, J. G., C. G. Igo, A. Heald, D. Hubert, and J. Yamamoto. (1998). *A Guide to Food and Fiber Systems Literacy.* Stillwater, OK: W. K. Kellogg Foundation & Oklahoma State University.

McDermott, T. J., and N. A. Knobloch. (2005). A comparison of national leaders' strategic thinking to the strategic intentions of the agricultural education profession. *Journal of Agricultural Education, 46*(1), 55–67. doi: 10.5032/jae.2005.01055.

Moore, G. E. (Ed.). (1998). A primer for agricultural education. *The Agricultural Education Magazine, 71*(3).

Moore, G. E. (Ed.). (2000). Reinventing agricultural education for the year 2020. *The Agricultural Education Magazine, 72*(4).

National FFA Organization. (2011). *Official FFA Manual.* Indianapolis: Author.

National Research Council. (1988). *Understanding Agriculture: New Directions for Education.* Washington, DC: National Academy Press.

Phipps, L. J., E. W. Osborne, J. E. Dyer, and A. L. Ball. (2008). *Handbook on Agricultural Education in Public Schools* (6th ed.). Clifton Park, NY: Thomson Delmar Learning.

Powell, D., D. Agnew, and C. Trexler. (2008). Agricultural literacy: Clarifying a vision for practical application. *Journal of Agricultural Education, 49*(1), 85–98. doi:10.5032/jae.2008.01085.

16
Middle School Agricultural Education

Bob Stewart was coming out of the university library when he ran into Jan Davis, a fellow senior in teaching methods class.

"Hey, Jan, did you have a good visit to the school where you are student teaching?" Bob asked.

"I certainly did, and it's going to be great!" replied Jan.

"You seem pretty excited about it, even though it is a middle school. I don't know how you can teach that crazy bunch," Bob said.

"Have you visited a middle school recently? I did. What I saw and what I thought I would see were very different. I thought I would see loud, rude children running the hallways. I thought I would see hyperactive human packages of young people trying to figure out what life was all about. And the teachers … I thought they would be frantically chasing about trying to establish order and that they would have few instructional resources."

"Yeah, that sounds about right. Those kids are out of …" said Bob.

Jan interrupted, "Was I surprised! The hallways were quiet. The students were pleasant and cooperative. When the class period I was observing ended, they changed classes in an orderly manner and without lagging behind. The teachers are the best! They were businesslike and professional. There was no downtime—everyone was engaged in the learning process. This settles it for me. I am sure that student teaching at the middle school is a great idea. See you in class this afternoon!" With a wave, Jan entered the library.

"Hmmm. Maybe I have it all wrong when it comes to middle schools. I'll have to check into the possibility of teaching at a middle school when I graduate from college." Bob thought to himself as he walked toward his next class.

objectives

This chapter discusses middle school agricultural education. It has the following objectives:

1. Explain the concept of middle schools in American education
2. Discuss the role of middle school agricultural education
3. Identify stages of student maturation and other developmental factors influencing methods of instruction in middle schools
4. Describe instructional strategies in the middle school environment
5. Describe the function of FFA and supervised experience

terms

career exploration
Erik Erikson
Food for America
horizontal articulation

student-centered instruction
vertical articulation

16–1. Middle school students may be involved with school garden plots. (Courtesy, Education Images)

MIDDLE SCHOOLS IN AMERICAN EDUCATION

For many adults, middle school represents that awkward period in their lives when they were caught between the relatively safe environment of elementary school and the socially challenging environment of high school. For many people, middle school is characterized as that bewildering period when they experienced phenomenal changes physically, socially, and emotionally. Students entering middle school for the first time find that lessons focus on developing competency and on performance goals—a significant change from elementary school. The students who felt they had mastered elementary school now find themselves in a whole new world—middle school (Anderman & Midgley, 1997).

A middle school is a school administrative unit between the primary or elementary grades and the high school grades. The grades included are typically 6 through 8, though other variations can be used.

The middle school is a useful by-product of the collision between educational theory and practice. Adolescents between the ages of 11 and 14 are experiencing the joys and frustrations of puberty and experiencing a mental metamorphosis that gives them the tools for solving problems in ways they never imagined before. It stands to reason that students in this age group would need a special and unique environment for learning. Thus, the middle school was born.

To meet the needs of young adolescents, there are essential elements that must be provided at the middle school level (Association for Middle Level Education, 2010). These are as follows:

- A combination of student-centered and teacher-centered instructional methods that encourage and develop critical thinking.
- A rigorous, age-appropriate curriculum that uses technology to bring the world into the mind of the learner.
- Learning experiences that allow for intellectual exploration, including education in the arts and humanities.
- Flexibility in organizing instruction for maximum benefit to students.

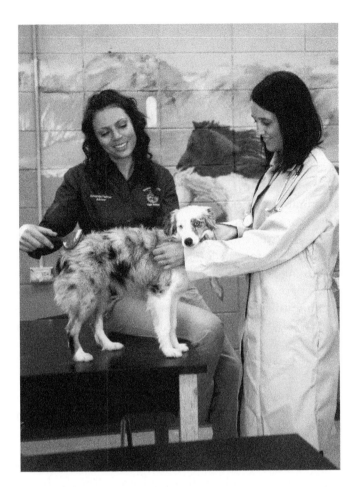

16–2. Middle school students respond to hands-on learning with animals and plants.
(Courtesy, Education Images)

- Learning experiences that foster the development of the whole person.
- Academic and career guidance opportunities that encourage personal wellness and stewardship of a sustainable environment and economy.
- Appropriate assessment strategies that adequately measure student progress.
- Learning experiences that foster lifelong learning along with a desire to contribute to the growth of the community.
- Educational experiences that demonstrate citizenship and encourage students to be civic-minded learners.

Middle schools generally comprise grades 6 through 8, but there is much variation as to which grades are appropriate for the "middle grade" label. In some cases, grades 5 through 9 are considered appropriate for inclusion at a middle school, and these grades probably define the boundary of middle school grade levels. Public school systems have shifted away from the junior high school model of instruction and toward middle schools. There are currently more than 67,100 middle schools in the United States (U.S. Department of Education, 2011).

MIDDLE SCHOOL AGRICULTURAL EDUCATION

There are about 10,600 agriculture teachers in the United States. Of these, approximately 446 teach exclusively in middle schools. Overall, approximately 1,520 agriculture teachers teach in both high school and middle school settings. Twenty-seven states report some type of middle grades agricultural education programming (Kantrovich, 2010). These instructors generally teach a mixture of agricultural technology and career exploration topics. According to the National FFA Organization, there are approximately 37,826 middle school students enrolled as FFA members (National FFA Organization, 2012a).

One major advantage to the high school agricultural education program is that it exposes students to a wide variety of learning experiences related to the agricultural sciences. The learning experiences are coupled with opportunities for career exploration, thus enabling students to begin the process of formulating career goals based upon their occupational interests in the food, fiber, and natural resources industries.

Middle school agricultural education programs assist students in developing necessary social skills and decision-making skills while providing basic knowledge about agriculture and natural resources. This allows for practical application of academic disciplines. Furthermore, middle school programs also allow for the developmental needs of youth. To help students develop a sense of belonging and self-worth, the agriculture teacher arranges for career interest surveys that help the students identify their strengths. Learning activities that develop citizenship and teamwork skills provide outlets for growth of social skills. Lessons that focus on international agriculture build a sense of community in the classroom among students of differing cultures. To promote problem solving and critical thinking, the teacher provides instructional activities that require these skills to solve authentic problems.

The Agriculture Curriculum

The agricultural education program in middle school focuses on two major areas: agricultural literacy and career exploration. The purpose of the program is to provide the fuel that ignites the spark of interest in a student for agricultural occupations. Because a large number of agricultural occupations are available to high school and college graduates, it is essential that students receive a concentrated and directed study of agricultural careers at some point before graduating from high school (National FFA Organization, 2012b).

Career exploration is the use of career-related investigative activities carried out in and outside the classroom. The goal is to help students identify future potential career interests and gain information to help as they plan their higher levels of education. Career exploration may involve reading about careers, watching videos,

16–3. Resource people can be used to enrich instruction with realia. (Courtesy, NRCS, USDA)

participating in job shadowing, volunteering, and gaining information in other ways. Overall, the best career exploration involves "hands-on" experiences that provide vivid real-world opportunities.

Through student-centered instructional methods, the middle school agricultural education program offers students the opportunity to learn about agricultural careers and about some of the skills needed in those careers. Some basic things a student might learn about a given agricultural career could be the following:

- Vocabulary and nomenclature specific to an agricultural occupation
- The impact of agriculture on the local, state, national, and global economy
- Basic skills in agricultural sciences, biotechnology, and engineering
- Leadership and human relations skills
- The overall importance of agriculture to humanity

Since the middle school is an excellent place for students to begin developing personal and social skills for life after high school, the middle school agricultural education program should incorporate experiences that expose students to the world they will soon enter as adults. Through exploratory activities, students can learn about ethnic and social diversity within the workplace, as well as glimpse the technological and intellectual demands placed on workers in the industry (Kerka, 1994).

Agricultural education has always gone beyond learning about "cows and plows." With the exponential increase in technology in the agricultural sciences, there is a greater need for highly skilled workers in the industry. Career preparation should begin in a middle school agricultural education program.

Interdisciplinary Methods of Instruction

A unique advantage of middle school education is the opportunity for interdisciplinary methods of instruction. Agricultural education teachers should grasp the chance to work with other teachers in developing meaningful learning experiences for students. For instance, an agriculture teacher could partner with a biology teacher to study the aquatic life in a pond near the school. The biology teacher's goals would be met through an interesting lab activity, while the agriculture teacher's goals would be met through a realistic activity that demonstrates the tasks associated with an agricultural career area, such as wildlife biology.

An essential phase of planning a middle school agricultural education curriculum is to articulate it with the curriculum at the high school level. Articulation is achieved when educational activities are organized to facilitate the continuous and efficient progress of students from one grade or class to another. Courses of study are sequenced so that prerequisites are developed in the lower grades. Two types of articulation are vertical and horizontal.

Vertical articulation means that articulation has focused on the upward movement to higher grades. For example, students in grade 7 learn skills that prepare them for grade 8, those in grade 8 learn skills that prepare them for grade 9, and so on.

Horizontal articulation refers to articulation within the same grade level among the various courses that are being taken. For example, the science and language arts curriculums may be articulated and demonstrate continuity from one class to another.

A prime example of how vertical and horizontal articulation plays out in the school setting can be found in school gardens and outdoor learning labs. For instance, research studies have demonstrated that middle school outdoor learning labs in plant sciences can

- Increase a student's knowledge and interest in plant science.
- Provide awareness about the environment.
- Provide knowledge about food production and food systems.
- Encourage teamwork and social skill development.

There are more than 10,000 school gardens and outdoor learning labs on middle grade campuses in the United States. Students who are actively engaged in gardening on the school campus are more likely to be engaged in science education, and participate more effectively in other school disciplines, too. The excitement of learning in the garden often transfers into learning in the science classroom (Skinner, Chi, & Group, 2012).

ADOLESCENT DEVELOPMENT AND THE MIDDLE SCHOOL

The maturity and cognitive level of middle graders make it a challenge to teach specialized content in the agricultural sciences. For instance, an 11-year-old middle grade student might be quite successful at understanding the rudiments of soil fertility, but he or she might not be successful at understanding the concepts associated with calculating the universal soil loss equation. A number of theories attempt to explain the cognitive and social development of adolescents. Perhaps these theories can shed some light on the types of learning experiences appropriate for middle school students.

Stages of Social–Emotional Development

Erik Erikson proposed a psychological theory of development. He described human socialization as an eight-stage process. Each stage is the result of a felt internal need that must be suitably addressed to reach the intended outcome—a normal, well-balanced, emotionally stable child. The first stage encompasses the period from birth to age 1 and is characterized by the individual learning to trust others. If the child is embraced by a nurturing environment, then he or she develops a basic optimism and trust of others. Jean Piaget identified this stage as a child's sensory-motor period, during which the child develops physical coordination of limbs and moves from reflexive action to synthesized action.

In the second stage, the 2 to 3 age bracket, the individual begins to develop a need for autonomy. Many parents know this stage as the "terrible twos," and it is characterized by a child's need to be assertive.

Erickson identifies the third stage as the "play" stage. In this stage, the child learns to imagine and plays cooperatively with other children. It is in this stage that the child first begins to develop skills in leading and following. Piaget's preoperational period coincides with Erickson's third and fourth stages, in that the child develops transductive reasoning and more complex speech patterns.

Erickson's fourth stage begins sometime around the age at which the child goes to school. In this stage, the child begins to master peer relations, cognitive skills in reading and math, and the complex rules associated with formal play and organized recreation. The child is likely to enter middle school during this stage and progress from it to the next stage before going on to high school.

Erickson's fifth stage is just beginning upon the child's departure from middle school. It is characterized by a need to develop a unique identity. Piaget identified this stage as the concrete operations period, in which the child develops organized and logical thinking skills. Before leaving middle school, the child will have entered Piaget's period of formal operations and begun to develop abstract thought and prepositional logic ("if-then" processes). Furthermore, the middle grade years see the beginning of the physical maturation process and puberty.

The transition from middle school to high school is a challenge for most adolescents. The environment in high school is significantly more impersonal and grade oriented. The competition for academic scholarships begins in high school, along with physical competition in organized athletics. The high school years are the beginning of part-time jobs and serious talk about colleges and careers. In agricultural education, the curriculum narrows to specialized occupations in agriculture, and students are required to choose the classes that most nearly match their career goals (Mizelle & Irvin, 2000). The high school environment is very different from the middle school environment, and preparing for the transition from one to the other begins in middle school.

The first five stages of Erikson's model provide the foundation for the three remaining phases of human development. Students who have successfully resolved conflicts in the first five stages will be prepared for young adulthood, which is stage six. In stage six, young adults hone their sense of identity in preparation for developing love relationships with others. Young adults who find it difficult to trust and commit to others will experience difficulty in developing intimate relationships and may feel socially isolated.

As individuals move into middle adulthood, they experience the seventh stage by building upon intimate relationships and become caregivers through parenting and mentoring. In Erikson's final and eighth stage of human growth and development, adults come to terms with their existence by reflecting on their lives in a positive manner. Adults in this stage of maturity understand that they cannot change their past and accept responsibility for what their lives have become. Upon death, the maturation process is complete.

Educators should create a positive, secure environment for students. This will help students progress through these natural stages successfully.

Transitioning from Elementary School to Middle School

Belonging is a major concern to middle school students. They have to feel like they belong in the community, that is, school. The most effective teachers are those who empathize with middle school students and help students respond to the pressures of homework, middle school culture, and social relationships. Middle school students are developing, but have not fully developed, planning and decision-making skills and ethical reasoning skills. They are learning how to control impulsive behavior. It may prove helpful for the agriculture teacher to employ techniques that help middle school students survive the transition from elementary school. Figure 16–4 below provides tips for helping students transition to middle school.

Arrange a school visit day for elementary school students to the agricultural program at the middle school. Have tours of the labs and greenhouse facilities, and teach a mini-unit of instruction on an agricultural subject.

Meet with elementary school teachers of rising middle grades students to gather information on new students in your program. Information about particular student strengths, and their personal or career interests are useful.

Send a letter of congratulations to rising middle grades students on their graduation from elementary school.

Continue to make connections with parents. Invite them to open house programs, and provide regular feedback on student progress.

Remember, the teacher sets the tone and morale in the classroom. A healthy dose of humor keeps the stress levels down and encourages students to develop a good professional working relationship with the teacher.

16–4. Tips for teachers in helping students transition from elementary school to middle school.
(Wormeli, 2011)

INSTRUCTIONAL STRATEGIES IN THE MIDDLE SCHOOL

Because adolescents are changing socially, emotionally, and physiologically in the middle school period, it is important that the agriculture teacher design the curriculum and instruction in a manner that recognizes the specific needs of these students. Adolescent students may have difficulty engaging in the lesson from time to time. Disengaged students need more exposure to differentiated instruction, as delivered by a competent and caring teacher. In middle school agricultural education, the teacher can engage reluctant learners by (Turner, 2011)

- Focusing student attention on the lesson and eliminating or curtailing distractions.
- Frequently explaining the importance of the lesson being learned.
- Using the personal experiences of students as part of the instructional process.
- Providing several methods or opportunities for students to learn the lesson content.
- Conducting diagnostic tests to determine how students are constructing meaning from the lesson content.
- Paying careful attention to the cultural backgrounds of students, and providing an atmosphere of acceptance in the learning environment.

Student-Centered Methods

Middle grades agricultural education programs offer early adolescents an opportunity to benefit from experiential learning (Fritz & Moody, 1997). ***Student-centered instruction*** provides learning experiences that most nearly match the maturity level of the students. The teacher becomes a facilitator and helps the students learn the material through processes designed by the students. Because the students create the processes that take them to the material to be learned, the students are self-motivated to acquire the knowledge. Examples of student-centered methods are represented in Table 16–1.

When student-centered methods are used, it is important that the teacher effectively facilitate instruction. The teacher's role is to determine which method is appropriate for the content and for the developmental level of the students and to teach the students how to participate in the method. The teacher must be well organized and have thoroughly planned each use of the method so that there are few distractions from the learning process. Student-centered methods are by no means voluntary participation methods. All students must participate in the manner requested by the teacher.

FFA and Supervised Experience in the Middle Grades

FFA   The National FFA Organization offers a number of membership benefits for members in the middle grades. Middle grade students who join FFA are afforded the same rights and responsibilities as other members of the organization when it comes to local chapter governance. Middle grade students receive the FFA *New Horizons* magazine and can attend leadership programs, conventions, and meetings when appropriate and when allowed by the school system. (National FFA Organization, 2012b)

The primary disadvantage to FFA membership for the middle grades rests with the age and maturity level of middle grade students. Middle grade students are often prohibited from overnight travel by local school boards. Overnight travel creates special liability issues for younger students, and some school boards and administrators are unwilling to take the risk of letting middle graders attend FFA functions

Table 16–1
EXAMPLES OF STUDENT-CENTERED TEACHING STRATEGIES

Strategy	Description
Phillips 66 method	Six students are given six minutes to develop a solution to a specific problem. The first minute of the group work is devoted to organizing the group and choosing a reporter. The next four minutes are devoted to solving the problem. The last minute is devoted to preparing the group report.
Cooperative learning	A small group of students is given a problem to solve. All students must actively participate in the problem-solving process. Students learn the basics of accountability and important social skills in addition to the content of the lesson.
Inner circle/outer circle	A group of students discusses a problem or works on solving a problem using psychomotor skills. The remainder of the class observes the process and takes notes. At the end of the allotted time, the work of the group is discussed by the whole class.
Role-playing and dramatic skits	Students act out a particular problem and its solution. At the end of the students' presentation, the problem and its recommended solution are discussed by the whole class.

overnight. This limits many middle grade students to day travel to conventions and camps.

Career Development Events, Leadership Programs, and FFA Degrees   The National FFA Organization is also concerned about the liability issues surrounding middle school students in some Career Development Events. Is it acceptable to have 12-year-old students perform agricultural mechanics skills or participate in livestock events? Another question that arises out of middle grade student participation in Career Development Events is, "Can students acquire enough content knowledge in the middle school agricultural education classroom to prepare them to compete alongside high school juniors and seniors?"

The National FFA Organization has recognized the special circumstances of competition involving middle school students and has made some changes in the Career Development Event structure to accommodate these students. Most notably is the Agriscience Fair. This event has a middle school division, and students can compete against other students with similar learning experiences. The goals of the Agriscience Fair are as follows:

- Provide students with a real-life experience in the scientific process.
- Recognize students with exceptional ability in the agricultural sciences.
- Reinforce material learned in agriscience courses.

In the Agriscience Fair, students have the opportunity to perform experimental research in biochemistry and food science, environmental science, animal science, botany, and agricultural engineering.

FFA offers a number of leadership programs at the state and national levels for middle grade students. Participants explore common leadership and career interests with others, and the conference provides introductory training in service learning and community development.

The Discovery FFA Degree is the first degree a student can obtain as a member of FFA. This degree was created for the purpose of encouraging middle grade students to learn more about FFA and the agricultural industry. The specific requirements for this degree can be found in Chapter 23.

Supervised Experience and Career Exploration

Supervised experience in agriculture should begin in middle school agricultural education programs. Middle grade students can create exploratory experiences based upon their own learning experiences in the agricultural education classroom.

According to the expanded model for supervised experience developed by Gary Moore and James Flowers at North Carolina State University, exploratory supervised experience can lead to a more sophisticated understanding of the agricultural industry (Moore & Flowers, 2012). If the middle grade student stops taking agricultural education courses in the eighth grade, he or she should have learned enough about the agricultural industry to have a significant measure of appreciation for it. If the student chooses to continue to take agriculture courses in high school, the exploratory experiences he or she began in middle school can blend into an entrepreneurship or placement supervised experience.

If the student was active in the FFA Agriscience Fair program, then he or she might continue exploratory work by developing an experimental or analytical supervised experience. The middle grade student need not wait until enrollment in a high school agricultural education program to begin work on an experimental or analytical supervised experience.

It is never too early for a student to learn about food, fiber, and natural resources. For some students, their first significant experience with agriculture happens during a Food for America program at their school. ***Food for America*** is a program sponsored by FFA that provides instruction on a variety of agricultural literacy topics to elementary and middle school students. Students who have interest in the agricultural sciences can investigate them more thoroughly when they enter

16–5. Note the prominence of exploratory experience in this model of supervised experience. (Developed by Gary Moore and James Flowers, North Carolina State University. Reprinted with permission.)

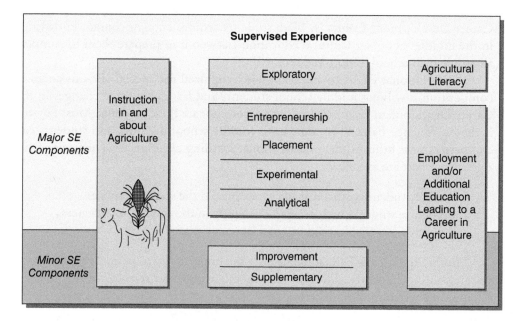

16–6. Middle school exploration may involve observing others at work. (Courtesy, National FFA Organization)

middle school and enroll in an occupational exploration course. The middle school agricultural education course can provide appropriate learning experiences that encourage a student to seek more knowledge about agriculture and perhaps one day become a useful and productive participant in the agricultural industry.

REVIEWING SUMMARY

A middle school is a school between the primary or elementary grades and the high school grades. With grades 6 through 8 typically included, the students are at a time of great personal development. In the next twenty years, most jobs in the United States will require some form of higher education, but not necessarily a baccalaureate degree. Many students will be able to find valuable careers with only an associate's degree, and most of these students will need skills developed through career and technical education in order to be successful. Considering that most middle school students see themselves attending college, there is certainly room for agricultural education in middle schools (Symonds, Schwartz, & Ferguson, 2011).

Agricultural education in the middle school meets two major goals: agricultural literacy and career exploration. Agricultural literacy focuses on developing an appropriate understanding of food, fiber, and natural resources and of appropriate areas, such as horticulture, wildlife, and the environment. Career exploration is the investigation of various career areas, using a wide range of approaches.

Articulation is an important concept in middle school curriculum development. The curriculum should facilitate vertical and horizontal movement of students as well as instructional content. Adolescent development has been studied and described as a stage-based process. Erik Erikson is well known for his eight-stage human socialization process as related to middle school students.

Instructional strategies may be student-centered and interdisciplinary. Student-centered instruction is intended to match the maturation levels of the learners. Teachers determine and use a range of approaches in teaching.

FFA has an important role in middle schools. Career Development Events, leadership programs, and related activities promote student development at the middle school level. Supervised experience beginning in middle school can lead to a more sophisticated understanding of the agricultural industry.

QUESTIONS FOR REVIEW AND DISCUSSION

1. What is student-centered learning, and how does it differ from traditional classroom methods?
2. What are some concerns associated with middle grade student participation in FFA activities, and what are some possible solutions to these concerns?
3. What types of supervised experiences can a student have in a middle school agricultural education program?
4. What is a middle school?
5. What is career exploration? What is the goal of career exploration?
6. What is articulation? Distinguish between vertical and horizontal articulation.
7. What are the eight stages in the human socialization process developed by Erik Erikson?

ACTIVITIES

1. Visit a middle school agricultural education program. Ask the agriscience teachers the following questions. Then, write a brief summary of your interview and use it as a basis for a classroom discussion.
   - Why did you choose to become a middle school teacher? Why did you choose agricultural education as your field of expertise?
   - What do you enjoy most about teaching in middle school?
   - What do you enjoy least about teaching in middle school?
   - How have middle schools changed since you first entered the teaching profession? How has agricultural education changed as well?
   - How do you handle the stress of teaching?
   - How is your program funded?
   - What effect does FFA have on your program? Does it help or hinder your instruction?

   - Generally speaking, what teaching methods work best for your students?
   - How do you manage your time as a teacher? (Alternate question: How do you accomplish all that you must do in order to be an effective teacher? What are some methods you use to utilize your time better?)
   - How much assistance do you get from school administrators and parents?
2. Investigate the Association for Middle Level Education (AMLE). Begin with the association's Web site. Determine the mission and goals of AMLE, how it supports the professional development of middle school teachers, and its advocacy role with policy makers. Prepare a report on your findings.

CHAPTER BIBLIOGRAPHY

Anderman, E. M., and C. Midgley. (1997). Changes in achievement goal orientations, perceived academic competence, and grades across the transition to middle-level schools. *Contemporary Educational Psychology, 22*(3), 269–98. doi: 10.1006/ceps.1996.0926

Association for Middle Level Education. (2010). *This We Believe: Keys to Educating Young Adolescents.*

Westerville, OH: Association for Middle Level Education.

Fritz, S., and L. Moody. (1997). Assessment of junior high/middle school agricultural education programs in Nebraska. *Journal of Agricultural Education, 38*(1), 61–65.

Kantrovich, A. J. (2010). *The 36th Volume of a National Study of the Supply and Demand for Teachers of*

Agricultural Education 2006–2009: American Association for Agricultural Education. Accessed on November 5, 2012.

Kerka, S. (1994). Vocational education in the middle school. *ERIC Digest No. 155.* Columbus, OH: ERIC Clearinghouse on Adult, Career, and Vocational Education.

Mizelle, N. B., and J. L. Irvin. (2000). Transition from middle school into high school. *Middle School Journal, 31*(5), 57–61.

Moore, G., and J. Flowers. (2012). *The North Carolina SAE Model.* Raleigh: North Carolina State University.

National FFA Organization. (2012a). *FFA Statistics.* Retrieved April 13, 2012.

National FFA Organization. (2012b). *National FFA Agriscience Fair 2012–2016.* Indianapolis, IN: National FFA Organization.

Skinner, E. A., U. Chi, and the Learning-Gardens Educational Assessment Group. (2012). Intrinsic motivation and engagement as "active ingredients" in garden-based education: Examining models and measures derived from self-determination theory. *Journal of Environmental Education, 43*(1), 16–36.

Symonds, W. C., R. B. Schwartz, and R. Ferguson. (2011). *Meeting the Challenge of Preparing Young Americans for the 21st Century.* Report issued by the Pathways to Prosperity Project: Harvard Graduate School of Education.

Turner, S. L. (2011). Student-centered instruction: Integrating the learning sciences to support elementary and middle school learners. *Preventing School Failure, 55*(3), 123–31. doi: 10.1080/10459880903472884

U.S. Department of Education. (2011). *America's High School Graduates: Results of the 2009 NAEP High School Transcript Study.* Washington, DC: U.S. Department of Education.

17

High School Agricultural Education

objectives

This chapter explains the offering of high school agricultural education in the educational system of the United States. It has the following objectives:

1. Explain high schools in American education

2. Discuss the role of high school agricultural education

3. Identify student maturation and developmental factors influencing instruction

4. Describe instructional strategies appropriate for high school students

What is high school agricultural education like today? Consider these examples:

- Lauren Wise teaches in a one-teacher agriculture department at a small school with grades kindergarten through 12 all in the same building.
- Jennifer Smart teaches in a five-teacher agriculture department in a large ninth- through twelfth-grade comprehensive high school.
- Octavius Brown teaches horticulture and landscape management at a tenth- through twelfth-grade area career and technical education center.
- Cato Cantrell rides the bus to school, has four classes on block scheduling (one is veterinary science), and gets off the bus each afternoon for supervised experience at a veterinary clinic near his home.
- Jasmine Robles goes to her local high school for first-period agriculture, then returns to her house where her mother home schools her.

terms

area career and technical education center
charter school
cognitive development
comprehensive high school
development
elementary school
home school
Jean Piaget

Lev Vygotsky
magnet school
scaffolding
scheme
secondary school
Seven Cardinal Principles
tertiary system
voucher
zone of proximal development

17–1. A current high school agriculture facility in California bears little resemblance to a high school agriculture facility from the mid-1900s.
(Courtesy, Education Images)

HIGH SCHOOLS IN AMERICAN EDUCATION

Formal schools have existed in the United States since its founding. The majority of these schools in the 1600s and 1700s were religious in nature, whether publicly or privately funded.

The 1800s brought structure to education, and eventually a laddered system of elementary, secondary, and tertiary schools took form. An ***elementary school*** was typically a school with grades 1 through 6. A ***secondary school*** was usually a school with grades 7 through 12. The ***tertiary system*** was the colleges and other institutions after high school. This structure served as a good foundation on which today's education was built.

The twentieth century saw an ever-increasing percentage of American adolescents attending at least some high school. In the early 1900s, the junior high school developed as an entity separate from the high school. The junior high school contained grades 7 through 9, and the high school contained grades 10 through 12. Although in its beginnings the junior high school was designed to meet the unique needs of preadolescents, it quickly came to follow the high school in structure and organization.

Beginning in the 1960s and continuing throughout the remainder of the century, the middle school concept took hold and spread. The middle school was designed for students in grades 5 or 6 through grade 8. Ninth-grade students were moved to the high school so the curricular, physiological, and social needs of middle school students could be better met. (Middle school agricultural education was discussed in Chapter 16.)

The early years of the twenty-first century have seen continued changes. The number of students who are being home schooled has increased. ***Home school*** consists of teaching by parents and others outside of traditional school buildings and typically in the home. Home school may include face-to-face instruction, distance education, or a combination of the two. Students may be home schooled until the start of middle or high school or until completion of high school.

17–2. Two agriculture teachers stand at the front of a combination junior and senior high school.
(Courtesy, Education Images)

Another kind of school is the charter school. A ***charter school*** is a publicly supported school given freedom to operate outside of state and local bureaucratic control. Charter schools are able to experiment and try educational innovations. However, they are still required to meet local and state performance standards for their students.

Early college high schools are a newer form of secondary education. High school students earn both high school graduation credit and college credit for their course work. Students typically graduate with an associate's degree or completion of two years of college credits. Early college high schools may be aligned with a community college, other two-year college, or a four-year university. Students may take dual credit courses from their high school teachers, travel to the college campus, or take distance-based college courses.

A voucher approach has received considerable debate. With ***vouchers***, parents can choose the school their children attend, whether public or private, and receive public funds through a direct payment or tax credit arrangement. Public school educators often fear that a voucher system will erode support for the public schools.

Purposes of a Secondary Education

Education in the United States has been organized according to three guiding principles: local control, federalism, and professionalism (Chubb, 2001).

In the principle of local control, public schools are controlled and administered by local boards of education. These entities make curricular, staffing, and other educational decisions. A local body may control the collecting of taxes for school use and typically must approve a budget of expenditures.

The federalism principle gives the states the responsibility for education. This ensures that no one group can exercise its force on a national level. At the same time, it provides flexibility among the states for innovation and experimentation.

The principle of professionalism states that education is best delivered by professional educators who make decisions based on the best interests of the students. Among others, this is one reason that teachers are held to high standards of credentialing.

17–3. A teacher educator is discussing local program plans with agriculture teachers.
(Courtesy, Education Images)

Committee of Ten Report   Over the years, groups, committees, and commissions have been formed to determine the purposes of a secondary education. In 1893, the National Education Association (NEA) funded a study that came to be called the Committee of Ten Report. The committee recognized that a secondary education should be for all students, not just those going to college. However, the nine subjects recommended by the Committee of Ten were traditional liberal arts in nature, thus continuing the focus of secondary education as primarily preparation for college. The nine subjects were Latin, Greek, English, modern languages, physics, astronomy and chemistry, natural history, history, and geography. The work of the Committee of Ten influenced other committees that established a standard unit of credit for high school subjects called the Carnegie unit. The Carnegie unit, which is defined in Chapter 8, is still used today to determine high school graduation and college entrance requirements. The Committee of Ten Report directed secondary education for the next twenty-five years and continues to influence educational practices to this day.

Cardinal Principles Report   Secondary education was dramatically shaped by another NEA committee. The Cardinal Principles Report of 1918 changed the focus of secondary education and shaped it for the remainder of the 1900s. Instead of making subject area recommendations, the Cardinal Principles Report gave objectives to be met. These objectives are known as the *Seven Cardinal Principles*. They include command of the fundamental processes (basic skills, such as reading, writing, and mathematics), worthy home membership, health, vocation, citizenship, worthy use of leisure time, and ethical character. For the first time, vocational, citizenship, and personal development areas were to be a part of secondary education for everyone. To meet this broadened purpose, the comprehensive high school was developed.

Reports from the 1980s to the 2000s   In 1987, the Southern Regional Education Board (SREB) and several state partners began an initiative called *High Schools That Work* (SREB, 2012). The initiative has ten key practices including high expectations and continuous improvement to prepare students for the world of work and further education (SREB, 2012). The philosophy of the program is to create a learning environment that allows students to master a rigorous academic and technical curriculum. More than 1,200 middle school, high school, and higher education sites were using the High Schools That Work framework in thirty states and the District of Columbia.

In the 1990s, the U.S. Secretary of Labor appointed a commission to find the skills students need to be successful in the workforce. The Secretary's Commission on Achieving Necessary Skills (SCANS) concluded: All high school students must be competent, all companies/employers must be characterized by high performance, and all schools must become high performing. The SCANS report identified five competencies (resources, interpersonal skills, information, systems, and technology) and three foundational skills (basic skills, thinking skills, and personal qualities) essential for job performance (SCANS, 1991).

The National Governors Association (NGA) was formed in 1908 to provide the nation's governors a vehicle to promote leadership, share best practices, and speak with a unified voice on public policy. Over the past century, the NGA has issued numerous reports regarding education. The 1989 National Education Summit led to *Goals 2000*, which was signed into law by President Clinton. The 2005 National Education Summit had as its goal to redesign high schools, particularly to improve graduation rates (NGA, 2008).

The NGA Center for Best Practices and the Council of Chief State School Officers (CCSSO) released *Common Core State Standards* in 2010 (Common Core State Standards Initiative, 2012). Two categories of standards, English-Language Arts and Mathematics, were developed to prepare K–12 students for college and career readiness. Most states have adopted the standards.

Organization of American High Schools

High schools in the United States are organized in various ways. All are intended to provide the amount of education associated with a specific diploma or certificate.

The Comprehensive High School   A *comprehensive high school* encompasses the full range of subjects and activities in academic, career and technical, citizenship, and personal development areas. The curriculum includes courses that all students take, such as English; courses that students select based on their career goals, such as sciences; and elective courses, such as art. In addition to the classroom subjects, a comprehensive high school has citizenship development activities, such as clubs and organizations, and personal development activities, such as sports and driver's education.

The number of career and technical education areas taught in comprehensive high schools is dependent upon their size and communities. Some may offer a minimum of technology education (formerly called industrial arts), business education, and consumer and family sciences (formerly called home economics). Others may offer a wide array, including automotive technology, building trades, metalworking, electronics, and health sciences. Agricultural education is offered in about 8,000 U.S. public high schools.

Specialized High Schools   An *area career and technical education center* is a specialized type of high school designed to prepare students for careers. Students may enter the workforce directly upon high school graduation or continue on to a trade, technical, community, junior, or four-year college. Area centers may draw from comprehensive high schools and teach only career and technical subjects, or they may be self-contained, teaching academic as well as career and technical subjects.

Area centers tend to offer a great variety of career and technical education areas. Teachers in these schools typically must document work experience in the areas in which they teach. Courses may be in two- or three-hour blocks, allowing students to

17–4. An agriculture teacher stands at the door to her agricultural education classroom and lab.
(Courtesy, Education Images)

complete detailed, complex assignments and more closely simulating the real world of work. These longer blocks of time also allow students to put classroom theory immediately into practice in the laboratory.

A ***magnet school*** is a specialized high school designed to attract students for a specific purpose. A magnet school is organized around a theme, such as the arts, science, mathematics, or even agriculture. To parents, magnet schools may be viewed as safer and more academically challenging. In most instances, students must apply to attend a magnet school, and competition for available spaces is intense. Magnet schools are most prevalent in urban areas. The Chicago High School for the Agricultural Sciences in Chicago, Illinois, and the STAR (Science and Technology of Agriculture and its Resources) Academy in Indianapolis, Indiana, are two examples of agricultural magnet schools.

HIGH SCHOOL AGRICULTURAL EDUCATION IN AMERICAN EDUCATION

Agricultural education emerged along with American education to serve an important role. The history of education and the history of agricultural education were covered earlier in the book. This section reviews and expands agricultural education in the high schools.

Before Smith-Hughes

For the first seventy-five years of the nineteenth century, agricultural education was not a part of public secondary education. However, during the period 1875 to 1900, many states established residential agricultural schools. Beginning in 1889, Alabama established secondary agricultural schools in each of its congressional districts. Many state agriculture colleges offered agricultural instruction of less than college grade either as separate secondary schools or as short, intensive courses.

The first decade of the 1900s brought efforts to establish secondary agricultural education as a course of study in county high schools. Many states provided funding for counties that taught and maintained agriculture courses. Other high schools added agricultural education even without the benefit of state monetary aid. Some high schools teaching agriculture had land and facilities available for experiments, demonstrations, and student work.

By 1915, agricultural education was taught in almost 4,000 secondary schools in the United States. The level and quality of instruction varied greatly. In some schools, agriculture was taught as a science by teachers who were not always trained at a state agricultural college. In other schools, the instruction was based on textbooks or covered the sociological aspects of the subject and was taught by teachers not trained in agriculture. In still other secondary schools, agriculture had a more vocational focus, included laboratory and hands-on instruction, and was taught by trained agriculture teachers.

From Smith-Hughes to the 1970s

In 1917, the Smith-Hughes Vocational Education Act provided federal funds for instruction in agriculture, home economics, and trades and industry. This act dramatically changed the scope and focus of secondary agricultural education. The number of agriculture teachers and secondary students of agriculture greatly increased.

Teacher education for agriculture teachers, although occurring before 1917, became more important to assure a supply of qualified teachers for all the new secondary agricultural education programs.

State supervisors of agricultural education were hired. Home projects became a required part of secondary agricultural education. Local extension agents, as a part of the 1914 Smith-Lever Act, and secondary agriculture teachers, as a part of the 1917 Smith-Hughes Act, had to work out agreements on how they were to work with both children and adults.

From 1917 to the early 1960s, agricultural education on the secondary level grew and expanded. Although certain aspects changed, such as instructional technologies and subject-matter content, the focus remained vocational, oriented on production agriculture primarily for males. Both societal and legislative changes of the 1960s affected agricultural education.

The Vocational Education Act of 1963 expanded agricultural education instruction to off-farm agricultural occupations and subjects. This opened the door for instruction in areas such as horticulture, natural resources, and agricultural business. The combination of the addition of courses in off-farm agricultural subjects and the acceptance, in 1969, of females into FFA greatly increased the number of females enrolling in secondary agricultural education.

The civil rights movement of the 1960s and concurrent civil rights legislation led to the desegregation of public schools and therefore of agricultural education. This change was epitomized by the merging of the New Farmers of America (NFA) and the Future Farmers of America (FFA) in 1965. Desegregation led to a decline in the number of African American agriculture teachers and over the years to a decline in the number of African American secondary agriculture students. (More information about the NFA is in Chapter 23.)

From the 1970s to 2000

The 1970s saw increasing enrollments in secondary agricultural education, peaking in 1976 at almost three quarters of a million students (Camp, Broyles, & Skelton, 2002). The Vocational Education Amendments of 1968 and 1976 brought about an increased emphasis on overcoming gender discrimination and gender stereotyping, which further enhanced the effort to increase enrollment of females in agricultural education. These amendments, along with other federal legislation, encouraged the inclusion of special needs students in agricultural education.

From the mid-1970s to the early 1990s, enrollment in secondary agricultural education declined. This decline has been attributed to numerous factors. In the 1980s, a major farm crisis accelerated the decline of families earning their living from farming. This was coupled with the long-term trend of overall declining farm numbers. The pendulum of educational reform swung toward increasing requirements in core academic subjects to the detriment of electives. In addition, the generation of baby boomers had worked its way through the secondary school system by the mid-1980s, and a smaller population of Generation Xers followed. The publication of the *Green Book* in 1988 by the National Academy of Sciences set the course for numerous changes in agricultural education over the next decade.

The 1990s saw increasing enrollments, greater emphasis on the recruitment and retention of female and minority students, and a greater diversity of curriculum offerings. This decade also saw the rapid increase in the use of microcomputers in the agricultural education classroom and eventually access to the Internet. During this time, many states greatly expanded the offering of agricultural education courses, especially non-sequenced, agricultural literacy–type courses. Although in existence in a few cities before this time, urban agricultural education received a renewed emphasis.

Beginning of the Twenty-First Century and Beyond

Agricultural education at the beginning of the twenty-first century is taught in approximately 8,000 U.S. middle and high schools to approximately 1 million students in grades 6 through 12. Classes include a diversity of students with regard to ethnicity, gender, geographic area, and ability level. This diversity extends to the reasons students enroll in agricultural education. Some still enroll for vocational reasons, whether they are planning to enter agricultural occupations immediately or continue on to further education after high school. Others enroll for avocational or personal-interest reasons. In some high schools, certain agriculture classes count toward science graduation requirements. Finally, the extension of agricultural education classes into middle schools has greatly expanded.

In 2000, the *National Strategic Plan and Action Agenda for Agricultural Education: Reinventing Agricultural Education for the Year 2020* was released. This document set out the vision, mission, and goals for agricultural education for the

17–5. Information technology has changed many practices in this Indiana high school.
(Courtesy, Education Images)

next twenty years. The vision of agricultural education was "a world where all people value and understand the vital role of agriculture, food, fiber, and natural resources systems in advancing personal and global well-being" (National Council for Agricultural Education, 2000, p. 3). The mission of agricultural education was set as preparing "students for successful careers and a lifetime of informed choices in the global agriculture, food, fiber, and natural resources systems" (National Council for Agricultural Education, 2000, p. 3). This mission envisions an expanding of agricultural education in both reach and impact. The National Council for Agricultural Education, through its strategic plan, worked to accomplish the vision, mission, and goals set forth in this plan.

ADOLESCENT DEVELOPMENT

Secondary students are going through mental, emotional, physical, hormonal, and social changes. This adolescent development affects what is taught and how it is taught. *Development* is the orderly, lasting change that occurs in humans as a result of maturation, learning, and life experiences. Many social scientists have studied and written about adolescent development.

Woolfolk (1998) gave the following three general principles of development:

1. People develop at different rates. Adolescents of the same age will be at varying stages of physical, mental, and social development.
2. Development is relatively orderly. We can predict that most people will go through the same stages of development in about the same order.
3. Development takes place gradually. Change and growth take time and, at least in mental and social development, may include periods of decline before advancement.

Cognitive Development

Cognitive development is the development of an individual's thought processes. It includes adaptation to the environment and the assimilation of information.

Piaget   In the mid-1900s, *Jean Piaget*, a Swiss psychologist, developed a theory of cognitive development that is still in use today. He proposed that we all have the tendency toward organization and adaptation. We organize our thinking into structures called schemes. *Schemes* are the mental representations of objects and events of the world. We then adapt our schemes as new information and experiences cause us to adjust our thinking.

Piaget is most famous for his "four stages of cognitive development" theory. Although psychologists have questioned whether the number and order of stages are correct, most agree that Piaget's observations about cognitive development are accurate. The first two stages, sensorimotor and preoperational, occur in children from birth to about 7 years of age. Although agriculture teachers rarely teach students of this age range, certain characteristics of the preoperational stage are important for teachers to consider. Students in this stage need visual aids to help them learn concepts. Demonstrations and student practice are important for student understanding. Also, students need real-world experiences, such as those provided by field trips, to build a foundation for learning concepts.

Piaget's third stage is called concrete operational and generally covers children ages 7 to 11. It is at this stage that students begin to think logically. They still need concrete examples, but these can increasingly take the form of symbols as opposed to real objects. Students continue to need clear instructions and opportunities to experiment. Students can now understand more complex ideas but benefit when the concept presented is connected to a familiar idea.

The fourth stage in Piaget's theory is called formal operational and includes age 11 to adult. Students at this stage are able to think hypothetically and abstractly and to combine information into new concepts. Teachers should involve students in solving problems, in developing hypotheses, and in testing and exploring complex concepts. Teachers should not expect all high school students to be in Piaget's fourth stage. In fact, many psychologists debate whether adults operate at the formal operational stage in all areas.

Vygotsky   *Lev Vygotsky*, a Russian psychologist, had a view of cognitive development different from Piaget's view. The sociocultural theory emphasizes the role people play in a child's cognitive development. This is particularly true with regard to the child's culture and the child's tools, such as language. Children develop through their interactions with adults, especially as adults provide assistance in the form of scaffolding. *Scaffolding* is the help, prompts, and questions an adult gives a child to assist the child's understanding.

Vygotsky proposed the concept of the "zone of proximal development." For any particular range of tasks, the *zone of proximal development* is that level of proficiency in which students cannot accomplish the tasks alone but can accomplish them with assistance (Woolfolk, 1998). Above the zone, students are capable of accomplishing the tasks by themselves. Below the zone, students cannot accomplish the tasks even with assistance. Only within the zone can learning take place and instruction be successful.

Personal, Social, and Emotional Development

Erikson's psychosocial theory of human development was introduced in Chapter 16. It is expanded here as a base for high school students.

Erikson based his theory on his observation that people have the same basic needs and that a purpose of society is to provide for those needs. He proposed that all

17–6. Personal development that begins in middle school can extend through high school. (© Shutterstock/Andy Dean Photography)

people go through eight stages of psychosocial development. Each stage is characterized by a crisis or conflict between a positive alternative and a potentially unhealthy alternative. How a person resolves each stage has an impact on later stages. The first three stages cover birth to approximately 6 years of age. The final two stages cover middle and late adulthood. The middle three stages occur during the time a person is in school or just out of school.

Erikson's fourth stage occurs during middle childhood and involves the conflict of industry versus inferiority. This conflict is characterized by the learning of new skills. Children who master the skills and become competent develop a sense of accomplishment. Those who do not master the skills risk developing a sense of failure and inferiority. Teachers can encourage industry by making sure students can set and work toward realistic goals, show independence and responsibility, and recognize their progress and improvements (Woolfolk, 1998).

Erikson's fifth stage occurs during adolescence and involves the conflict of identity versus confusion. At this stage, roughly ages 12 to 18, peers become much more important. Adolescents must develop their identity in terms of personal, social, occupational, and sexual roles. This stage involves experimentation and clarification of roles. Teachers can support identity formation by providing role models, resources to help work out personal problems, acceptance of experimentations that are not harmful and do not interfere with learning, and realistic feedback (Woolfolk, 1998).

17–7. A Minnesota student gains hands-on experience with a guinea pig in the school lab.
(Courtesy, Education Images)

Table 17–1
ERICKSON'S STAGES OF HUMAN GROWTH AND DEVELOPMENT SUMMARIZED

Stages 1-4
Life Stage: Infancy—Childhood
Approximate Ages: Birth—12
Conflicts: Trust, Autonomy, Initiative, Industry

Stage 5
Life Stage: Adolescence
Approximate Ages: 13—18
Conflict: Identity vs. Role Confusion

Stage 6
Life Stage: Young Adulthood
Approximate Ages: 19—25
Conflict: Intimacy vs. Isolation

Stages 7-8
Life Stage: Middle Adulthood—Late Adulthood
Approximate Ages: 26—Death
Conflicts: Generativity, Integrity

Source: Bee, H. L., and S. K. Mitchell. (1984). *The Developing Person: A Life Span Approach* (2nd ed.). New York: Harper & Row, p. 19.

Erikson's sixth stage typically occurs during young adulthood but may have impact on some high school students. At this stage, the conflict is between intimacy and isolation. Young adults need to be willing to relate to others and to develop intimate relationships. Teachers need to watch for students who isolate themselves and have few friendships.

Just before high school, students in grades 5 through 8 (middle school) have gone through rapid developmental changes. This age group demonstrates a great range of individual developmental levels. Students of the same chronological age will exhibit vastly different levels of physical, cognitive, and social development.

Physically, middle school students are developing in weight, height, strength, and endurance. Many middle school students will be awkward and uncoordinated as their bones grow faster than their muscles. Students become self-conscious about how they look as their bodies change in size and structure. Students at this age, especially females, will begin puberty and may not understand all the changes their bodies are going through. Students may grow tired more easily, have periods of restlessness, and eat more in both frequency and quantity.

Emotionally, middle school students are developing their self-identity. They are beginning to define themselves independent from their families. They may have exaggerated emotions, experiencing high "highs" as well as low "lows." Middle school students may have unwarranted fears and at the same time have feelings of invulnerability.

Socially, peer groups become increasingly important for middle school students. They compare themselves with peers and follow group norms in dress, speech, and behavior. A middle school peer group may become exclusive, territorial, and even cruel to those not in the group. On the other hand, middle school students are becoming increasingly conscious socially and may take on causes of justice and assist those who are less fortunate. Parents and authority are still important, but this age group may also seek out other adults as mentors and role models.

INSTRUCTIONAL STRATEGIES APPROPRIATE FOR SECONDARY STUDENTS

Agriculture teachers should organize their classrooms according to the developmental levels of students. For example, with middle school students, some will be able to think and work abstractly, whereas many others will still be in the concrete operational stage. For the benefit of both groups of students, agriculture teachers should include problem-solving experiences but make sure to give clear instructions and expectations.

Agriculture teachers of middle school students must realize that these students need social opportunities in order to learn effectively. Teachers should plan activities that allow students to talk and move while accomplishing educational objectives. Problem-solving exercises that involve justice, righting wrongs, or helping those less fortunate can actively engage middle school students in the learning situation.

High school students, more so than middle school students, will be on similar developmental levels. However, students may have trouble transferring learning from one situation to another. For example, they may have learned concepts in geometry and biology that are applicable in agriculture but not be able to use the learning in agriculture. If possible, the agriculture teacher should individualize scaffolding so that each student receives only the help and prompts needed for advancement to the next step. The teacher should design tasks and problems so each student can operate within his or her own zone of proximal development.

High school students are very much in Erikson's fifth stage of identity development. Agriculture teachers should use SAE and FFA activities to help students in this task. Assistance in occupational identity development should be provided to all agricultural education students. The opportunities within FFA for social and personal identity development are numerous. Individualizing instruction within the classroom and laboratory can help students safely experiment with different roles, especially in the occupational area.

REVIEWING SUMMARY

Secondary education in the United States has gone through numerous transformations during the past 300-plus years. What was once reserved for the rich and elite is now a universal right for all American children. In the transformations, secondary education has included everything from purely academics for the training of the mind, to purely vocational for the working classes, to a mixture of academic and vocational designed to meet individual student needs.

In recent years, greater emphasis has been placed on early secondary education. The middle school concept recognizes that students in grades 5 through 8 are in a unique developmental stage and should be educated accordingly. This concept also takes into account that middle school students are developing physically, emotionally, and socially in addition to intellectually.

Agricultural education has been a part of secondary education since the late 1800s. With the passage of the Smith-Hughes Act in 1917 and the provision of federal support, agricultural education gained a larger role in public secondary education. The 1960s brought social and legislative changes to agricultural education. The 1980s saw declining enrollments and a reexamination of the role of agricultural education. At the beginning of the twenty-first century, agricultural education is working to set its own vision and chart its own course.

Our understanding of adolescent cognitive development has been greatly influenced by Jean Piaget and Lev Vygotsky. Piaget's theory has had substantial impact on what we do in the classroom and how we do it. Vygotsky's theory emphasizes the role language plays in cognitive development.

Erikson's psychosocial theory of development explains personal, social, and emotional development. Three of Erikson's eight stages occur during the years within secondary education. Erikson's theory has implications for teaching, especially middle school students.

QUESTIONS FOR REVIEW AND DISCUSSION

1. What is the difference between a comprehensive high school and an area career and technical education center?
2. Explain the rationale for charter schools.
3. What are the Seven Cardinal Principles?
4. What was the impact of the Smith-Hughes Act on secondary agricultural education?
5. Describe the legislative and social forces that shaped agricultural education in the 1960s.
6. What are the characteristics of agricultural education at the beginning of the twenty-first century?
7. How can an agriculture teacher's knowledge of Piaget's four stages of cognitive development enhance the agricultural education learning experience for students?
8. What are examples of how the zone of proximal development can be used in agricultural education instruction?
9. How can Erikson's psychosocial theory of development influence the total agricultural education program?
10. What variations may be made between teaching middle school students and high school students?

ACTIVITIES

1. Investigate what the Cardinal Principles for a secondary education should be for the twenty-first century.
2. Investigate current legislation from your state and the federal government regarding secondary education. On the federal level, you will want to investigate the latest Perkins Act, the Individuals with Disabilities Act, the Elementary and Secondary Education Act, and the Higher Education Act, among others. On the state level, you will want to search for information regarding home schooling, charter schools, school vouchers, and other alternatives to the traditional public secondary education.
3. Using the *Journal of Agricultural Education, The Agricultural Education Magazine*, and the proceedings of the National Agricultural Education Research Conference, write a report on research about an aspect of secondary agricultural education. Current and past issues of these journals can be found in the university library. Selected past issues can be found online at the American Association for Agricultural Education and the National Association of Agricultural Educators Web sites.

CHAPTER BIBLIOGRAPHY

Association for Career and Technical Education (ACTE). (2003). *ACTE Web Site.* Accessed on August 22, 2003.

Bee, H. L., and S. K. Mitchell. (1984). *The Developing Person: A Life Span Approach* (2nd ed.). New York: Harper & Row.

Binkley, H. R., and R. W. Tulloch. (1981). *Teaching Vocational Agriculture/Agribusiness.* Danville, IL: The Interstate Printers & Publishers, Inc.

Boyer, E. L. (1983). *High School.* New York: Harper & Row.

Camp, W. G., T. Broyles, and N. S. Skelton. (2002). *A National Study of the Supply and Demand for Teachers of Agricultural Education in 1999–2001* (online). Accessed on October 3, 2003.

Chubb, J. E. (2001). The system. In T. M. Moe (Ed.), *A Primer on America's Schools* (pp. 15–42). Stanford, CA: Hoover Institution Press.

Common Core States Standards Initiative. (2012). *Common Core States Standards Initiative Web Site.*

Eggen, P., and D. Kauchak. (1997). *Educational Psychology: Windows on Classrooms.* Upper Saddle River, NJ: Prentice Hall.

Goodlad, J. I. (1994). *Educational Renewal: Better Teachers, Better Schools.* San Francisco: Jossey-Bass.

Herbst, J. (1996). *The Once and Future School: Three Hundred and Fifty Years of American Secondary Education.* New York: Routledge.

Hlebowitsh, P. S. (2001). *Foundations of American Education* (2nd ed.). Belmont, CA: Wadsworth/Thomson Learning.

Moe, T. M. (Ed.). (2001). *A Primer on America's Schools.* Stanford, CA: Hoover Institution Press.

National Association of Agricultural Educators (NAAE). (2003). *NAAE Web Site.* Accessed on August 22, 2003.

National Council for Agricultural Education. (2000). *National Strategic Plan and Action Agenda for Agricultural Education: Reinventing Agricultural Education for the Year 2020.* Alexandria, VA: Author.

National FFA Organization. (2003). *Local Program Resource Guide: 2003–2004* [CD-ROM]. Indianapolis: Author.

National FFA Organization. (2011). *Local Program Success* (online). Access on November 30, 2011.

National FFA Organization. (2011). *Official FFA Manual.* Indianapolis: Author.

National Governors Association (NGA). (2008). *National Governors Association Web Site.* Accessed on August 14, 2008.

Newcomb, L. H., J. D. McCracken, J. R. Warmbrod, and M. S. Whittington. (2004). *Methods of Teaching Agriculture* (3rd ed.). Upper Saddle River, NJ: Prentice Hall.

Phipps, L. J., and E. W. Osborne. (1988). *Handbook on Agricultural Education in Public Schools* (5th ed.). Danville, IL: The Interstate Printers & Publishers, Inc.

Pulliam, J. D., and J. J. Van Patten. (1999). *History of Education in America* (7th ed.). Upper Saddle River, NJ: Prentice Hall.

Secretary's Commission on Achieving Necessary Skills (SCANS). (1991, June). *What Work Requires of Schools: A SCANS Report for America 2000.* U.S. Department of Labor. Accessed on August 14, 2008.

Southern Regional Education Board (SREB). (2012). *High Schools That Work Web Site.* Accessed on August 14, 2008.

True, A. C. (1929). *A History of Agricultural Education in the United States, 1785–1925.* U.S. Department of Agriculture, Miscellaneous Publication No. 36. Washington, DC: GPO.

Wiles, J., and J. Bondi. (2001). *The New American Middle School: Educating Preadolescents in an Era of Change* (3rd ed.). Upper Saddle River, NJ: Prentice Hall.

Woolfolk, A. E. (1998). *Educational Psychology* (7th ed.). Needham Heights, MA: Allyn & Bacon.

18

Adult and Postsecondary Education

Beyond secondary school classes, agricultural education takes on different forms. Two examples are described here. How do you feel that these appeal to adults and meet their needs?

- The Willis River Young Farmers chapter will conduct machinery safety demonstrations at the high school agriculture facilities next Tuesday at 7:30 P.M. Demonstrations include safety with PTO (power takeoff), augers, and grain heads and the prevention of tractor rollover. The public is invited.
- The Horticulture Department at Slate Valley Community College is offering continuing education classes in floral design. Classes will meet from 6:30 to 8:30 P.M. on Mondays beginning January 15 and ending February 26. As this is a hands-on class, enrollment is limited to twenty. Please preregister by January 10. A materials fee of $25 is due at the first class meeting. Call (555) 555-1234 between 8:00 A.M. and 4:00 P.M. to preregister, or visit the college's Web site for more information.

objectives

This chapter explains the importance of understanding the needs of adult students and the organization of adult agricultural education. It has the following objectives:

1. Explain the meaning of adult, adult education, postsecondary education, and continuing education
2. Describe the nature of adults as learners
3. Describe the organization of adult/young adult agricultural education
4. Discuss the role of postsecondary education
5. Discuss adult education in agriculture offered through secondary schools
6. Explain the laws of adult learning
7. Discuss teaching approaches for adults and young adults
8. Explain the role of student organizations with adult and postsecondary students

terms

adult
adult agricultural education
adult basic education
adult education
andragogy
chautauqua movement

lyceum
National Postsecondary Agricultural
  Student Organization
National Young Farmer Education
  Association
postsecondary education

18–1. Adult instruction in agricultural education covers a variety of topics and includes a wide range of audiences.

(Courtesy, Education Images)

ADULTS AND ADULT EDUCATION

Who is an adult? Is the definition based on age, life status, independence from parents, or some other condition? What is adult education? This section clarifies these two concepts.

Being an Adult

Throughout time, many cultures have developed rites of passage that clearly demonstrate when a child has passed into adulthood. In the United States, it is a much more difficult issue. A person can vote at age 18 but cannot legally purchase alcoholic beverages until age 21. Most states provide for a person to get a driver's license at or near age 16 but have a higher age limit for a person to get a commercial driver's license.

The age at which an alleged criminal can be tried as an adult is less than 18 in most states; however, most states prevent a youth under 18 from being a party to a contract or other legal instrument. Does marriage or having children define adulthood? Does holding a full-time job? Are 18-year-old college students adults because they live in apartments away from their parents? Are 30-year-olds living at home with their parents not adults?

All this background is provided to illustrate how the word *adult* is defined by our culture. An **adult** is an individual who has reached the stage in life to be personally responsible for himself or herself and who has assumed a productive role in society. This definition complies with the legal and cultural definitions we often use for being an adult in the United States. In short, an adult works to gain income for support and to support other individuals. Adults assume productive roles in society. Adults are often said to be men and women as opposed to children and youth.

Educating Adults

Adult education is much like *adult*—it is hard to define. People think of many things when they hear the term *adult education*. **Adult education** is the process of providing learning opportunities to improve adults in a wide range of areas of their lives. Many people associate adult education with adult basic education, such as instruction in English and mathematics in order to receive a GED (general equivalency diploma).

18–2. Field trips for first-hand observations are often used in adult agricultural education.

(Courtesy, Education Images)

Adult basic education (ABE) is instruction for adults in basic and specific skills needed to function in society. It is for individuals who did not achieve or did not participate in education as a child. Most often, the instruction focuses on reading, mathematics, and other basic areas. However, adult education is much broader than ABE and includes a wide range of workshops, seminars, classes, tours, self-study, and other means of improving the education of adults.

Adult agricultural education is education for men and women who are engaged in the ordinary business of life. They typically have jobs in an area of the agricultural industry. They have families and assume many responsibilities with their families and the communities in which they live.

The Cooperative Extension Service, through non-formal education, offers workshops, seminars, and classes on a variety of topics. Most religious organizations have adult education programs in the tenets of the faith as well as life topics such as finance, relationships, and personal growth. Businesses and agencies conduct training for employees to develop new skills related to their work. Colleges sponsor continuing education classes and activities. Apprenticeship programs exist for many of the trades. Local high schools and career centers offer adults instruction in languages, computers, construction, horticulture, welding, and much more.

According to the National Center for Education Statistics (NCES), about 40 percent of American adults participate in one or more adult education activities. The NCES placed these activities into six categories:

1. English as a Second Language, which includes language instruction for learners whose first language is not English, as well as instruction in basic American culture
2. Adult basic education, which includes instruction designed to assist adults in obtaining a high school diploma or its equivalent
3. Credential programs, which include college or technical degrees as well as certificates for jobs or licenses
4. Apprenticeship and formal training for trades occupations

5. Courses related to work and advancing in one's career

6. Personal development courses, with hobbies, religion, health, and sports being examples

In agricultural education, adult education has experienced periods when it has thrived and others when it has faded. In some states, agriculture teachers are employed to work exclusively or primarily with adults engaged in production agriculture. These teachers may work out of local public secondary school systems or community/junior college systems. In other instances, high school agriculture teachers offer night classes in areas related to their expertise and facilities, such as welding, construction, horticulture, landscape management, or financial management. Young Farmers, FFA Alumni, or collaborations with the Cooperative Extension Service or Farm Bureau are additional adult education avenues for agriculture teachers.

Postsecondary Education

Postsecondary education is typically defined as education offered in grades 13 and 14 at technical schools, community colleges, or junior colleges. In some cases, it includes college- and/or university-level education.

Education beyond high school may or may not have an "adult focus." Some postsecondary education is merely a continuation of high school education. It may be designed to prepare students for entering a skilled occupation or for gaining additional education. Other postsecondary education focuses on the needs of adults. The needs of the individuals served often depend on the philosophy of those in charge of planning the educational programs at the postsecondary school.

Continuing Education

Continuing education is part-time adult education. It is education beyond normal levels of formal educational attainment, such as high school or college. Continuing education also refers to education that enhances skills and keeps people's knowledge and skills current. For example, schoolteachers take continuing education. So do veterinarians, agronomists, floral designers, and many others.

18–3. Continuing education usually has a specific, hands-on focus and provides training in skills that can be immediately used.
(Courtesy, Education Images)

Continuing education units may be offered for participation in programs of continuing education. These units are not college credits but, in some cases, are used in certification or recertification much as college credits are used.

CHARACTERISTICS OF ADULTS IN EDUCATIONAL SETTINGS

Some adult educators feel that it is wise to discuss characteristics of adults in educational settings rather than dwell on what makes an individual an adult. Malcolm Knowles, known as the father of adult education, stated that individuals should be treated as adults educationally if they behave as adults by performing adult roles and if their self-concept is that of an adult.

Four broad categories of adult learner characteristics are used. These are as follows:

- Educational—how adults learn
- Motivational—why adults use adult education
- Cultural, physical, medical—what limits or enhances adults' learning
- Personal—what makes an adult learner different from a child

Educational Characteristics

Adults bring a wealth of experiences into the educational setting. The adult educator can draw on these experiences to make the learning more relevant.

Because of their careers and life responsibilities, adults tend to want the knowledge and skills they are learning to be immediately applicable and useful. They may prefer visual and hands-on learning as opposed to auditory and note-taking—in other words, learning by doing. An old saying in adult education circles is that adult learners "vote with their feet." They want value and appropriate instruction, and if they do not perceive this, they do not return for the next class.

Motivational Characteristics

Adults are motivated to participate in education for both external and internal reasons. They are often motivated by a real, crisis-driven need, such as lost job, career change, retirement, or financial problems. The motivation can also come from requirements such as certifications, job entry, or court orders.

Adults may be motivated by something they want (e.g., doing a training program to get higher pay) or something they do not want (e.g., taking a defensive driving course to reduce the impact of a traffic violation). Internally, adults may be motivated by a desire to enrich themselves or to learn a new skill. They may be pursuing a dream or may be motivated to change careers. Perhaps they now have the time to learn a new hobby. They may demonstrate anxiety, fear, or curiosity for things such as using new technologies.

Cultural/Physical/Medical Characteristics

Adults tend to be independent and make their own choices, which can greatly enhance the educational environment. Adults, as well as children, bring many cultural aspects into the educational setting. These include language, socioeconomic status, religious and cultural norms, and gender roles.

Older adults may have special needs related to the loss of stamina, hearing, eyesight, memory, dexterity, and/or mobility that require adaptation of the learning environment and elimination of learning barriers. Adults may have medical needs or limitations that require retraining or rehabilitation.

Personal Characteristics

With adults, timing is important. Adults, as opposed to children, are more likely to have full-time jobs, their own children or other dependents, community commitments, social obligations, and time restraints. With certain age groups or socioeconomic groups, transportation may be an issue.

As opposed to children who have compulsory education, adults engage in education for various reasons. Some do so for the social aspects—a chance to meet other people. Others, maybe through retirement, have extra time on their hands or want to learn a hobby. For others, technological changes and the ever-expanding knowledge base may necessitate their seeking further education and training.

ORGANIZATION OF ADULT EDUCATION

The delivery of adult education has evolved over many years. Several important approaches have shaped how adult education is carried out in the United States, particularly as related to agriculture.

History of Adult Education in Agriculture

True (1929) gives a detailed history of agricultural education in the United States. Adult agricultural education in the United States can be traced back to the colonies. Fairs, primarily for the sale of agricultural products, were held as early as 1644.

In 1785, societies were formed to promote and improve agricultural production. The Philadelphia Society for Promoting Agriculture had as its purpose to "print memoirs, offer prizes for experiments, improvements, and agricultural essays, and encourage the establishment of other societies through the country" (True, 1929, p. 7). The South Carolina Society for Promoting and Improving Agriculture and Other Rural Concerns was also organized in 1785. This movement grew so that in 1852 the United States Agricultural Society was organized.

Lyceums and Chautauquas   In 1826, Josiah Holbrook founded the first American lyceum in Millbury, Massachusetts. *Lyceums* were to be societies where the arts, sciences, agriculture, and public issues were presented and discussed. "In 1831, about 900 towns had lyceums, and for the next 20 years most public lectures were delivered before such organizations" (True, 1929, p. 32). After the Civil War, the lyceum movement faded.

About the time the lyceum movement diminished in importance, the *chautauqua movement* began. A Methodist Episcopal camp meeting in Chautauqua, New York, had held a summer Sunday-school institute for several years. In 1874, secular as well as religious instruction was included. The institute was such a success that local chautauquas sprang up over the country, providing lectures and entertainment. These continued until the middle 1920s, although the original chautauqua site still draws summer visitors (History Channel, 2003a).

Cooperative Extension Service   Federal legislation designed to promote adult agricultural education was enacted in the early twentieth century. Extension work had been conducted in the state of New York as early as 1894. In the early 1900s, Seaman Knapp, known as the founder of farm demonstration work, was traveling the South, teaching better practices to farmers through demonstrations. These and other events led to the Smith-Lever Act of 1914, which established the Cooperative Extension Service. The act set the purpose of agricultural cooperative extension work as "the giving of instruction and practical demonstrations in agriculture and home economics . . . through field demonstrations, publications, and otherwise" (True, 1929, p. 288).

18–4. The USDA complex in Washington, DC, is the headquarters for a number of national efforts in adult education in agriculture.

(Courtesy, U.S. Department of Agriculture)

Public School Adult Agricultural Education   Public school–based adult agricultural education was recognized in the Smith-Hughes Act of 1917. The Vocational Education Act of 1963 broadened the scope of adult agricultural education to include nonfarm occupations in areas such as horticulture; agricultural mechanics; agricultural business, processing, sales, and services; and natural resources. The Carl D. Perkins Vocational Education Act of 1984 placed a greater emphasis on the education of adults. It focused federal funding on training and retraining adults, individuals who are single parents, those with limited English proficiency, those with disadvantages or disabilities, and men or women entering nontraditional occupations. The Perkins Act was amended in 1990 and 1998. The Carl D. Perkins Vocational and Technical Education Act of 1998 emphasized developing not only vocational and technical skills of secondary and postsecondary students but their academic skills as well. The 1998 Act was reauthorized in 2006 by the U.S. Congress, extending through 2012.

POSTSECONDARY PROGRAMS OF STUDY

Adult agricultural education on the postsecondary (after-high-school) level can be categorized into three areas: programs leading to an associate's degree, programs leading to transfer to a four-year baccalaureate degree, and short-term programs leading to certification. These programs tend to be offered at community or junior colleges; however, some four-year colleges and universities also have postsecondary programs. Many community and junior colleges tend to be for commuting students. Residence facilities may not be provided. The campuses easily accommodate both part-time and full-time students.

Two-year associate's degree programs concentrate instruction on specialized courses designed to prepare graduates for specific careers. Instruction tends to be career-focused and hands-on. Classes may be smaller than the typical university first-year and second-year courses. Many of these programs have internship requirements that students complete within their areas of interest. Admission requirements tend to be more flexible than those for baccalaureate degree admission. Associate's degree students may also take some courses along with baccalaureate-degree-seeking

18–5. Adults like to be recognized for their educational accomplishments.
(Courtesy, Education Images)

students. Individual institutions and states have guidelines for how many, if any, of the courses taken in these programs can transfer into a four-year program. Career preparation programs may have agreements with major employers that provide facilities, up-to-date equipment, and internship opportunities.

Junior and community colleges also provide courses for students to transfer into baccalaureate degree programs at state and land-grant universities. These courses may include general education courses taken by all first- and second-year college or university students, such as English, mathematics, humanities, and basic sciences. They can also include basic lower-level courses in animal science, agronomy, or agricultural economics that are applicable to several majors. The junior or community college may have an articulation agreement with four-year colleges and universities to assure students that courses taken will transfer and count toward degree requirements.

Junior and community colleges typically also offer courses for workplace training or retraining, personal growth, and lifelong learning. Some of these may be intensive, lasting less than one year, and lead to certifications. An employer may work directly with a junior or community college to tailor a course or series of courses specifically for that employer.

ADULT EDUCATION IN SECONDARY PROGRAMS

Secondary public schools may offer adult agricultural education in both career centers and comprehensive high schools. The instruction may be career-focused or interest-based.

Funding Programs

Funding for adult education on this level may take several forms. Secondary teachers may teach part-time adult agricultural education in addition to their in-school classes. In most cases, such agriculture teachers will receive a specific salary supplement for teaching adults in addition to their secondary student teaching load. This supplement may come from state funding, local funding, or user fees. Some states employ agriculture teachers specifically for adult instruction. Their salaries may be funded through the school systems or be supported by user fees.

Nature of Instruction

The majority of adult instruction on this level is short term and topic specific. The adult educator must balance the time required to meet the instructional objectives adequately with the ability and interest of adults to devote time to the class.

Many agriculture teachers have found that the winter months are a time of year when adults are more willing to participate in adult educational activities. Also, if classes can be structured to occur just before application is required, adults are more likely to participate. For example, classes on gardening, flowers, and landscaping are best conducted just before the spring gardening season begins.

Role of Secondary Agriculture Teachers

Should secondary agriculture teachers teach adults? This question is more relevant now than at any time in the past thirty years. With increasing class size in secondary classes, greater emphasis being placed on middle school agricultural education, and increased accountability pressures, do agriculture teachers have the time and resources to conduct quality adult education programs?

Many would argue that by engaging adults in agricultural education, greater support will be built for the secondary program. This could lead to increased resources or even the hiring of additional agriculture teachers. Others, however, espouse the view that agriculture teachers are employed to teach secondary students first and foremost. They believe that anything that takes away from this focus is a detriment to the secondary students.

This debate must be resolved at the local level. The optimum adult education program is organized to meet the needs of the community, using the resources available, with benefit to both the adult learners and the agriculture teacher.

LAWS OF LEARNING FOR ADULT EDUCATION

Kahler et al. (1985) summarized eight laws of learning for adult education. These laws are based on psychological and educational research and have been confirmed by recent brain-based research. The laws provide a foundation for supporting adults' learning and meeting their interests and needs.

The Desire to Learn

This law states that the learner must be motivated in order for effective learning to take place. The interest may stem from the learner wanting to delve into the subject or from the need to gain new knowledge. The desire to learn can be influenced by the teacher and can spread from one learner to others.

Understanding of the Task

Learners need to know what is expected of them. This includes knowing the goals and objectives of the instruction, the degree to which performance is expected, and the activities required to master the task. The instructor needs to structure the learning situation so that the task and content are at the educational level of the learner.

The Law of Exercise

Practice, practice, practice! Adults need the opportunity to learn by doing. This can occur through experiments, problem solving, or even home or workplace implementation of what was learned in the classroom. Breaking up the repetition cycle with periods of reflection or analysis tends to increase learning effectiveness. This law is closely associated with the principle that tasks not performed on a regular basis or information not used is forgotten.

The Law of Effect

This law states that learning is affected by whether something provides satisfaction or annoyance. Results that are satisfying tend to strengthen learning, whereas results that are annoying tend to weaken learning. This law can be applied in the adult education classroom by making instruction interesting, having a comfortable environment, and structuring learning so that learners see progress in their work.

The Law of Association

Learning is enhanced when new information can be associated with previous experiences or grouped with similar information. The law of effect is important here, as an adult educator should draw on learners' life and work experiences that resulted in satisfaction to form connections with new content.

Interest, Vividness, and Intensity

The greater the vividness and intensity of something, the more likely the learner is to show interest in and retain the learning. Vividness can be enhanced through the use of color, smells, emotions, movement, and clarity of words. Intensity can be obtained through unexpected results, shock, emotions, tasks requiring concentration, and the finding of learners' passions.

Mind-Set

Learners bring with them a mind-set that can either block learning or ease learning. Many adults have memories of past educational failures, fears of less-than-acceptable performance, prejudices, and biases. All these can hinder current learning. The adult educational environment needs to be one in which the learner has the freedom to explore new ideas and ways of thinking. The educational setting needs to be inviting and open.

Knowledge of Success and Failure

Learners need feedback in order to grow and develop. The adult educator provides feedback, but the learner also needs to be taught to conduct self-evaluations. Adult educators must be careful that any negative feedback is provided in such a way that it promotes growth rather than discouragement.

TEACHING APPROACHES

With some adaptation, many of the same teaching approaches and methods used with youth are also used with adults. However, certain teaching approaches work specifically with adults. Malcolm Knowles (1989) used the term *andragogy* to describe characteristics of adult education.

Andragogy is the study of processes and practices in adult education. It is based on the needs and readiness of the adult learners. As such, the instruction is more learner-focused, collaborative, and problem-centered than traditional pedagogy. The instruction takes advantage of learner experiences and recognizes that adults need to know the purpose of the learning and have responsibility for the learning.

Methods of Teaching Adults

How does the adult educator decide what teaching methods to use? Through knowledge of the characteristics of adults and the laws of learning, the adult educator can best organize instruction to meet student needs. However, before the teaching method is selected, the first determination is deciding what should be taught.

Getting Organized   As detailed in Chapter 7, advisory committees and citizen groups are one means of determining subject matter, recruiting adult students, and otherwise assisting with an adult program. Advisory committees are to provide guidance regarding the total agricultural education program, which includes the adult education component. Figure 18–6 shows a community survey, another means that can be used to gather information on adult educational needs and interests. The survey can be placed online (if the school Internet use policy allows online surveys), mailed to randomly selected community members, mailed to selected persons who have previously expressed interest in the program, placed in the community newspaper, or placed in selected local businesses. Depending on the outlet chosen, the survey will need to be modified appropriately.

Once the class topics have been mutually determined, appropriate teaching methodologies can be selected. Chapters 8 and 13 will be useful when selecting teaching methodologies. If the objective of the topic is on a knowledge/cognitive level, then some form of lecture supplemented with instructional materials, visuals, demonstrations, and group discussion is appropriate. You, as the adult educator, do not have to be the expert on every topic. Utilize resource persons, such as extension educators and local persons experienced in the topic, as guest presenters. Give the adult learners frequent opportunities to apply the knowledge to real-life situations through scenarios, case studies, or discussions of problems experienced by class members.

Teaching Strategies   Adults respond best if active teaching strategies are used that show the usefulness of the information being taught. They want teachers to be well prepared and use formal methods of teaching, though often in informal ways.

The Slate Valley Horticulture Department would like to offer adult education classes to the community. We want to plan programs that meet your interests and needs. Please take a few moments to complete this survey and return it by November 15. Thank you!

* * *

Rank your interest in learning about these topics from 1 to 5, with 1 being the topic you would most want to learn about and 5 being the least.

_____ Arranging cut flowers
_____ Exploring commercial greenhouse production and its developing markets
_____ Growing indoor plants successfully
_____ Growing roses, shrubs, and ornamentals successfully
_____ Using a small, backyard greenhouse effectively

If the classes you ranked number 1 and number 2 were offered, what is the likelihood you would take the classes this January and February. (Check one only, please.)

_____ Definitely _____ Probably _____ Maybe _____ No

Each class is designed for 8 to 10 sessions of 2 hours each. Please indicate the class structure that best fits your schedule. (Check one only, please.)

_____ Two nights per week for 4 or 5 weeks
_____ One night per week for 8 to 10 weeks
_____ Saturday 8:00 A.M. to noon for 4 or 5 weeks

18–6. Sample adult interest survey form.

Presentations are often used to provide new information. The presentations may include realia to show the relationships, step-by-step procedures, or comparisons. Resource persons are used to assure that information is current, valid, and reliable.

Discussions are used to help assimilate the information and solve problems. Discussions should be carefully guided so that students stay on the topic. A presentation may be followed with discussion time to clarify points and help class members assimilate the information.

Field trips and tours are used to introduce new technologies and practices. Planning to assure maximum learning is essential. Such events should be scheduled at times when adults are available.

When topics on the psychomotor level are involved, adults want to put them into action as soon as possible. Demonstrations followed by guided practice, experiments, or exercises are appropriate. Frequent feedback, including instructor evaluations and self-evaluations, is critical for the learners to develop their skills and achieve progress. Laboratory activities are used to develop specific skills. These are useful with some adults, especially those who want to learn the skills and who will find the new skills useful.

Newer technologies available through computers and the Internet are appropriate with some adults. Some background instruction in computers and the Internet may be needed.

STUDENT ORGANIZATIONS

Adult students in agricultural education are served by student organizations much as students at the secondary level are served by FFA.

NYFEA

The *National Young Farmer Education Association* (NYFEA) is the official adult student organization for agricultural education as recognized by the U.S. Department of Education. Though the initial focus was on younger individuals or those just beginning agricultural careers, the Young Farmers serves all ages at the local, state, and national levels. NYFEA is the organization that provides for the coordination of programs, the delivery of educational opportunities, and the recognition of members on a national level. NYFEA members are engaged in leadership development, business skill development, and community service. The NYFEA mission is stated in its constitution. NYFEA supports agricultural leaders by providing education, training, and opportunities in leadership, agricultural education, and community service (NYFEA, 2012).

Young Farmer clubs in Ohio began as early as 1927 (Phipps & Osborne, 1988). Over the years more clubs started and state associations were organized. These state associations collaborated on Winter Institutes, where host states invited young farmers from around the country to participate in educational seminars, take part in field trips to agricultural and cultural components of the host states, and socialize. Young Farmer clubs, activities, and state collaborations grew such that in 1982 NYFEA was incorporated.

About two thirds of the states have young farmer associations and affiliation with NYFEA. NYFEA sponsors two major national meetings per year. The NYFEA Institute continues the tradition of the Winter Institutes that have been held in recent years in such diverse settings as Cheyenne, Wyoming; Indianapolis, Indiana;

Oklahoma City, Oklahoma; Baltimore, Maryland; Monterey, California; and Kansas City, Missouri.

A local Young Farmer chapter should be organized to meet the educational, social, and community service needs of its members. The agriculture teacher is the advisor and facilitator for the Young Farmer chapter. The Young Farmer chapter should not take an inordinate amount of the agriculture teacher's time. As adults, Young Farmers are able to organize and conduct most of their activities on their own initiative.

PAS

The **_National Postsecondary Agricultural Student Organization_** (PAS) is one of the eleven student organizations approved by the U.S. Department of Education as integral parts of career and technical education. PAS (for students beyond high school) is an organization associated with agriculture, agribusiness, and natural resources in approved postsecondary institutions offering associate or baccalaureate degrees or vocational diplomas or certificates.

Draft bylaws for a national PAS organization were formulated in 1979. PAS was officially founded in March 1980 in Kansas City, Missouri. PAS has membership from sixty-two chapters located in eighteen states. It is available to students in postsecondary programs of agriculture/agribusiness/natural resources in approximately 550 institutions in all fifty states.

The motto of PAS is "Uniting Education and Industry in Agriculture" (PAS, 2012). PAS provides opportunities for its members in individual growth, leadership, and career preparation. Activities are organized around these three areas. At the organization's national meetings, PAS members compete in the various career and leadership areas.

FFA Alumni

The National FFA Alumni Association is affiliated with the National FFA Organization. It serves as one of the four types of membership in FFA. The FFA Alumni is organized at local, state, and national levels. It involves adults in educational ways that support FFA. (More about the National FFA Alumni Association is presented in Chapter 7 of this book.)

REVIEWING SUMMARY

Adult education in agriculture is an important component of a total agricultural education program. It takes many forms. It ranges from formal degree programs at colleges and universities to topic-specific presentations by the local agriculture teacher.

A clear definition of when a person becomes an adult is difficult to obtain. However, there are educational characteristics that determine when a learner should be treated as an adult. Although some of these characteristics can also be true of youth, they most appropriately apply to adults.

The eight laws of learning for adult education should be used to guide educators in structuring learning environments for adults. Selecting subject matter to teach adults should be a collaborative decision involving the teacher, the adult students, and the community. The agricultural education advisory committee, along with data obtained from community surveys, should provide input into the decision. Teaching methods should match the subject matter and objectives of the instruction.

Organizations such as NYFEA, PAS, and FFA Alumni are able to assist agriculture teachers in organizing and delivering instruction to adults. NYFEA is primarily designed for out-of-school adults, whereas PAS is designed for students of postsecondary agricultural programs. Both organizations are recognized and supported by the U.S. Department of Education.

QUESTIONS FOR REVIEW AND DISCUSSION

1. How do adults learn, and how does this differ from how youth learn?
2. Who is an adult? Defend your answer.
3. What today serves the role of lyceums and chautauquas?
4. Educationally, are college students more like adults or youth?
5. How did federal legislation influence adult agricultural education in the early 1900s?
6. How have the various Perkins Acts made changes for adult agricultural education?
7. What is the definition of postsecondary education?
8. How is adult agricultural education funded on the secondary level?
9. What does *andragogy* mean?
10. How are the eight laws of learning for adult education used in agricultural education?
11. What are the differences between the two student organizations for adult agricultural education?
12. What is continuing education?

ACTIVITIES

1. Investigate adult agricultural education in your state. Topics to cover include Young Farmers, PAS, and postsecondary agricultural colleges. Are secondary agriculture teachers paid for adult education, and if so, how?
2. Investigate current federal legislation regarding adult agricultural education. In addition to the latest Perkins Act, investigate acts for secondary and higher education.
3. Using the *Journal of Agricultural Education*, *The Agricultural Education Magazine*, and the proceedings of the National Agricultural Education Research Conference, write a report on research about adult agricultural education. Current and past issues of these journals can be found in the university library. Selected past issues can be found online at the American Association for Agricultural Education and the National Association of Agricultural Educators Web sites.
4. Review the book entitled *Effective Adult Learning* (Birkenholz, 1999). Note the attributes of adults as learners and how educational programs are planned and delivered to maximize adult learning. Prepare a written report on your observations. Make an oral presentation in class.

CHAPTER BIBLIOGRAPHY

Association for Career and Technical Education (ACTE). (2003). *ACTE Web Site.* Accessed on August 22, 2003.

Bailey, J. C. (1945). *Seaman A. Knapp, Schoolmaster of American Agriculture.* New York: Columbia University Press.

Binkley, H. R., and R. W. Tulloch. (1981). *Teaching Vocational Agriculture/Agribusiness.* Danville, IL: The Interstate Printers & Publishers, Inc.

Birkenholz, R. J. (1999). *Effective Adult Learning.* Danville, IL: Interstate Publishers, Inc. (Available from Prentice Hall, Upper Saddle River, NJ.)

Cravens, M. (2003). *Lyceum, Chautauqua, and Magic.* Accessed on August 21, 2003.

Hettinger, J. (1999). The new Perkins . . . finally. *Techniques, 74*(1), 40–42.

History Channel. (2003a). *Chautauqua Movement.* Accessed on August 21, 2003.

History Channel. (2003b). *Lyceum.* Accessed on August 21, 2003.

Kahler, A. A., B. Morgan, G. E. Holmes, and C. E. Bundy. (1985). *Methods in Adult Education* (4th ed.). Danville, IL: The Interstate Printers & Publishers, Inc.

Knowles, M. S. (1989). *The Making of an Adult Educator: An Autobiographical Journey.* San Francisco, CA: Jossey-Bass.

Merriam, S. B., and R. G. Brockett. (1997). *The Profession and Practice of Adult Education: An Introduction.* San Francisco, CA: Jossey-Bass.

Merriam, S. B., and P. M. Cunningham. (1989). *Handbook of Adult and Continuing Education.* San Francisco, CA: Jossey-Bass.

National Association of Agricultural Educators (NAAE). (2003). *NAAE Web Site.* Accessed on August 22, 2003.

U.S. Department of Education, National Center for Education Statistics. (2007). *The Condition of Education 2007 (NCES 2007-064)*. Washington, DC: U.S. Government Printing Office.

National FFA Organization. (2003). *Local Program Resource Guide: 2003–2004* [CD-ROM]. Indianapolis: Author.

National FFA Organization. (2011a). *Local Program Success* (online). Accessed on November 30, 2011.

National FFA Organization. (2011b). *Official FFA Manual*. Indianapolis: Author.

National Postsecondary Agricultural Student Organization (PAS). (2012). *PAS Web Site.* Accessed on February 12, 2012.

National Young Farmer Educational Association (NYFEA). (2012). *NYFEA Web Site.* Accessed on November 12, 2012.

Newcomb, L. H., J. D. McCracken, J. R. Warmbrod, and M. S. Whittington. (2004). *Methods of Teaching Agriculture* (3rd ed.). Upper Saddle River, NJ: Prentice Hall.

Phipps, L. J., and E. W. Osborne. (1988). *Handbook on Agricultural Education in Public Schools* (5th ed.). Danville, IL: The Interstate Printers & Publishers, Inc.

Seevers, B., D. Graham, J. Gamon, and N. Conklin. (1997). *Education Through Cooperative Extension.* Albany, NY: Delmar.

Tippett, D. (2003). The learners we teach. In V. M. Chamberlain and M. N. Cummings (Eds.), *Creative Instructional Methods for Family and Consumer Sciences; Nutrition and Wellness.* New York: Glencoe/McGraw-Hill.

True, A. C. (1929). *A History of Agricultural Education in the United States, 1785–1925.* U.S. Department of Agriculture, Miscellaneous Publication No. 36. Washington, DC: GPO.

VocEd. (1985). The Carl D. Perkins Vocational Education Act of 1984. (Entire issue). *60*(4).

Wilson, A. L., and E. R. Hayes (Eds). (2000). *Handbook of Adult and Continuing Education.* San Francisco: Jossey-Bass.

19
Evaluating Learning

Mr. Spears, student teacher at Bayside High School, is frustrated. On the last test on animal digestion, the class average was not very high. He was sure the students would do well on the next test. Two weeks were spent covering the material. The class had examined the effect of stomach acids on foods, made models of types of teeth, researched methane gas released by cows, and took a field trip to a beef cattle research farm.

And yet, the test scores were not much better on the next test. Mr. Spears could not figure out why the students did not do better on the test on digestive system parts. He had tried to teach for mastery.

What do you think Mr. Spears needs to do? Maybe his test needed improvement as a tool for assessing student learning.

terms

achievement evaluation
authentic assessment
bell-shaped curve
constructed-response question
criterion-referenced test
fixed-response question
formative evaluation
grade
mastery learning

norm-referenced test
performance-based evaluation
portfolio
pretest
product scale
scoring rubric
summative evaluation
table of specifications

19–1. Observation of student performance on learning activities is often used in evaluation in agricultural education. (This shows students using a Quad/Sect Kit to a one square meter quadrat-rectangular plot used for population studies.)
(Courtesy, Education Images)

THE MEANING OF EVALUATION

Evaluation is a process to analyze educational effectiveness (student achievement) by using measurement tools. These tools include paper-and-pencil tests, observations, performances, and portfolios.

Evaluation serves multiple purposes. It can (1) evaluate academic achievement of individual students, (2) diagnose academic problems or deficiencies of individual students or classes, (3) evaluate the effectiveness of a curriculum or other educational product, and (4) evaluate the educational progress of large groups in order to inform public policy decision makers (Ahmann & Glock, 1981). Although evaluation is often viewed as teacher-driven, an overarching goal of evaluation should be to hold students accountable for their own learning. In this respect, evaluation is a feedback tool to assist students in advancing their learning.

Evaluation occurring before or during instruction is called **formative evaluation**. The purpose of formative evaluation is to determine the starting point for instruction and to give feedback to students and the teacher during the instruction. Formative feedback occurs during the process so changes can be made and plans modified to address deficiencies or problems.

Evaluation occurring after instruction is called **summative evaluation**. The purpose of summative evaluation is to determine how well students achieved and in many cases to provide a grade. Summative evaluation can provide feedback to determine the next level at which a student should be placed and changes for the next time a topic is taught.

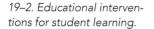

19–2. Educational interventions for student learning.

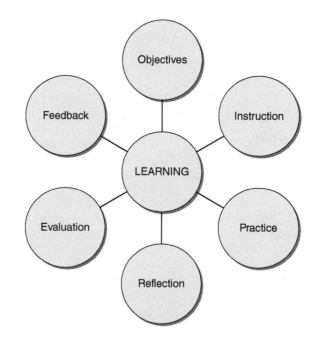

Many people wrongly view evaluation (or grades) as an end in and of itself rather than as part of a larger process. As shown in Figure 19–2, evaluation is just one of many educational interventions for facilitating student learning. When viewed in this respect, evaluation becomes one of several tools the agriculture teacher uses to assist students in their education.

The next section will describe the two most common categories of evaluation. Achievement evaluation tells how much a student knows. Performance-based evaluation tells what a student is able to do.

ACHIEVEMENT EVALUATION

Achievement evaluation is the practice of measuring how much students know about a subject or how well they are able to use that knowledge cognitively. Tests to measure student achievement can be standardized, such as state competency tests, or teacher made. These tend to be paper-and-pencil tests, though computer-based and Internet approaches to testing are being used.

A ***pretest*** is a test given before instruction is begun to determine previous student learning in the subject area. Knowledge or skills in which students already have competence can be either lightly reviewed or omitted altogether. Achievement evaluations are further broken down into norm-referenced and criterion-referenced.

Norm-Referenced Testing

A ***norm-referenced test*** is a test that compares the achievement of an individual student with an established group norm. Norm-referenced tests are constructed so most individuals score in the average range and a few score in the high and low ranges. This yields a ***bell-shaped curve*** in which the number of students scoring at a particular level is plotted such that the number of scores at each point yields a line in the shape of a bell. It is a graphic depiction of how a large number of student scores would be distributed. The bell-shaped curve is known as a normal distribution curve. Norm-referenced tests are used to compare individuals with the group and with other individuals.

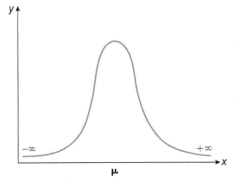

19–3. A bell-shaped curve.

Very few standardized tests have existed in agricultural education at the national level. It is difficult to determine what all agricultural education students should be knowledgeable in with such diversity in growing regions, cultural practices, and products. Some states have developed standardized tests for agricultural courses taught in those states.

Criterion-Referenced Testing

A *criterion-referenced test* is a test that compares the score of an individual student with established criteria or standards. This type of test is designed to determine how students compare with established criteria rather than how they compare with others.

Mastery learning is often associated with criterion-referenced tests. In *mastery learning*, students are given frequent opportunities to practice and multiple opportunities to improve their scores. An agriculture teacher who requires 100 percent correct on a safety test is using a criterion-referenced test with a mastery approach.

An agriculture teacher who uses well-written behavioral objectives, as described in Chapter 13, to both teach the content and develop the evaluation is using a criterion-referenced approach. This is how most teacher-made tests and chapter tests in textbooks are developed. The agriculture teacher is most concerned with knowing what individual students have learned rather than how students compare with each other.

PERFORMANCE-BASED EVALUATION

Performance-based evaluation is the use of evaluation approaches that measure how well students can do something under given conditions, such as weld a flat bead on a piece of mild steel. Typically students demonstrate a skill or process, construct a product, or provide some other tangible evidence of their learning. Teachers observe and judge the evidence using tools such as rubrics, checklists, rating scales, or product scales.

Using Performance-Based Evaluation

Performance-based evaluation is perceived to have several advantages over achievement evaluation. These include clarity, authenticity, subjectivity, allowance for learner differences, and engagement of learners (Tanner, 2001).

Under performance-based evaluation, students have a greater knowledge of what is expected of them and the level of performance required for success. This type of evaluation also has more meaning for the students in the real world. Although subjectivity can be a detriment, performance-based evaluation allows teachers to judge student performance in areas of effort, improvement, timeliness, and care. Performance-based

19–4. The ability of a student to weigh a given amount of chemical efficiently and precisely is an indication of performance.
(Courtesy, Education Images)

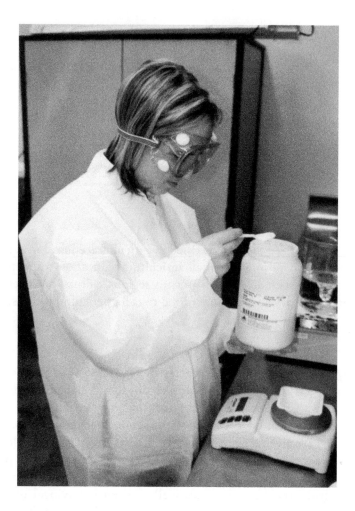

evaluation can be tailored for an individual student, thereby accounting for learner differences and ideally engaging the student in the assessment. This makes the assessment more a part of the overall learning process.

Agriculture teachers have numerous opportunities to use performance-based evaluation. Most laboratory situations require a teacher to evaluate how well students can apply and demonstrate what they know. Career-directed instruction lends itself to performance-based evaluation.

Supervised agricultural experience (SAE) and FFA Career Development Events are other examples of performance-based evaluations.

Authentic Assessment

Authentic assessment is a type of performance-based assessment that places the student in a situation or simulation that closely mimics the workplace or other life setting. In many career and technical education areas, the school setting is designed to match the workplace setting as closely as possible. The tools, equipment, and tasks completed all model those of the workplace. Most supervised experience carries this a step further and places the student into the actual workplace.

Reproducing actual workplaces may not be practical. Schools may not have the budgets to purchase or build replicas of workplaces. Besides, what is state of the art one year may be obsolete three years later. Finally, space and time may limit how closely workplaces can be reproduced. Regardless, authentic assessment can still be used. Simulations, such as computer programs, models, and learning stations, can provide enough realism to evaluate students adequately.

19–5. The quality of a hanging basket is an indication of the proficiency of a student in producing such a product. (Courtesy, Education Images)

DEVELOPING ASSESSMENT INSTRUMENTS AND TEST ITEMS

The purpose of assessment instruments is to measure accurately the knowledge, skills, and dispositions of students. A well-designed assessment instrument provides the agriculture teacher vital information regarding student learning in a particular subject area. However, assessment instruments can also measure other student variables. For example, sleepiness, hunger, bad mood, or unpleasant home life can negatively affect a student's performance.

Tests may be prepared "on paper" or administered with a computer. Computer tests may be teacher made or obtained from a provider. Some state departments of education prepare tests that are administered to all students enrolled in particular classes. Commercially prepared CD-based and online tests are available, which may be completed by using a computer keyboard or by using a pencil after printing them out on paper. If the tests are taken electronically, scores are determined and maintained by the computer or an online server.

Students with lower-than-grade-level ability in reading, writing, math, or thinking skills tend to have difficulty with assessments regardless of their knowledge of the subject area. Students may also be unmotivated toward the subject matter or have negative attitudes toward the class, the school, the teacher, or themselves. Educational problems such as poor study habits or failure to be attentive during instruction, practice, and review can also negatively affect students' assessment performance.

Phipps and Osborne (1988) described possible causes of problems in student achievement (Table 19–1). When developing the assessment instrument, the agriculture teacher should consider the level of objectives and instruction. In Chapter 13, the six levels of cognitive objectives in Bloom's taxonomy were discussed. Some evaluation techniques are best for certain levels of objectives. For example, essay questions are better than true/false questions for synthesis-level objectives. Some subject matter is best evaluated using one evaluation technique over others. For example, the proper

19–6. A teacher prepares an online test to be completed by students in the school computer lab.
(Courtesy, CAERT, Inc.)

Table 19–1
POSSIBLE CAUSES OF PROBLEMS IN STUDENT ACHIEVEMENT

Level One Causes	Level Two Causes
Inappropriate evaluation techniques	Type of evaluation is not parallel to objectives and instruction. Type of evaluation is not suited to subject matter.
Poorly designed evaluation	Evaluation is graded unfairly. Evaluation has unclear or inappropriate questions. Evaluation questions are not parallel to objectives and instruction.
Lack of student understanding and skill	Objectives are not clear to students. Student motivation is low. Students do not have the background knowledge or skills to perform as expected. Students are not mentally or physically capable of performing at expected levels. Teaching is ineffective.

Source: Adapted from Phipps & Osborne (1988).

welding of a joint is best evaluated through the performance of the task, not through a paper-and-pencil test.

Evaluation instruments should not be graded arbitrarily. Teachers must take care to construct questions so they are clear and appropriate. A lesson taught using knowledge- and comprehension-level objectives should have an assessment instrument written to test the students on those levels only.

The teacher needs to make sure that students know what is expected of them, are motivated to learn, and have been taught in the most appropriate manner. It is sometimes difficult to determine whether poor student performance is a result of causes stemming from the teacher or from the student.

Table of Specifications

Teachers develop evaluation instruments weekly if not daily. Examples of written instruments include quizzes, tests, and exams. Teachers need to ensure that the evaluation instruments adequately and accurately address the amount and level of the subject matter reflected in the objectives and the instruction.

A *table of specifications* is a procedure used to chart test content, its relative emphasis, and its level. The table can also chart the types of questions, such as true/false, matching, multiple choice, and so forth.

An example of a table of specifications for a lesson on parliamentary procedure is presented as Table 19–2.

Fixed-Response Question

A common type of test question is the *fixed-response question*. This requires the student to select the correct answer from given choices. Fixed-response questions include multiple choice, true/false, and matching. Examples and guidelines for each are given in the next sections.

Multiple Choice   Multiple-choice test items are the most frequently used and are viewed as objective. To be effective, multiple-choice items need to be correctly constructed. Good multiple-choice questions can measure low-level to high-level objectives, be just as effective as essay questions, and measure a specific content focus (Tanner, 2001).

A multiple-choice question is constructed with a stem followed by four or five choices. Only one choice is the correct or best answer; the rest are called distractors. The stem is written as a sentence and should allow the student to know what is required before reading the choices. A stem should be written in simple, clear language and contain one idea. Negative wording should be avoided in the stem. The stem should not contain vague terms, such as *sometimes*, *usually*, and *may*, or absolute terms, such as *always*, *never*, *none*, and *only*.

Writing good distractors is the key to a good multiple-choice question. Distractors should be written as students who have a general sense of the topic might think. Distractors should be logical, credible, and even attractive to the students. Distractors should be similar to the correct response in terms of length, grammar, and complexity. They should not give clues to correct answers in other questions. It is all right occasionally to make a distracter humorous. This is especially appropriate for questions at the beginning of a test to relax students.

Using the phrases "all of the above" and "none of the above" as choices should usually be avoided. Using these phrases as distractors is not effective if one of them is never the correct choice. On the other hand, if students can easily guess when one of these phrases is the correct answer, the question loses its ability to

Table 19–2
TABLE OF SPECIFICATIONS FOR A FIFTY-ITEM TEST ON PARLIAMENTARY PROCEDURE

Instructional Content	Relative Emphasis	Levels of Bloom's Taxonomy*							
		1	2	3	4	5	6	7	8
Reasons for holding a meeting	2%	1							
Steps to planning a meeting	4%		2						
Order of business	8%	2	2						
Purposes of a properly arranged meeting room	8%	2	2						
Symbols of FFA officer stations	12%	6							
Definition of parliamentary procedure	2%	1							
Purposes of parliamentary procedure	4%		2						
Uses of the gavel	8%	4							
Methods of voting	8%			4					
Definition of main motion	8%	4							
Four motion classifications	12%			6					
Definition of amendment	4%	2							
Conducting a meeting using parliamentary procedure	20%					10			
	100%	22	8	10		10			

Type of Questions	Number of Questions
True/false	10
Matching	10
Multiple choice	10
Fill in the blank	5
Short answer	3
Essay	2
Role play	10
Total number of questions	50

*1 = knowledge, 2 = comprehension, 3 = application, 4 = analysis, 5 = synthesis, 6 = evaluation, 7 = affective, 8 = psycho-motor

discriminate between students who do and do not know the content. Teachers also need to avoid patterns, such as the correct answer being choice "c" more than any other choice.

The following example is based on the table of specifications presented as Table 19–2.

Multiple-Choice Example

1. Two taps of the gavel are used to
   a. have everyone be seated.
   b. have everyone stand up.
   c. announce a vote decision.
   d. call a meeting to order.

True/False   True/false questions have many advantages. They are viewed as easy to construct and grade. They are objective in that a given question is either correct or incorrect. True/false questions require less reading than multiple-choice questions, so the test appears shorter. However, true/false questions also have disadvantages. Students are prone to guess if they do not know an answer, since they have a 50 percent chance of being correct. Poorly constructed questions that are ambiguous or contain more than one idea can lead to student confusion and to arguments about whether a question is graded correctly. Also, poorly constructed questions may give too many clues, allowing students to answer the questions correctly without really knowing the content. The teacher may also be tempted to write true/false questions on trivial facts, as these questions are the easiest to write.

Every true/false item should be written as a statement sentence about one important concept. Both true and false statements should be the same length. Patterns should be avoided, such as three true followed by three false items. There is no clear guideline for the percentage of true items versus false items. In general, the test should have an equal number of true and false or a greater number of false items. To indicate the choice of answer, a student should either circle a preprinted "T" or "F" or write "true" or "false" in a blank.

There are two common methods to use for reducing the effect of guessing on true/false items. The first method, a formula, computes the student score by subtracting the number of incorrect responses from the number of correct responses ($S = C - I$). Items left blank do not count against the student, so if he or she is not sure of an answer, it is better not to guess. The second method rewards the student for explaining why an item is false. For each item the student determines is false, he or she writes a statement telling why the item is false or what would make the statement true. This second method requires more grading time for the teacher but encourages the student to think more deeply about the item.

The following example is based on the table of specifications presented as Table 19–2.

True/False Example

T   F   1. To amend a main motion, a member rises, gains recognition from the chair, and says, "I wish to amend the motion by . . ."

Matching   Matching questions are popular on tests for evaluating students on lists of related information. They are easy for the teacher to write and quick for the students to answer. Matching questions do not take up much space on a page, so the test appears shorter than a multiple-choice test.

Matching does have some disadvantages. Matching is best suited for knowledge-level objectives, such as recall, define, and label. Instructions are more important for matching than for other types of questions. Will answers be used more than once? Will there be more than one correct response to an item? Will some answers not be used?

The following guidelines should help when constructing matching items: Place the words on the left and the list with the definitions or descriptions on the right. Each set of matching items should deal with a single concept. One list should have choices that will not be used so students will not be able to guess the last answer by the process of elimination. If possible, put one of the lists in alphabetical or chronological order or in some other logical pattern. Place the entire matching set on the same page. The matching set should have between five and twelve items.

The following example is taken from the table of specifications in Table 19–2.

Matching Example

a. Parliamentarian   _____ 1. Emblem of Washington
b. President   _____ 2. Ear of corn
c. Reporter   _____ 3. American flag
d. Secretary   _____ 4. Handclasp of friendship
e. Sentinel   _____ 5. Plow
f. Treasurer   _____ 6. Rising sun
g. Vice President

Constructed-Response Question

Another common type of test question is the constructed-response question. The *constructed-response question* is the kind of item that requires the student to develop his or her own answer to the question. Constructed-response questions include fill in the blank, short answer, and essay. Examples and guidelines for each are given in the next sections.

Fill in the Blank Fill-in-the-blank questions are similar to the fixed-response questions described in the previous sections except the student supplies the word or words to make each statement correct. Because of this, guessing is reduced or eliminated. Questions of this type are easy to write but rarely measure above the recall level. They are best suited for definitions, concepts, and key terms.

Fill-in-the-blank questions tend to be the most problematic of all the types. Although the teacher may believe a question is properly worded, students will invariably give numerous answers that are technically correct yet not what the teacher expected. The teacher also must decide whether to accept misspellings and synonyms. Teachers can eliminate some of these problems by providing the students a list of words or terms to be used. However, this adds the problems of student guessing, extra test length, and the supplying of clues for the answers.

These guidelines should help when constructing fill-in-the-blank items: Construct each statement so the blank is at the end of the sentence. If too many words are left out, the statement loses its meaning and students are confused. There is no need to make the length of the blank fit the size of the answer or put the same number of blanks as words in the answer. It is necessary only to provide enough space for the student to write in the longest correct answer. Avoid using statements directly from textbooks or other curriculum materials. Doing this results in the statement being out of context and ambiguous.

The following example is based on the table of specifications presented as Table 19–2.

Fill-in-the-Blank Example

1. The type of motion used to provide proper and fair treatment to all members is called _____.

Short Answer Short-answer questions are similar to fill-in-the-blank questions but typically require slightly more writing and thought for the student. These questions are designed to test a student's knowledge of definitions, series, names, and relationships. Short-answer questions allow the student more freedom in type and length of response.

Just as short-answer questions have advantages similar to those of fill-in-the-blank questions, they also have similar disadvantages. The teacher must decide how to grade misspellings, partially correct answers, and technically correct answers. The teacher must also give clear directions regarding how much to write. Does the size of the white space give an indication of the response expected? Are one-word answers sufficient, or are explanations necessary?

The following example is based on the table of specifications presented as Table 19–2.

Short-Answer Example

1. List the three ways to amend a motion and give an example of each.

Essay Essay questions are capable of measuring levels of objectives from simple to complex. This is both their strength and their weakness. A teacher may write an essay question expecting to measure students' higher-order thinking abilities but instead receive answers on the recall level. Also, essays take a substantial amount of teacher time to grade. Other types of questions may be sufficient for testing lower-order objectives.

An essay question needs clear instructions. Is the expected answer to be one paragraph or multiple paragraphs? Are some topic areas required for full credit? Are students graded on spelling, grammar, and sentence structure?

These guidelines should help when constructing essay items: The teacher should decide before writing a question what objective the question is covering and what cognitive level the answer should reflect. The teacher should also develop a scoring rubric before administering the test. The rubric should include relevant content the teacher will look for in grading and the relative importance placed on each element of the answer. The teacher should grade all students on the same essay question before grading them on the next essay question. The pile of student tests should be reshuffled between questions, as some research has shown that teachers tend to grade papers at the beginning and end of the pile differently.

The following example is based on the table of specifications presented as Table 19–2.

Essay Example

1. Explain how parliamentary procedure maintains order in an FFA meeting. Make sure to address the minority, the majority, and outside guests.

Useful guidelines in preparing a teacher-made written test are as follows:

- Cover in the test only what has been taught based on educational objectives.
- Provide instructions on how to complete the test.
- Use wording appropriate for the grade level.
- Spell all words correctly.
- Place a title and a date at the top of the first page.
- Provide a line on which the student can write his or her name.
- Use appropriate punctuation and grammar.
- Write test items in a straightforward manner—no tricky questions.
- Number all items.
- Provide sufficient space for the student to give answers.
- Proof the test before making copies.

19–7. How to prepare a good written test.

ALTERNATIVE ASSESSMENT

Quizzes, tests, and exams are valid measures of what a student knows at a particular point in time. However, they are not the most effective tools for measuring progress over time or demonstrating creativity. Written tests are also not good at measuring affective and psychomotor objectives. Other assessment tools, such as portfolios and projects, are better suited to these purposes.

Portfolios and projects are examples of performance-based evaluation, discussed in an earlier section of this chapter. Depending on how they are structured, portfolios and projects are also examples of authentic assessments. Portfolios and projects are discussed in greater detail in the following sections.

Portfolios

A *portfolio* is a collection of materials that represent a student's work. It may be contained in a notebook or a folder or be on a CD-ROM or a Web site. The portfolio usually contains a variety of samples of the student's work. These samples can be assignments (both graded and ungraded), tests, photographs of projects or activities, drawings, video and audio clips, reports and computer projects, certificates and letters of commendation, and other items that document the student's progress and accomplishments.

How a portfolio is organized and what items are included depend on the purpose of the portfolio. Portfolios typically serve three basic purposes: documentation of growth, documentation of best efforts, and compilation for assessment.

A portfolio provides information on the whole student. This follows the theory of multiple intelligences and the brain-based learning theory discussed in Chapter 12. Portfolios provide students the opportunity to use creativity in demonstrating all aspects of their learning (cognitive, affective, and psychomotor) together rather than in separate parts. Growth portfolios provide students the opportunity to reflect on their efforts. This can lead to deeper learning and greater development of students' critical-thinking abilities.

19–8. A teacher is reviewing a portfolio with a student.
(Courtesy, Education Images)

Teachers using student assessment portfolios for grading purposes must develop clear, detailed criteria before students begin their portfolios. It is also critical that scoring rubrics be developed early in the process. Teachers need to take care that the rubrics are fair, consistent, and detailed. Criteria and rubrics should not be too prescriptive, as this will stifle student creativity.

Agricultural education teachers use student portfolios in the classroom, in supervised experience, and in FFA. This may come as a surprise to someone who believes portfolios are a new educational innovation. Supervised experience record books, proficiency award applications, and degree applications are examples of portfolios. Although portfolio examples are less numerous in the classroom, some agriculture teachers have students develop portfolios for drawings, experiments, and projects.

Projects

A project is a task or problem completed by a student during or outside of class time. A project supplements instruction and allows a student to apply the information that was studied.

Examples of projects are numerous in agricultural education. Almost every area of agricultural mechanics uses student projects. The areas of natural resources, animal sciences, plant and soil sciences, horticulture, and agricultural business are replete with project examples.

Planning by both the teacher and the student yields a successful student project. Figure 19–10 shows a project planning sheet that is adaptable to a variety of agricultural subjects. If appropriate, project drawings, schematics, flow charts, and other planning tools can be added. The agriculture teacher may also want to add an "Estimated Arrival Time" column to the "Materials Needed" section. A key material that takes weeks to arrive after it is ordered means the student must plan far in advance, alter the plan, work on other aspects of the project, or arrange for alternative class activities in the interim. Project procedures should be detailed, step by step, and in

19–9. Projects provide students opportunities to utilize many of their intelligences.
(Courtesy, Education Images)

```
┌─────────────────────────────────────────────────────────────────────┐
│                                                                     │
│   Student's Name: _____ Date: _____      │
│                                                                     │
│   Title of Project: _____ Project Description: _____ │
│                                                                     │
│   Materials Needed: (Include dimensions, source, quantity, cost.)   │
│                                                                     │
│   Item and                                                          │
│   Description    Source    Quantity    Unit Cost    Total Cost      │
│                                                                     │
│                                                                     │
│                                                                     │
│                                                                     │
│                                                                     │
│   Project Procedures:                                               │
│                                                                     │
│                                                                     │
│                                                                     │
│                                                                     │
│   Evaluation Criteria:                                              │
│                                                                     │
│                                                                     │
│                                                                     │
│                                                                     │
└─────────────────────────────────────────────────────────────────────┘
```

19–10. Sample project planning sheet.

logical order. Many students, especially younger ones, will need help in this area. Depending on the purpose and type of project, the evaluation criteria may be student or teacher developed. It will also vary as to whether these are individualized or the same for all students.

Figure 19–11 is an example of a daily documentation sheet to be used when students are working on projects. It is designed for students to be self-accountable for use of class time, progress toward goals, and quality of work. The "Daily Grade" section can be used for self-assessment by the student or as a grading mechanism for the teacher. It can accommodate various grading scales. A check mark (✓), minus (–), zero (0), or smiley face (☺) / sad face (☹) system is quick and gives students broad feedback. A numerical system (0, 1, 2; 0–4; or 0–5) can give more focused feedback and is more easily adapted to grade-book systems.

GRADING

Virtually all American public school systems require teachers to assign grades to student work on a regular basis. A *grade* is a mark or descriptive statement that rates the extent to which a student has achieved a standard. The most common arrangement requires, at a minimum, assigning grades for the grading period, semester, and course. More frequent grading is often required. Students are typically assigned either letter (A–F) or numerical (0–100) grades, with numerical grades often later converted to letter grades. The format and structure of grading are typically specified in school system policy manuals.

Student's Name: _____ Date: _____

Title of Project: _____

Materials, Tools, and Equipment to Be Used Today:

Project Steps or Procedures to Be Accomplished Today:

Materials to Be Ordered:

Daily Grade:

- Project progress (keeps on task, meets goals, etc.)
  Comments: _____

- Work quality (follows specifications, uses materials wisely, etc.)
  Comments: _____

- Safety (follows safety rules, protects self and others, etc.)
  Comments: _____

- Work habits (begins and ends on time, cleans up area, puts tools away, refrains from horseplay, etc.)
  Comments: _____

- Other:
  Comments: _____

- Total points _____

19–11. Sample daily documentation sheet for a project.

The agriculture teacher has several grading decisions to make. How will individual assignments be graded? What weight will be given to each assignment? How will grades be reported to students and parents? What about qualitative assessments that do not easily convert to letter or numerical grades? Should SAE and FFA be graded? These and other questions will be addressed in the following sections.

Scoring Rubrics

A *scoring rubric* is a set of guidelines for judging student work. For a rubric to be effective, the agriculture teacher needs to make several decisions before and during construction of the rubric. How much precision is required to discriminate between products of differing quality? Sometimes a 3-point scale (high, middle, low) is sufficient, while other times maybe a 10-point scale is required.

The teacher should use language that describes each level on the scale selected. Consideration needs to be given to the highest point on the scale and to the point at which below is unacceptable. Defining these two points is critical and will make constructing the remainder of the scale easier. The difference between any two points on the scale should be equivalent to the difference between any other two points. For example, the difference in quality between level 4 and level 5 performances should be the same as the difference in quality between level 2 and level 3 performances.

Scoring rubrics must give the grading criteria or categories, levels of performance (the scale), and descriptions of what constitutes performance under each level. The more specific and detailed the descriptions, the better students understand what is expected of them. Precise descriptions also reduce or eliminate arguments over assignment of grades and provide feedback for improvement.

A special type of rubric is a product scale. A **product scale** is a series of actual objects representing each level of quality along the scale. Each object is given a numerical or letter grade indicating its position on the scale. A student's product is then compared with the product scale and receives the number or letter of the scale item most closely matching the student's product. Product scales in agricultural education are often used in agricultural mechanics for evaluating welds. The agriculture teacher can develop a board containing welds of varying qualities, from exceptional to inferior. If a mastery learning approach is being used, a student can self-evaluate his or her weld and continue working until a weld of the desired quality is produced.

Qualitative Assessments

The mission of agricultural education is to prepare the student for a successful career and a lifetime of informed choices in the plants, animals, and natural resources systems. This implies that agricultural education is concerned with developing the total student. To accomplish this mission, agriculture teachers must provide their students with qualitative assessments on attitudes, behaviors, and efforts. Qualitative assessments include rating scales, checklists, and essays.

Rating scales and checklists are useful for assessing attitudes and behaviors. A checklist, as the name implies, is a listing of characteristics, qualities, attributes, or behaviors that the teacher places a check mark beside if an item is observed. Checklist items should be worded in a parallel manner—for example, all items stated in a positive fashion. This avoids confusion about whether a check mark is desirable. If the checklist is used for determining a grade, directions must include the number of check marks required for a certain grade and must specify whether particular items must be checked to obtain a certain grade.

Essays allow students to provide background information and demonstrate higher-order thinking skills. Students can give examples and explanations more easily

	Habitually late	Sometimes late	On time	Early
Punctuality:	1	2	3	4

19–12. Sample rating scale item for the quality of punctuality. A circle is placed around the number of the quality that best represents a student's performance.

Table 19–3
SAMPLE WEIGHTINGS FOR ASSIGNING GRADING-PERIOD GRADES

Category	Weighting (%)*
Quizzes	10–15
Tests	25–35
Homework	10–15
Daily grades	10–15
Laboratory exercises	25–35
Projects	25–35
Grading-period exam	10–15
Notebook	5–15
SE	5–15
FFA	5–15

*Actual weightings used will add to 100%. Not all categories will be used each grading period.

with essays. Finally, essays give students practice in writing skills that are applicable to numerous occupations.

Calculating Grades

Course grades are indicators of student accomplishment, growth, and level of understanding of course content. Fair, consistent, and accurate calculating of student grades is important.

The variety of grading periods at schools is astounding. There are seven-period days, trimesters, full-year blocks, semester blocks, and variations on these. Grading periods may occur at four-week, six-week, nine-week, or other intervals. Grades are typically calculated at the end of the grading period, and many times at the midpoint as well. School systems may require that student grades be reported to students and parents at the midpoint and at the end of the grading period. More frequent reporting, such as weekly, is common when an online grade book accessible by students and parents is used.

Many teachers use a system of contacting a different group of parents each week to update them on their students' performance and progress. Some schools require FFA members to verify passing grades in their classes before they are allowed to participate in out-of-school activities.

How does an agriculture teacher determine a student's grading-period grade? Electronic grade books simplify the task of calculating grades; however, teachers must still assign values and weights to each assignment. Sample weightings are presented in Table 19–3.

Grading SAE and FFA

Many questions arise related to grading SAE and FFA participation. Should agriculture students receive grades for supervised experience? Many factors enter into answering this question. Do students receive academic credit on their high school transcripts for supervised experience? Are all agricultural education students required to conduct supervised experience? What is the philosophy of the local program? Are

19–13. Should "bringing home a trophy" from FFA competition be considered in grading?
(Courtesy, Education Images)

there state requirements concerning supervised experience? Do school administrators, parents, and advisory boards support the grading of students on activities conducted outside of regular school hours?

If academic credit is given, then SAE must be graded as pass/fail at a minimum. Otherwise, the agriculture teacher should discuss the preceding questions with local administrators, the Agricultural Education Advisory Board, the State Agriculture Teachers' Association, the Agricultural Education Teacher Education Program, and the State Department of Education Agricultural Education Staff.

Should agriculture students receive grades for participation in FFA activities? This question also requires consideration of many of the same factors as supervised experience. The agriculture teacher should consult with the same entities as described in the preceding paragraph in making this decision. Additional questions to be considered regarding grading participation in FFA activities include the following: Are all students FFA members, even those in semester courses? Do students have equal opportunity to participate in activities? Will some students be excluded from activities because of gender or financial constraints? Will students participating in sports or other organizations be able to participate in enough FFA activities to earn the highest possible grade? Can a grading system be developed that is not biased in favor of FFA officers to the detriment of students who are not officers?

REVIEWING SUMMARY

Evaluation is an integral part of the education process. It involves more than tests and grading and ideally leads to students taking responsibility for their own learning. Evaluation should occur before, during, and at the end of instruction.

Achievement evaluations are common in today's educational environment. Both standardized and teacher-made achievement tests are used throughout the curriculum. Norm-referenced tests are useful in determining how much individual students know in comparison to the group or to other individuals. Criterion-referenced tests are useful in determining how much students know in comparison to standards and to established criteria.

Performance-based evaluations have been used in career and technical education for a long time. The use of authentic assessments, portfolios, projects, and qualitative assessments provides a more nearly complete picture of the total student. Although these may be more difficult to score, they yield rich data about the student.

Teacher-made assessment instruments are used on an almost daily basis in schools. It is critical that teachers take the time and effort required to construct valid instruments. The type of question, the wording of question items, and the scoring rubric must all be carefully selected.

QUESTIONS FOR REVIEW AND DISCUSSION

1. What are examples of formative evaluations used in agricultural education?
2. Why are norm-referenced standardized tests rarely used in agricultural education?
3. Does a mastery learning approach encourage students to put forth less than their best effort?
4. How are performance-based evaluations used in agricultural education?
5. What is the relationship between authentic assessment and SAE?
6. What are possible causes of students performing poorly on a teacher-made test?
7. What are examples of appropriate uses of true/false questions in agricultural education assessment instruments?
8. What are the advantages and disadvantages of constructed-response questions?
9. What are examples of uses of student portfolios in agricultural education?
10. When is it appropriate for students to receive grades for SAE or FFA participation?

ACTIVITIES

1. Investigate the debate over the use of standardized testing in U.S. public schools. Prepare a written report on your findings. Be sure to include the pros and cons of standardized testing in your report.
2. Develop a table of specifications for a unit of instruction you might teach. From the table, develop the assessment instrument.
3. Using the *Journal of Agricultural Education, The Agricultural Education Magazine*, and the proceedings of the National Agricultural Education Research Conference, write a report on research about evaluating learning within agricultural education. Current and past issues of these journals can be found in the university library. Selected past issues can be found online at the American Association for Agricultural Education and the National Association of Agricultural Educators Web sites.
4. Explore an online assessment system. MYcaert is one integrated online system in agricultural education that includes an assessment tool. For additional information, go to the MYcaert Web site. Prepare a report on your findings.

CHAPTER BIBLIOGRAPHY

Ahmann, J. S., and M. D. Glock. (1981). *Evaluating Student Progress: Principles of Tests and Measurements* (6th ed.). Needham Heights, MA: Allyn & Bacon.

Binkley, H. R., and R. W. Tulloch. (1981). *Teaching Vocational Agriculture/Agribusiness*. Danville, IL: The Interstate Printers & Publishers, Inc.

Center for Agricultural and Environmental Research and Training (CAERT), Inc. (2003). *Indiana AgriScience Lesson Plan Library* [CD-ROM]. Danville, IL: Author.

Cole, D. J., C. W. Ryan, and F. Kick. (1995). *Portfolios Across the Curriculum and Beyond*. Thousand Oaks, CA: Corwin Press.

Enger, S. K., and R. E. Yager. (2001). *Assessing Student Understanding in Science: A Standards-Based K–12 Handbook*. Thousand Oaks, CA: Corwin Press.

Finch, C. R., and J. R. Crunkilton. (1999). *Curriculum Development in Vocational and Technical Education: Planning, Content, and Implementation* (5th ed.). Needham Heights, MA: Allyn & Bacon.

Joint Committee on Standards for Educational Evaluation. (2003). *The Student Evaluation Standards: How to Improve Evaluations of Students.* Thousand Oaks, CA: Corwin Press.

Kahler, A. A., B. Morgan, G. E. Holmes, and C. E. Bundy. (1985). *Methods in Adult Education* (4th ed.). Danville, IL: The Interstate Printers & Publishers, Inc.

National FFA Organization. (2003). *Local Program Resource Guide: 2003–2004* [CD-ROM]. Indianapolis: Author.

National FFA Organization. (2011a). *Local Program Success* (online). Accessed on November 30, 2011.

National FFA Organization. (2011b). *Official FFA Manual.* Indianapolis: Author.

Newcomb, L. H., J. D. McCracken, J. R. Warmbrod, and M. S. Whittington. (2004). *Methods of Teaching Agriculture* (3rd ed.). Upper Saddle River, NJ: Prentice Hall.

Office of Educational Research and Improvement. (1997). *Studies of Education Reform: Assessment of Student Performance.* Washington, DC: U.S. Government Printing Office.

Phipps, L. J., and E. W. Osborne. (1988). *Handbook on Agricultural Education in Public Schools* (5th ed.). Danville, IL: The Interstate Printers & Publishers, Inc.

Rolheiser, C., B. Bower, and L. Stevahn. (2000). *The Portfolio Organizer: Succeeding with Portfolios in Your Classroom.* Alexandria, VA: Association for Supervision and Curriculum Development.

Tanner, D. E. (2001). *Assessing Academic Achievement.* Needham Heights, MA: Allyn & Bacon.

Vacc, N. A., L. C. Loesch, and R. E. Lubik. (2001). *Writing Multiple-Choice Test Items.* (ERIC Document Reproduction Service No. ED457440).

20

Meeting the Needs
of Diverse Students

Welcome to Ms. Kim's fifth-period Natural Resources class. She has twenty-four students who are tenth through twelfth graders—fifteen females and nine males. The students are rather diverse in many ways. Dexter is going to be the valedictorian of his class. Alisha is the star hitter on the volleyball team and eleventh-grade class president. Jose is the first member of his family born in the United States, and Spanish is the language spoken at home. Ten of Ms. Kim's students have Individualized Education Plans (IEPs). She has one autistic student, one visually impaired, and one emotionally disturbed student who is prone to violence.

Today's lesson is on ecosystems, and Ms. Kim is excited and ready to get every student engaged in the topic. How do you think Ms. Kim has prepared? How would you prepare to teach this class?

objectives

This chapter explains the importance of understanding the diversity of the student population. It has the following objectives:

1. Discuss the meaning and importance of diversity

2. Explain multicultural education and describe useful approaches

3. Relate the meaning of prejudice and discrimination in education

4. Summarize the meaning and practice of the least-restrictive environment in the education of students with special needs

5. Identify approaches that are appropriate with academically gifted students

6. Explain considerations with students who have physical disabilities

7. Describe teaching variations with students who have mental or emotional disabilities

terms

cultural pluralism
disability
discrimination
diversity
handicap

learning disability
least restrictive environment
multicultural education
prejudice
stereotype

20–1. All students respond to meaningful, hands-on instruction. (Shutterstock © Goodluz)

DIVERSITY

Diversity is the variety of differences within a category or classification. For example, a school described as ethnically diverse may have within its student body African Americans, Hispanics, Hmong, Vietnamese, and people of other ethnic backgrounds.

Although ethnicity, gender, and socioeconomic status are the most discussed forms of diversity, others are critical for the agriculture teacher to bear in mind. Geography, family situation, religious and value beliefs, mental and physical abilities, age, and language also need to be considered.

20–2. Students learn best in a comfortable environment that provides needed learning resources. (Shutterstock © Monkey Business images)

Many people stereotype agricultural education students as white male farmers; however, this image is no longer true. Students whose families make 100 percent of their income from farming are a small percentage of agricultural education students. Today, agricultural education students are also from urban and suburban backgrounds, rural backgrounds not connected to production agriculture, and farms where most family income is nonfarm. Nationally, females make up about two fifths of agricultural education students. Greater than 75 percent of agricultural education students are white.

Just as students may be stereotyped, certain attributes may also be ascribed to teachers. An example is the notion that all agriculture teachers are white males. Both males and females teach agriculture. These individuals represent a wide range of ethnic, social, and cultural backgrounds. Diversity among teachers should be respected and appreciated just as it is among students.

Diversity brings unique talents, experiences, and learning styles to the agricultural education classroom. A teacher should use these differences as a basis for enhancing instruction. Some differences do present definite challenges to a teacher. These differences, however, provide the opportunity for the teacher to gain personal satisfaction from seeing students achieve and be successful in their schoolwork.

MULTICULTURAL EDUCATION AND APPROACHES

"*Multicultural education* incorporates the idea that all students—regardless of their gender and social class, and their ethnic, racial or cultural characteristics— should have an equal opportunity to learn in school" (Banks, 1989, p. 2). Concurrent with this is a belief by the teacher that all students can learn. Banks explained that multicultural education is an idea, an educational reform movement, and a process.

The ideology of multicultural education is that the purpose of schools is for all students to have the opportunity to learn. Unfortunately, some schools have been configured so that students with certain characteristics have a better opportunity to learn than others. This disparity requires a reform movement to change the educational system. Finally, multicultural education is a process that takes time, effort, and a willingness to change.

A major goal of multicultural education is the improvement of academic achievement for all students. Multicultural education strives for all students to acquire the reading, writing, and math skills necessary to compete in the global economy (Banks, 2008). A second goal is "to foster unity with diversity" (Banks, 2009, p. 10). All students should appreciate what is the common American identity while also being respected and able to express their unique cultural identity.

Sleeter and Grant (1994) described five approaches to multicultural education. These are as follows:

1. Teaching the exceptional and the culturally different
2. Human relations
3. Single-group studies
4. Multicultural education
5. Education that is multicultural and social reconstructionist

Single-group studies, such as women's studies, Jewish studies, or African American studies, are typically on the collegiate level and will not be discussed here. All the following approaches have valid teaching advice as well as potential problems. In all cases, the agriculture teacher is encouraged to treat all students as individuals who desire to learn and are capable of learning.

20–3. Realia promote student interest and learning.
(Courtesy, Education Images)

Teaching the Exceptional and the Culturally Different

This approach views the dominant American society as basically good. The purpose of education is to ensure that all students are capable of becoming productive citizens who fit into the culture. Differences should be viewed simply as differences, not as deficiencies. Teachers who follow this approach will take students where they are and work to build them up to where they need to be. For students who are English language learners, teachers may use learning materials in the students' first language until the students are able to learn in English. These teachers will try to structure their instruction to accommodate students of varying learning styles and cognitive abilities. They will use examples and visuals that students relate to and understand. A criticism of this approach is that it values the mainstream culture to the exclusion of all others.

Agriculture teachers following this approach may structure their classes in this way: They work with special education teachers to give students with special needs the extra help they need to succeed. Step-by-step instructions for projects, assignments, laboratories, and such are given in both oral and written form. Students who speak limited English may be paired with students who are bilingual. Not only do agriculture teachers screen visuals and materials to eliminate their being stereotypical, racist, or sexist, but they also make an effort to show females and people of color in agricultural careers.

Human Relations

The human relations approach aims to reduce stereotypes, facilitate positive relationships, and promote tolerance and acceptance. Whereas the first approach targeted those students needing assistance, the human relations approach targets all students. Teachers using this approach will combine students into varying work groups. These teachers will make sure that all students are involved in classroom activities and that everyone gets along. They will not tolerate name-calling or hateful speech. They will work to break down barriers between groups and to refute stereotypes.

Teachers using the human relations approach structure the classroom so that every student can succeed; that is, they make sure that the success of one student is not at the expense of another student. The human relations approach has received several criticisms. Some criticize it for emphasizing relationships at the expense of academics. Others see it as a simplistic answer to deep social problems. They see the human relations approach as merely concentrating on "Can't we all just like each other?" without addressing the issues of poverty, discrimination, and hate.

An agriculture teacher following this approach may structure his or her classes in this way: The first week of school, do activities that demonstrate how much in common the students have with each other. Posters around the room are motivational and value the uniqueness of individuals and groups. The FFA chapter has activities that are attractive to a variety of students and does not hold activities that exclude students. The chapter emphasizes community service and personal development of all members over competition or development of positional leaders only.

MULTICULTURAL EDUCATION

The multicultural education approach aims to change education and society such that equality of all and cultural pluralism are promoted. ***Cultural pluralism*** is the social philosophy that the culture of countries such as the United States is a product of the cultures of the various immigrant groups to the countries. This approach targets not only all students but also the school and the community.

In a pluralistic society, all groups are free to maintain and develop their own cultural systems. Teachers using this approach are proactive in eliminating equality concerns. They will advocate making all facilities and activities accessible to students and parents with physical disabilities. They will work to change disciplinary procedures that punish one group unfairly.

Teachers following this approach would support students wearing items of religious significance. In addition, they would make accommodations for students observing religious holidays that are not school break days. They would voice their concerns in favor of faculty and administrator demographics reflecting the gender, ethnic, and disability diversity of the community.

The multicultural education approach also has its criticisms. Some criticize it for enhancing divisiveness by encouraging differences. By emphasizing educational content from a multitude of cultures, students may miss out on needed content from the mainstream culture. Others argue that the multicultural education approach changes cognitive knowledge but not affective behaviors. If this is true, students would know more about other cultures yet still have the same attitudes, prejudices, and biases. Teachers can structure their classrooms to address these criticisms.

Agriculture teachers following this approach may structure their classes using fundamentals of multicultural education. They would evaluate their agriculture program to identify and eliminate any inequality and discrimination. They would actively recruit students from minority groups and advocate that new faculty hires reflect diversity. Teachers would also encourage students to celebrate their differences and to accept differences in others. Students would be taught to analyze content critically for biases and work to eliminate any found. Instruction would be student-centered, addressing all learning styles. Students would be taught about the agricultural accomplishments of people from minority groups. As an example, not only would students learn the history of New Farmers of America, the African American student organization until 1965, but they would also explore the aspects of the NFA/FFA merger that were detrimental to African American teachers and students. (More about the NFA is in Chapter 23.)

20–4. FFA participation should promote diversity. (Courtesy, National FFA Organization)

Multiculturalism and Social Reconstructionist Approach

The multicultural and social reconstructionist approach is similar to the multicultural education approach. It also aims to change education and society such that equality and pluralism are promoted. However, it goes a step further and advocates teaching students how to work toward changing society. Teachers using this approach organize their instruction around current social issues, such as racism, ageism, sexism, and classism. They explore oppression and power and guide students to work toward alleviating oppression and equalizing power. Students are a part of the democratic decision-making process of the classroom. Diversity is not only accepted but expected.

An agriculture teacher following this approach may structure his or her class in this way: A unit on agricultural lending may be taught from the perspective of African American farmers who were discriminated against. After exploring the issues of discrimination, power, social class, and poverty, the agriculture teacher would lead the class to analyze lending practices in the community. The students would then actively work to reverse any inequities. Another unit might look at fruit and vegetable production from the perspective of the migrant worker. The same procedure would be used as described above. A third example would be exploring the importation of cut flowers. Students would look at issues such as environmental impact and worker exploitation in the exporting country.

PREJUDICE AND DISCRIMINATION

Prejudice is a preference or idea formed without having complete information. It is closely related to a stereotype. A ***stereotype*** is a conception resulting from the assignment of oversimplified characteristics to a whole group.

Everyone has prejudices. A person who does not eat vegetables because of a dislike for the taste of broccoli is prejudiced against vegetables. They have taken a stereotype of broccoli and applied it to all vegetables. Others are prejudiced against certain colors

20–5. Animals in school labs provide opportunities for students to develop and meet psychological needs.
(Shutterstock © alexnika)

on automobiles because "they show dust too easily." These prejudices may be harmless; however, prejudices against people tend to harm human relationships and serve to drive groups apart. Prejudices against people or groups tend to be irrational and hurtful. Everyone deserves to be treated as an individual worthy of respect and dignity. This is not possible if prejudices are maintained. Stereotypes keep people from learning about others as individuals and can lead to irrational judgments and actions based on generalizations.

Discrimination is the act of treating one group differently than another group. Prejudices many times lead to discriminatory practices. For example, a male superintendent may be prejudiced, believing that all women are weak disciplinarians and indecisive in their decision making. This could lead him to discriminate against qualified women applicants for administrative roles in the school system.

Most forms of discrimination are illegal in the school system. Title IX of the 1972 Education Amendments prohibits discrimination in educational programs or activities on the basis of gender. The Civil Rights Act of 1964 prohibits discrimination in voting, public accommodations, schools, and employment on the basis of race, color, religion, gender, or national origin. The No Child Left Behind Act of 2001 upholds the right to a public education for children of migrant workers. Public Law 94-142, the Education for All Handicapped Children Act of 1975, requires a free, appropriate public education for all handicapped children. This law and the use of the word handicapped are discussed in detail later in this chapter.

Over the years, agricultural education has worked to eliminate discriminatory policies. In 1965, during the time of the Civil Rights Act, New Farmers of America (NFA) and Future Farmers of America (now FFA) were merged. This brought the African American student organization, NFA, and the predominately white student organization, FFA, together under the same umbrella. Over the years, females had enrolled in small percentages in agricultural education. The opening of national FFA membership to females in 1969 greatly expanded the number of females enrolling in agricultural education. Beginning in earnest in the 1970s and continuing today, students with disabilities and students with educational handicaps have been a part of agricultural education classrooms.

STUDENTS WITH SPECIAL NEEDS

Students with certain characteristics have been identified as needing special instruction within the school system. In 1975, PL 94-142, the Education for All Handicapped Children Act, was passed. PL 94-142 is also known as the Individuals with Disabilities

Education Act, or IDEA. This act had far-reaching implications for education. It required that students with handicaps be educated in the *least restrictive environment* or the closest-to-normal educational setting, meaning that students with handicaps and students without would be in the same classrooms. Students had to be evaluated before they could be placed in special education programs. The law also required that special education students have Individualized Education Plans (IEPs). PL 94-142 greatly opened the doors of education to students with handicaps and the channels of communication with their parents.

The 1997 IDEA was designed to enhance and expand the original 1975 act. It raises academic expectations for students with disabilities, increases parental involvement in the educational process, involves regular classroom teachers more in the planning process, includes all students in assessment and public reports, and supports greater professional development (USDE, n.d.). The Individuals with Disabilities Education Improvement Act (IDEIA) of 2004 took effect on July 1, 2005, and more closely aligns the requirements of IDEA with the No Child Left Behind Act of 2001.

Changes in the IEP process will have much impact on agriculture teachers. Agriculture teachers will have greater IEP input and will be expected to participate more fully in the IEP development process. They may have more students with disabilities in their classrooms because the financial incentive for placing these students in special classrooms was removed. However, agriculture teachers should have greater access to support services to assist them in teaching students with disabilities. Finally, agriculture teachers may have greater interaction with parents of students with disabilities than ever before.

The 1997 IDEA also changed the terminology from "handicapped children" to "children with disabilities." This is a subtle change but an important one. In changing the order of the words the focus changes to the child rather than the condition. According to Woolfolk (1998), a *disability* is the inability to do something specific, such as hear or see. A *handicap* is a disadvantage under certain situations. For example, being paraplegic is a disability for movement of the legs, but is a handicap only if the situation requires walking. However, it is not a handicap for tasks that require speaking or writing. The modification of using a wheelchair can eliminate the handicap from many situations.

The IDEIA of 2004 listed thirteen categories of disability for which children who have one or more need special education and related services. The categories are preschool disabled, mental retardation, hearing impairment, speech or language impairment, visual impairment, deaf-blindness, emotional disturbance, orthopedic impairment, autism, traumatic brain injury, other health impairment, specific learning disability, and multiple disabilities.

STUDENTS WHO ARE GIFTED

Students are classified by schools as gifted if they show above-average intelligence (typically defined by IQ) and/or show superior talent in areas such as music, art, or mathematics (Santrock, 2008). They may demonstrate above-average intellectual ability, creativity, and/or motivation to achieve (Woolfolk, 1998). Although completing class assignments may come easily to them, that is not a defining characteristic of students who are gifted. Instead, students who are gifted tend to master an area earlier than their peers, learn in ways different than their peers, and have a high internal motivation to excel in the area for which they are gifted (Santrock, 2008). Educational interventions for these students include special classes or programs, accelerated classes, alternative assignments, and outside-of-school opportunities.

Agriculture teachers can provide challenges and opportunities to students who are gifted. With any agricultural education course that is project-based, these students can choose a more demanding project—one that requires greater abstract thinking and creativity. The possibilities for independent study also exist. Many universities

20–6. Students bring a wide range of backgrounds into public schools.
(Shutterstock © Omer N Raja)

now offer college-level agriculture courses via distance education. Also, with the close working relationship that agriculture teachers have with Cooperative Extension educators and their state's agricultural university(ies), students who are gifted could be involved in field-based agricultural research projects.

FFA and SAE also provide opportunities to students who are gifted. The forty-nine proficiency award areas allow students to specialize in their areas of interest and challenge themselves to excel. The Agriscience Student Recognition and Scholarship Program provides opportunity for students to conduct research projects. The entrepreneurship aspects of agricultural education can be attractive to students who are gifted, as entrepreneurship allows students to exercise fully their creativity, abilities, and motivation to achieve.

STUDENTS WITH PHYSICAL DISABILITIES

The mainstreaming requirements of PL 94-142, IDEA, IDEIA, and the Carl D. Perkins Act and reauthorizations have led to more students with physical disabilities being in agricultural education courses. The tasks of the agricultural education classroom necessitate modifications for students with physical disabilities. Agriculture teachers also need to be aware that socialization may also be a need of students with physical disabilities.

Students who use orthopedic devices, such as wheelchairs or crutches, need access to buildings and facilities. Ramps may need to be added for classrooms or greenhouses. Outdoor laboratories and natural resources trails may need hardened paths, bridges over wet areas, and signage at a readable height. Agricultural mechanics equipment may need to be height-adjusted to allow students in wheelchairs to fit under worktables and to use tools.

Communications disabilities include hearing, vision, and speech impairments. Students with communications disabilities require classroom modifications also.

20–7. All students can be active participants in agricultural education. (Courtesy, National FFA Organization)

Many of these modifications are good teaching methods for all students. Table 20–1 outlines recommended modifications.

STUDENTS WITH MENTAL OR EMOTIONAL DISABILITIES

Students with mental disabilities have received their education by various means over the years. Before PL 94-142, the majority of these students, especially those with severe mental disabilities, were placed in special public or private facilities. Often these facilities provided inadequate educational and social training.

Table 20–1
TEACHING MODIFICATIONS FOR STUDENTS WITH COMMUNICATIONS DISABILITIES

Speak clearly.

Speak to the class, not to the chalkboard/whiteboard. Keep your mouth uncovered so lip readers can see your mouth movements.

Write plainly and in large letters when using the chalkboard/whiteboard or overhead projector.

If necessary, provide printed copies of notes for students with visual impairments. The school may provide a special magnification reader.

Teach other students to be understanding, accepting, and helpful. Do not tolerate belittling or hurtful comments.

If a task involves multiple steps, give instructions in both written and oral form.

Beginning in the 1970s and accelerating during the 1980s, students with mental disabilities were educated in public schools but in self-contained special education classrooms. This still did not provide adequate socialization and "real-world" experiences. Gradually these students were mainstreamed into more and more classes. This process was accelerated by IDEA 1997. Each successive rewrite of the Carl Perkins Act encouraged greater mainstreaming of students with disabilities into career and technical education programs, including agricultural education.

Intellectual Disability

Intellectual disability, sometimes still referred to in legislation and other documents as mental retardation, is defined as considerable limitations in both intellectual functioning and adaptive behavior (American Association on Intellectual and Developmental Disabilities, 2012). Limitations in adaptive behavior may include the inability to communicate, take care of self, gain employment, socialize with others, or provide for personal health and safety. Intellectual disability presents itself before the age of 18. Many school systems divide stages into mild, moderate, severe, and profound.

Students with intellectual disability need varying levels of support (Santrock, 2008). The lowest level is intermittent, in which support is needed only at intervals. The next level is limited, in which support is needed on a more consistent basis over time. The third level is extensive, in which regular or day-to-day support is needed. The highest level is pervasive, in which high-intensive support is needed in all life situations, including those that are life-sustaining.

An agriculture teacher is likely to have students requiring intermittent and limited support. These students will probably be below grade level in reading, mathematics, and general knowledge. They may also have difficulty in reasoning and understanding abstract concepts. Table 20–2 provides guidelines for teaching students with intellectual disability.

Table 20–2
TEACHING MODIFICATIONS FOR STUDENTS WITH INTELLECTUAL DISABILITY

Assess readiness. However little a student may know, he or she is ready to learn the next step. State and present objectives simply. This helps students know what they are to do and learn.

Base specific learning objectives on an analysis of a student's learning strengths and weaknesses. Present material in small, logical steps. Practice extensively before going on to the next step. Work on practical skills and concepts based on the demands of adult life.

Include all steps. Students with average intelligence can form conceptual bridges from one step to the next, but students with intellectual disability need every step and bridge made explicit. Make connections for the students. Do not expect them to "see" the connections.

Be prepared to present the same idea in many different forms.

Go back to a simpler level if you see a student is not following.

Be especially careful to motivate students and maintain attention.

Use materials that are appropriate for the students. A junior high student may need the low vocabulary of a Dr. Seuss book but will be insulted by the age of the characters and the content of the story.

Teach for success. Focus on a few target behaviors or skills so you and the students have a chance to experience success. Everyone needs positive reinforcement.

Be aware that these students must overlearn, repeat, and practice more than students of average intelligence. They must be taught how to study, and they must frequently review and practice their newly acquired skills in different settings.

Pay close attention to social relations. Simply including students with intellectual disability in a regular class will not guarantee that they will be accepted or that they will make and keep friends.

Source: Adapted from Woolfolk (1998).

Learning Disabilities

More than one third of the students with disabilities served under IDEIA are classified as having learning disabilities (National Center for Education Statistics, 2012). Of public school kindergarten through twelfth grade students, 5 percent were identified as having a specific learning disability. Although difficult to define, a *learning disability* is typically considered to be the failure or incapacity to function at the age-appropriate level in language, reading, spelling, mathematics, and/or other areas of learning. Most educators agree that learning disabilities include a heterogeneous group of disorders affecting students of average intelligence who have significant academic problems and perform well below what is expected.

Students with learning disabilities typically have their academic problems in a specific area, such as reading, writing, listening, mathematics, or reasoning, and perform at average or above-average levels in other areas. Students with learning disabilities need help overcoming frustrations, fear of failure, and other educational difficulties associated with their disabilities. A simple and first approach is to observe attendance patterns and study habits. Students who miss class frequently or have poor study habits tend to fall behind their peers and become discouraged. Changing these patterns and habits may be the only modification some students need. Other students will need additional support from special needs teachers in their area of learning disability—mathematics or reading comprehension, for example.

How does an agriculture teacher help students with learning disabilities? First, he or she should work closely with the special needs teachers. These experts will have suggestions and modifications for the agriculture teacher to use. Next, the agriculture teacher should observe the students for signs, such as frequent absences, frustration, or giving up. When these signs are present, the agriculture teacher needs to work with the student and the special needs teacher to remedy the situation. Finally, the agriculture teacher must use good, individualized teaching methods. What helps one student to be successful will be different from what helps another student.

Emotional or Behavioral Disabilities

Students with emotional or behavioral disabilities present problems in the classroom for both themselves and their peers. For a student to be classified with a behavioral disability, the behavior must be serious, persistent, and age-inappropriate (Eggen & Kauchak, 1997). Students with behavioral disabilities have trouble with self-control of behaviors and self-management of emotions.

Hallahan and Kauffman (1994) described two categories of behavioral disabilities. The first category is those behaviors that are external. These include aggression, uncooperativeness, and cruelty. Students with behavioral disabilities of this type may also have academic problems and are at high risk to drop out of school. The second category is those behaviors that are internal. Students with behavioral disabilities of this type are withdrawn or may suffer from anxiety or depression. They tend not to have social friends and may be quiet or shy in class.

Students with behavioral disabilities need positive reinforcement of their successes and accomplishments. Students with internal characteristics also need a teacher who will engage them as individuals and respect them without judging them. Students with external characteristics need firm and consistent rules with consequences that are spelled out. Students with behavioral disabilities may need a part of the classroom they can go to as a "gain control" spot. This allows them to remove themselves from the situation and defuse any escalating emotions or conflicts. The wise saying of "praise in public, discipline in private" is extra important for these students.

REVIEWING SUMMARY

The agricultural education classroom serves a diverse body of students. All students can benefit from the agricultural education program. It is the role of the teacher to structure the program to best meet student needs.

Gender, ethnic, language, cultural, socioeconomic, and geographic diversity exist within the agricultural education classroom. The agriculture teacher is encouraged to treat all students as individuals who desire to learn and are capable of learning. There are five approaches to multicultural education; however, they are not mutually exclusive. Awareness of the need for multicultural education and a desire to meet the needs of all students as individuals are key to educating students in today's diverse classroom.

Prejudices and stereotypes are harmful to teacher–student relationships as well as to learning. Teachers should work to identify and eliminate prejudices and stereotypes they hold. Discrimination is illegal and has no place in the agricultural education program. Policies and procedures should be put in place to ensure that all students have equal access to the agricultural education program.

More and more, students with special needs are included in the agricultural education classroom. Agriculture teachers should work with special needs teachers to develop appropriate modifications for these students. Agriculture teachers should be continually updating their knowledge of students with special needs and of recommended practices for facilitating their learning.

Students with learning disabilities are the largest group of students with disabilities. These students typically have academic problems in a specific area but perform at average or above-average levels in other areas. Agriculture teachers are encouraged to work individually with their students who have learning disabilities. What works for one student may not for another.

QUESTIONS FOR REVIEW AND DISCUSSION

1. What is multicultural education?
2. What are examples of prejudice and discrimination within agricultural education from the past?
3. Which of the five approaches to multicultural education best describes agricultural education within your state?
4. What is the ethnic, gender, and geographic (farm, rural, suburban, urban) diversity of agricultural education within your state?
5. What are examples of program changes and actions agriculture teachers could make to ensure gender, ethnic, and socioeconomic diversity and respect within their classroom?
6. What impacts have the 1975, 1997, and 2004 Individuals with Disabilities Education Acts had on agricultural education?
7. What is the difference between a disability and a handicap?
8. What are examples of modifications agriculture teachers could make for students who are gifted in their classes?
9. What are examples of modifications agriculture teachers could make for students with physical, mental, or emotional disabilities in their classes?
10. What are examples of modifications agriculture teachers could make for students with learning disabilities in their classes?

ACTIVITIES

1. Investigate students with special needs and agricultural education. Topics to cover include modifications in FFA and SAE. Are there national resource materials to assist students with special needs in agricultural education? Possible references include the National FFA Organization at its Web site, your state's teacher education institution(s), your state's agricultural education staff, and your state's agriculture teachers' professional association.
2. Investigate current federal legislation regarding students with special needs. You will want to include the latest Perkins Act, Individuals with Disabilities Education Act, Elementary and Secondary Education Act, and Higher Education Act, among others.
3. Using the *Journal of Agricultural Education, The Agricultural Education Magazine,* and the proceedings of the National Agricultural Education Research Conference, write a report on research about diversity in agricultural education. Current and past issues of these journals can be found in the university library. Selected past issues can be found online at the American Association for Agricultural Education and the National Association of Agricultural Educators Web site.

CHAPTER BIBLIOGRAPHY

American Association on Intellectual and Developmental Disabilities. (2012). *Definition of Intellectual Disability.* Accessed on April 20, 2012.

Association for Career and Technical Education (ACTE). (2012). *ACTE Web Site.* Accessed on November 5, 2012.

Banks, J. A. (1989). Multicultural education: Characteristics and goals. In J. A. Banks and C. A. McGee Banks (Eds.), *Multicultural Education: Issues and Perspectives* (pp. 1–26). Needham Heights, MA: Allyn & Bacon.

Banks, J. A. (2008). *An Introduction to Multicultural Education* (4th ed.). Boston: Pearson Education.

Banks, J. A. (2009). *Teaching Strategies for Ethnic Studies* (8th ed.). Boston: Pearson Education.

Banks, J. A., and C. A. McGee Banks (Eds.). (1989). *Multicultural Education: Issues and Perspectives.* Needham Heights, MA: Allyn & Bacon.

Banks, J. A., and C. A. McGee Banks (Eds.). (1995). *Handbook of Research on Multicultural Education.* New York: Macmillan.

Binkley, H. R., and R. W. Tulloch. (1981). *Teaching Vocational Agriculture/Agribusiness.* Danville, IL: The Interstate Printers & Publishers, Inc.

Colangelo, N., D. Dustin, and C. H. Foxley (Eds.). (1985). *Multicultural Nonsexist Education: A Human Relations Approach* (2nd ed.). Dubuque, IA: Kendall/Hunt Publishing Company.

Eggen, P., and D. Kauchak. (1997). *Educational Psychology: Windows on Classrooms.* Upper Saddle River, NJ: Prentice Hall.

Hallahan, D., and J. Kauffman. (1994). *Exceptional Children* (6th ed.). Needham Heights, MA: Allyn & Bacon.

Hettinger, J. (1999). The new Perkins . . . finally. *Techniques, 74*(1), 40–42.

Jackson, J. S. (Ed.). (1991). *Life in Black America.* Newbury Park, CA: Sage.

Kahler, A. A., B. Morgan, G. E. Holmes, and C. E. Bundy. (1985). *Methods in Adult Education* (4th ed.). Danville, IL: The Interstate Printers & Publishers, Inc.

McIntyre, A. (1997). *Making Meaning of Whiteness: Exploring Racial Identity with White Teachers.* Albany: State University of New York Press.

National Center for Education Statistics. (2012). *Fast Facts: Students with Disabilities.* Accessed on November 12, 2012.

National FFA Organization. (2011). *Official FFA Manual.* Indianapolis: Author.

National FFA Organization. (2012). *Local Program Success* (online). Accessed on November 12, 2012.

Phipps, L. J., and E. W. Osborne. (1988). *Handbook on Agricultural Education in Public Schools* (5th ed.). Danville, IL: The Interstate Printers & Publishers, Inc.

Santrock, J. W. (2008). *Educational Psychology* (3rd ed.). New York: McGraw-Hill.

Sleeter, C. E., and C. A. Grant. (1994). *Making Choices for Multicultural Education: Five Approaches to Race, Class, and Gender* (2nd ed.). New York: Macmillan.

U.S. Department of Education (USDE). (n.d.). *Building the Legacy: IDEA 2004 Resources.* Retrieved November 12, 2012.

U.S. Department of Education (USDE). (2003). *IDEA '97: General Information.* Retrieved November 12, 2012.

U.S. Department of Education (USDE). (2007). *Executive Summary OSERS 23rd Annual Report to Congress on the Implementation of the IDEA.* Retrieved September 12, 2012.

VocEd. (1985). The Carl D. Perkins Vocational Education Act of 1984 (Entire issue). *60*(4).

Woolfolk, A. E. (1998). *Educational Psychology* (7th ed.). Needham Heights, MA: Allyn & Bacon.

21
Using Laboratories

The Chestnut Grove High School agriculture department has a variety of laboratory facilities. An agri-biology laboratory has state-of-the-art equipment. A greenhouse is used for propagating and growing selected ornamental plants. A land laboratory is used by the agriculture and science departments and by the nearby elementary school. An animal sciences laboratory has Japanese quail, aquaculture tanks, and small animal cages.

Agricultural educators often talk about hands-on learning. We know that laboratory facilities should support the curriculum and promote mastery learning. Efficiently using laboratory facilities requires planning and instructional skill. How do agriculture teachers make the best use of laboratories?

objectives

This chapter explains the use of agricultural education laboratories to enhance student learning. It has the following objectives:

1. Identify the kinds and uses of laboratories in agricultural education

2. Discuss practices in laboratory management

3. Discuss how to develop laboratory activities

4. List and discuss safety considerations in laboratory teaching

terms

corrosive material
flammable material
grouping
laboratory
laboratory activity
laboratory instruction
laboratory superintendent
learning center

limiting factor
OSHA
oxidizer
protective clothing
reactive
readiness level
toxic material

21–1. The agricultural education laboratory provides a variety of hands-on activities such as small animal grooming. (Courtesy, Education Images)

LABORATORIES AND INSTRUCTION

Agricultural education programs may have several kinds of laboratory facilities. A **laboratory** is a space for individual or group student experiments, projects, or practice. In agricultural education, the activities may be in the areas of agriscience, agricultural mechanics, horticulture, plant and soil science, animal science, and natural resources.

Laboratories in agricultural education vary according to community, school, and state. They range from multipurpose large rooms, to single-purpose stations within a room, to outdoor land, pond, or greenhouse locations. Chapter 10 covered the space, equipment, and storage requirements for laboratory facilities. This chapter covers how best to use these facilities.

Laboratory instruction is organized instruction occurring in laboratories that, compared with classroom instruction, provides for greater freedom of movement for students, focuses on psychomotor skill development, uses special tools and equipment, and uses student self-directed learning (Phipps & Osborne, 1988). Laboratory instruction allows students to inquire into the content. The principles and concepts learned through classroom instruction are tested through application and hands-on learning. Students may conduct experiments, practice skills, and simulate real-world experiences.

A **laboratory activity** is an application of concepts and principles. The definition of a laboratory activity includes experiments, hypothesis testing, demonstrations, learning exercises, application exercises, and skills practice. For example, a lab in an agricultural mechanics laboratory may be wiring a circuit board with a single-pole light switch. A lab in an agriscience laboratory may be testing the effects of

21–2. The animal facility is one of the laboratories at Casa Roble High School in California.
(Courtesy, Education Images)

temperature on plant growth rates. A lab in a natural resources laboratory may be using a Biltmore stick to determine volume of standing trees.

Using Laboratories in Teaching Agricultural Education

The factors used in determining when and how to use agricultural education laboratories include curriculum, student, facilities, and teacher. Each of these categories will be discussed in more detail in the following sections.

Curriculum   The curriculum is a major determining factor in using agricultural education laboratories. As an example, a curriculum with a strong emphasis on agriscience will include instruction on plants. Instructional areas may include cells, plant functions, propagation, and various biotechnology processes. Following brain-based learning theory, discussed in Chapter 12, students might conduct a learning exercise to see the wholeness of a concept, such as how plants take up and transport water. They would then receive classroom instruction to learn the content and principles of plant water transportation. The instruction would end by using the laboratory for experiments on plant water transportation.

21–3. This student is extracting DNA from strawberries in an agriscience laboratory.
(Courtesy, Education Images)

21–4. An increment borer is being used to determine the age of a tree in a land laboratory.

(Courtesy, Education Images)

Another curricular factor is the degree of learning and the purpose of learning. In the example above, the exercises and experiments could be conducted in a classroom with tables and borrowed equipment. However, if the curricular focus is career-related, aimed at gaining the knowledge, skills, and dispositions required for employment in biotechnology fields, then a fully equipped agriscience laboratory is needed. A career-related focus would also require that more time be spent in the laboratory than an awareness-related focus. Students expected to gain competence in an area need additional practice in the skills connected with that area.

Student   The maturity and developmental levels of the students must be taken into account when deciding how to use laboratories, as was covered in Chapter 17.

Middle school students may uses laboratories frequently; however, their uses will probably be short in time and varied in focus. Laboratories provide this age group with concrete examples of abstract concepts. Activity is also involved, which can enhance learning.

High school students in advanced classes, on the other hand, may use laboratories a significant amount of time and focus on specific skills. Laboratories can be used to extend instruction into real-world settings using equipment found in the workplace. Although it is tempting to assign projects to advanced classes and use the laboratories all the time, this practice should be avoided. Students need to learn the "what" and the "why" of the content, as well as the "how." Problems and situations will arise that will require additional instruction.

The agriculture teacher must decide the ***readiness level*** of students for laboratory instruction. Do the students possess the requisite background knowledge in English, mathematics, or science? Has the requisite background knowledge in agriculture been learned to sufficient depth? Are the students physically capable of performing the required tasks? Are the students emotionally, mentally, and behaviorally ready? These are important questions for which the answers may be different among class periods during the same school year and between school years.

Facilities   Facilities are another factor in determining laboratory use. The first consideration is inventorying what facilities are available for the agriculture teacher to use. Obviously, in a single-teacher agriculture program, all laboratory facilities are

21–5. School animal laboratory facilities can be used by students for animal supervised experience.
(Courtesy, Education Images)

available for use. However, within a multiple-teacher program, a schedule must be developed for using the laboratories. There may be other laboratories within the school available for the agriculture teacher to use. Common shared facilities with science programs include small greenhouses, science laboratories, and outdoor nature laboratories. Common shared facilities with other career and technical education programs include mechanics laboratories, food science laboratories, and computer facilities. Computer and other mediated-instruction facilities are many times shared by the entire school and coordinated by the school's media specialist.

The condition of laboratory facilities will affect their usefulness. Facilities containing out-of-date or broken equipment lead to less than desirable laboratory instruction. In many cases, laboratories that were built for class sizes of twenty students now must accommodate class sizes of thirty or more. This strains both the space and the equipment. Students may not get the practice opportunities required for skill mastery.

In other instances, laboratory facilities built for one purpose can be converted to serve other purposes. Many aquaculture and hydroponics laboratories are housed in facilities once built for agricultural construction. Before converting facilities, the agriculture teacher should go through the program planning process described in Chapter 6. This will help the agriculture teacher determine whether the new use is the best for the program.

Teacher The comfort level of the teacher with agricultural laboratories influences how these laboratories are used. It is important that during teacher preparation students are exposed to and get experience with a variety of agricultural education laboratories. Some of this experience will occur in college classes, but much must be obtained by students in field experiences, in student teaching, and on their own initiative. Prospective agriculture teachers should reflect on their areas of need regarding laboratory instruction and develop plans for obtaining necessary knowledge and experiences.

21–6. A well-planned facility enhances student learning. (Courtesy, Education Images)

The care and maintenance of agricultural laboratories also affect the teacher. Does a laboratory require night, weekend, holiday, or summer care? If so, is this the responsibility of the teacher? The answer is yes, even if students do most of the work as part of SAE or other arrangements. The teacher has the ultimate responsibility for the health and well-being of the plants and animals in agricultural

21–7. A teacher individually provides instruction in preparing and using a slide for microscope observation. (Courtesy, Education Images)

education laboratories. Some teachers receive compensation or compensatory days for the outside-of-regular-hours work they do in connection with agricultural education laboratories.

MANAGING LABORATORIES

Managing agricultural laboratories requires the teacher to plan, organize, and be efficient. The teacher is in charge of assuring proper use and care of laboratories. The skills needed are in addition to the instructional skills of an agriculture teacher.

The agriculture teacher must purchase and maintain supplies, tools, and equipment for laboratory use. He or she must organize both the laboratory facilities and the instruction. Student behavior and safety are important laboratory considerations. Supplies, tools, and equipment; organization and cleanup; and student behavior are discussed in the following sections. Safety is discussed later in this chapter.

Supplies, Tools, and Equipment

Supplies, tools, and equipment of the proper type and amounts are critical for successful laboratory instruction. Among the duties of the teacher is to see that these are available when needed, correctly used, and properly stored when not in use.

Many agricultural education programs keep supplies, tools, and small equipment within an enclosed tool room. The advantages of this system include the agriculture teacher having greater control over item use, less chance for unauthorized use, and less laboratory space taken for item storage. Other programs keep these items in wall or floor cabinets within the laboratory space. The advantages of this system include having the items readily available in the areas where they are used, the ability to see quickly if items are missing, and the elimination of students grouping in a small storage room at the beginning and ending of the class period.

Regardless of which system is used, organization is a priority. Supplies must be ordered so they will be received by the time they are needed. Otherwise, instructional time is wasted waiting on materials. Supplies, such as chemicals, must be checked for viability, as their expiration dates may have passed since the last time the items were used. Instructions and material safety data sheets should be reviewed and filed for future reference. With the storage of all supplies, fire and health codes must be adhered to strictly.

Managing supplies also includes deciding what funds will be used to purchase them. An agricultural education program usually has a line item for supplies in its annual budget. Care must be taken to distribute the funds wisely across all courses and throughout the school year. Sometimes, a program operates on a fee system in which each agriculture course has a lab fee associated with it. Funds from this fee are used to purchase supplies only for that course.

In some cases, laboratories generate income that is used to purchase supplies. Examples include greenhouses, aquaculture ponds and tanks, and outdoor laboratories such as farm plots. Finally, some programs require students to bring in or purchase their own supplies. This is especially true of agricultural mechanics, particularly if students are doing individual projects.

Organization and Cleanup

Proper organization of laboratories and daily care are essential for efficient and safe use. Keeping a laboratory clean and neatly organized also reflects well in areas of public relations.

Learning Centers Many agricultural education laboratories do not have enough room or pieces of equipment for all students to do the same lab at the same time. For example, an agricultural mechanics laboratory may have only ten arc welding stations for a class of twenty-five students. Or a greenhouse may be too small for all students to be at the potting station at the same time. In these and other instances, a rotation schedule using learning centers is required.

A *learning center* is an area designed for one activity or lab and allows small groups of students to complete the activity or lab at the same time. A learning center is typically self-contained with all the equipment, tools, supplies, and instructions available in one place.

Learning centers allow for efficiency in both time and space. All students are engaged in active learning for the entire class period rather than having to wait for equipment to become available. Learning centers also require less space, as a smaller quantity of each item is needed. By working in groups at learning centers, students can observe and learn from peers.

Grouping, or placing students together to accomplish a common task or goal, is an important consideration in using learning centers. Students must receive prior instruction in the background content, concepts, and principles for all learning centers so the rotations will move efficiently. Because of this, the number of centers should be limited or designed around related content.

At certain points within the laboratory instruction, students may need additional classroom instruction or reteaching. At all times, the agriculture teacher should be alert to the need for instruction at a particular learning station. An example of learning centers used in an agriscience laboratory is given in Figure 21–8.

How should students be grouped? The first task is to design learning centers with activities that take approximately the same amount of time to complete. This reduces the chance for some students to get done early and become bored and for other students to feel rushed. The second task is to place students into the groups. If possible, groups should be of equal size so all students have equal opportunity to learn and participate. In most cases, heterogeneous rather than homogeneous grouping should be used according to student ability levels. This grouping strategy contributes to student achievement gains, reduces frustration and boredom, and provides the opportunity for peer teaching (Phipps, Osborne, Dyer, & Ball, 2008).

Cleanup Laboratories must be cleaned, with tools and supplies properly stored, every class period. The responsibility for this belongs to the students. Cleanup provides the opportunity to check the condition of tools and equipment as well as to restock supplies if needed. Students need instruction on why cleanliness and order are important. Several systems exist for assigning cleanup duties to students.

With laboratories that involve few tools and little debris, such as agriscience labs, students should be responsible for their own areas. Other laboratories, such as those for agricultural mechanics, have more tools used by multiple students and produce more debris. These require that the cleanup tasks be scheduled among the

Group 1	Group 2	Group 3	Group 4
Dicot seed dissection	Monocot seed dissection	Flower dissection	Stem dissection
Monocot seed dissection	Flower dissection	Stem dissection	Dicot seed dissection
Flower dissection	Stem dissection	Dicot seed dissection	Monocot seed dissection
Stem dissection	Dicot seed dissection	Monocot seed dissection	Flower dissection

21–8. An example of a student rotation schedule using four learning centers in an agriscience laboratory.

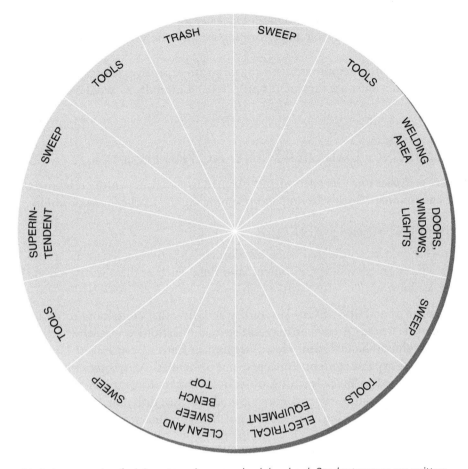

21–9. An example of a laboratory cleanup schedule wheel. Student names are written around the outside for each activity, or the students can be orally assigned to a group.

students. An example is shown in Figure 21–9. The number of students assigned to each task can be increased or decreased based on the number of students in the class and the particular requirements of the laboratory. The purpose of a cleanup schedule is to ensure that all tasks get accomplished and all items placed back where they belong.

One student can be designated ***laboratory superintendent*** to assist the teacher in making sure students do their tasks and to help out on tasks assigned to students who are absent. The laboratory superintendent can also check to make sure areas are clean, with tools and supplies properly put away. The agriculture teacher may assign one student the responsibility of tool manager. This person checks tools out and back into the tool storage room and checks the condition of tools when returned.

Student Behavior Issues

Laboratories involve special considerations in terms of student behavior. Inappropriate student behavior in a laboratory can cause accidents resulting in injury and harm. For this reason, great emphasis needs to be placed on students acting in a business-like, professional manner. Horseplay cannot be tolerated. Safety considerations are discussed in another section of this chapter. (Classroom management was covered in detail in Chapter 14.)

In laboratories, students must take greater responsibility for their own actions. Laboratory instruction tends to be more individualized, so most of the time the agriculture teacher is focusing attention on the instruction of other students. Students must be self-directed and capable of performing multiple steps without prompting or teacher guidance.

Laboratories often require students to work with expensive, sensitive equipment and tools. Students must use these items properly and carefully to avoid damaging or breaking them.

DEVELOPING LABORATORY ACTIVITIES

Laboratory activities must be selected and planned for efficiency in the teaching and learning environment. Such selection and planning are the teacher's responsibility.

Factors in Developing Activities

Five factors should be considered when developing lab activities. These are purpose, student readiness, methodology, resources, and time. Each of these is covered in the following sections.

- **Purpose of activity**—Every laboratory activity must contribute to efficient student learning. A teacher may use several questions to assess whether an activity should be included. How does it meet learning objectives? How does it teach or expand content, concepts, and principles? All lab activities should be structured to meet learning objectives and to accomplish educational purposes. Although students may view labs as fun, different from classroom instruction, or exciting, this should not be the primary reason for conducting a lab. A lab should teach a specific concept in itself or expand students' knowledge of the content.
- **Student readiness**—Efficient use of student and teacher time and school resources requires students to be at a stage of readiness. Do the students have the background knowledge, skills, and dispositions to conduct the lab? Most lab activities are conducted after students have received classroom instruction, although this is not always the case. Once students have learned the basic concepts and principles, they are better prepared to apply them in the laboratory. Another consideration is student maturity. Are the students physically capable of performing the gross and fine motor skills required by the lab? If not, continuing with the lab as planned may lead to student frustration and accidents.
- **Methodology**—Another factor to consider when developing a lab activity is lab methodology. What are the steps involved in the lab? Can the lab be finished in one class period, or will it involve several class periods? Does the teacher need to demonstrate, or can students successfully complete the activity without seeing it first? Does the lab involve a mastery of skills requiring repeated practice? Content in Chapter 13, "The Teaching Process," will be helpful in determining how best to structure the lab instruction.
- **Resources**—The proper resources must be available to carry out activities. These can be limiting factors. A *limiting factor* is a resource whose unavailability or time requirements may make conducting a lab impossible or require its modification. Each of the following should be considered carefully. If the agricultural education program does not have the resources, can resources be borrowed from other programs within the school? Would a field trip to a site with appropriate resources facilitate the laboratory instruction? Are there alternative labs that will yield equivalent student learning? Can some steps be done ahead of time to allow better utilization of class time during the lab activity?

- **Time**—As just mentioned, time is often a limiting factor in relation to resources. The time available for students to be engaged in a learning activity is important. School years and class periods are limited. Deciding whether an activity should be used requires consideration of what other learning may need to be left out of the teaching calendar. What is the best use of teaching and learning time? Is sufficient time available to carry an activity to a reasonable level of completion?

Planning Essentials

Laboratory instruction should follow a lesson plan just as should classroom instruction. The class period should be structured for maximum student learning and appropriate sequencing of activities. Because of the special characteristics of laboratory instruction, this type of lesson plan may be slightly different from one for classroom instruction. For example, a lab lesson plan should have notes reminding the teacher how far in advance to order and receive supplies, along with any special care instructions. A lab lesson plan may also contain group rotation and laboratory cleanup notes. Evaluations of student performance on laboratory activities are essential and must be included in the lesson plan. Finally, it is always recommended that the agriculture teacher actually do each lab before the students do it. This allows the teacher to see areas where questions will arise, problem areas, and areas where results may differ.

SAFETY CONSIDERATIONS IN LABORATORY TEACHING

Laboratory teaching requires attention to safety. All safety considerations must follow fire, health, and safety codes applicable to the school environment. Many of the safety standards are set by the federal Occupational Safety and Health Act, and managed through the Occupational Safety and Health Administration known also as **OSHA**. (Chapter 10 discussed safety in instructional environments.)

Safety rules, standards, and procedures are for the protection of both the student and the teacher. Not only must hearing and eye protection be provided, but its use must also be enforced. Students who do not use hearing and eye protection when it is required

21–10. Laboratory safety devices, such as this eye-wash station, need periodic checking.
(Courtesy, Education Images)

should not be allowed to continue in the laboratory activities. Eye protection provided must be of the appropriate type. The various types of welding, for example, require differing levels of tinting. As with all aspects of safety, the agriculture teacher must be a positive role model in the use of safety equipment. Always use appropriate (PPE) personal protective equipment when demonstrating or supervising laboratory activities.

Agricultural education laboratories may require **protective clothing**. Examples include gloves, headgear, chaps, and aprons. Students may need to bring in a set of work clothes or coveralls so their regular clothes do not get dirty or damaged. Footwear is also a concern. Sandals and tennis shoes do not provide the required protection for many agricultural education lab activities.

Students must also protect their respiratory functions. The appropriate dust mask should be worn when sanding, grinding, or spray painting. If students are involved in applying pesticides or other chemicals, special care should be taken to ensure that the appropriate respiratory protection is used.

HAZARDS IN LABORATORY SETTINGS

Several kinds of hazards may be present in an agricultural education laboratory. These vary with the nature of the instruction and the supplies or materials used. School boards may have regulations about the presence of some materials on the school grounds. Always know and abide by the regulations of the school board.

Corrosive Materials   A **corrosive material** is one that can cause chemical reactions that damage metal and can injure people. Corrosives are of three kinds: acids, bases, and other miscellaneous materials.

Some of the most corrosive acids in school laboratories are sulfuric acid, hydrochloric acid, acetic acid, nitric acid, and phosphoric acid. Common bases include sodium hydroxide (lye), potassium hydroxide, and aqueous ammonia. Miscellaneous materials include iodine, mercuric chloride, bromine, phenol, and ferric chloride. Besides being corrosive, materials containing mercury are toxic.

Fertilizers containing nitrogen, sulfur, and other materials can pose corrosion problems. Such fertilizers should be properly stored. Equipment for fertilizer application should be thoroughly cleaned after use.

Flammable Materials   A **flammable material** is a material that easily catches on fire and burns. Many factors, such as flash point and ignition temperature, should be taken into account with a flammable material. Flash point is the lowest temperature at which a chemical will give off vapors to form a flammable mixture in the air. Ignition temperature is the lowest temperature at which the vapor will ignite and continue to burn without heat from another source.

Gasoline, kerosene, diesel fuel, acetic acid, propane, and diethyl ether are examples of highly flammable materials in the agriscience and technology laboratory.

Organic solvents give off vapors that are flammable. Common examples of such solvents in the laboratory are acetone, benzene, ethanol, and toluene. Paints, varnishes, paint removers and thinners, and similar materials are often found in agricultural mechanics laboratories.

Toxic Materials   A **toxic material** is a substance that is poisonous to humans, other animals, and plants. Toxic materials are usually kept in containers marked with a skull and crossbones. Most pesticides are toxic. Not only should they be used properly, but they should also be stored properly.

It is a good idea to know how various substances can enter the body. Toxins can enter the body through inhalation. These irritate the nose and lungs and find their way into the bloodstream. Once in the bloodstream, they can cause injury in other parts of the body. Chemicals used as fumigants in greenhouses and other places can be particularly dangerous.

Some toxins are ingested. Most people would not think of purposefully eating a chemical; however, not washing their hands can lead to chemicals entering their bodies on food or in other ways.

Absorption occurs when chemicals enter the body through the skin. Rubber gloves, aprons, boots, and other protective devices should be worn to keep chemicals off the skin.

The effects of chemicals may occur immediately after exposure, or they may take a while to develop. They may range from mild nausea to cancer.

Teachers should be well aware of the school procedures regarding student exposure to toxic materials. The number of the nearest poison control center should be posted by the telephone.

Oxidizers and Reactives   Oxidizers and reactives are chemicals that can explode or react violently in other ways. An *oxidizer* is a material that can initiate and promote combustion. A *reactive* is a material that reacts with water or other materials. Many pyrophoric materials are reactives.

Some of the common fertilizers and chemicals used in agriculture are explosive. Ammonium nitrate fertilizer is an example. Other explosives include isopropyl ether, potassium nitrate, nitric acid, potassium chlorate, potassium, and sodium.

Some of these chemicals are very dangerous and become increasingly unstable with age. The agriculture teacher should obtain only those chemicals that will be needed and only in the amounts needed. Specialists in chemical disposal should be consulted when potential problems exist.

Electric Shock   Electric shock hazards are present in many agricultural education laboratories. Students and teachers may be exposed to these when using electricity to

21–11. Cleaning, storing, and inventorying is a regular responsibility in managing laboratories.
(Courtesy, Education Images)

heat, view, and prepare items in a laboratory. Only approved equipment should be used. Electrical service should be installed according to the electrical codes. Ground fault circuit interrupters (GFCIs) should be used if there is a possibility of water being in the work area.

Aquaculture laboratories often have a dangerous mix of water and electricity. All electrical work should be done by licensed electricians. All electrical devices should have GFCIs and be kept in good operating condition.

Learning activities in applied physical science may involve the study of electricity. Such activities should be carefully planned and monitored to ensure student and teacher safety. Whenever possible, low-voltage power sources should be used.

Safety with Specimens

Lab activities may involve animals, plants, and various microorganisms that pose potential hazards. Students and teachers should exercise safety. They should also assure the well-being of animals used in the activities.

Animals   Agricultural education courses may use live animals, animal tissues, and animal specimens for instructional purposes. Livestock may be kept in pens, barns, or pastures. Small animals may be kept in cages or pens. Fish may be kept in tanks or ponds. Preserved animal tissues and specimens may be purchased from supply houses.

With any use of animals, care needs to be taken for the safety and health of students. Live animals may injure students through kicks, bites, scratches, and other bodily harm. Students may also get diseases, parasites, or other contaminations from animals. Some students may be allergic to animal hairs or skin, and animal odors can affect students as well.

21–12. Land laboratories and school forests may have poisonous plants that can result in teacher and student injury. Carefully assess potential plant hazards and teach students to avoid them.

(Courtesy, Education Images)

Table 21–1
EXAMPLES OF TOXIC PLANT INJURIES

Symptom	Plants
Damage to internal organs (stomach, heart, liver, and kidneys)	Azalea, bird-of-paradise, castor bean, daffodil, foxglove, holly, hyacinth bulbs, mistletoe, nightshade, poinsettia, rhododendron, sweet pea, and wisteria
Skin rashes and blisters	Amaryllis; carnation; chrysanthemum; daffodil; geranium; iris; pencil cactus; poinsettia; poison ivy, oak, and sumac; tulip bulbs; and weeping fig
Swelling of mouth, upset stomach, and breathing difficulties	Calla lily, dieffenbachia, and philodendron

Note: Plant species listed here have been known to cause hazards to some individuals, companion animals, and other animals. Plants may be safe to some individuals and cause reactions with others. If an individual ingests any of these plant materials and appears ill, do not induce vomiting until instructed to do so by a poison control expert. If harmful plants come into contact with the skin, wash the area immediately with soap and water.

Source: Lee (2000).

All live animals must be properly fed, watered, housed, and maintained. Students should receive handling instructions before working with live animals and should wear proper clothing and protective gloves. If students are injured, they should report to the school nurse immediately. The school policies regarding reporting and caring for injuries should be followed.

Plants   Although plants cause fewer complications than animals, care should still be taken when working with them in the laboratory. Safety problems associated with plants include poisonous plants and allergic reactions. Poison ivy is the most common example of a plant that causes an allergic reaction. Table 21–1 gives examples of plants that are poisonous to humans.

Microorganisms   Bacteria, fungi, protozoa, and other microorganisms are often used in, or are present in, the laboratory. Some are kept or produced for useful purposes, such as in the biofilter of an aquaculture facility. Others are pests transported into the facility in some way, such as with the purchase of new plants for the school arboretum. Students should always follow appropriate procedures when working with microorganisms. These usually include wearing protective gloves and masks, washing hands thoroughly, and not having food or drink in the laboratory.

REVIEWING SUMMARY

Laboratory instruction in agricultural education is important for teaching applications of concepts and principles. In the laboratory, students gain needed skills and expanded knowledge regarding the content. In most cases, laboratory instruction involves the use of psychomotor skills, which are not used as often in classroom instruction.

Deciding when and how to use agricultural education laboratories involves looking at several factors. A major consideration is how a laboratory fits into the curriculum. The agriculture teacher must also decide to what extent students are ready mentally and physically for the laboratory instruction. Facilities also play a role in the decision process. A final factor is the teacher and his or her ability to manage the laboratory.

The management of laboratory space and supplies is an important responsibility. The agriculture teacher must decide management procedures, purchasing plans, organization, and care. As much as possible, the agriculture teacher should make students responsible for the maintenance of laboratory organization and cleanliness. Students should rotate among tasks and should be taught the reasons behind the tasks they are doing.

Safety is critical in agricultural laboratories. The agriculture teacher is responsible for following fire, health, and OSHA codes and regulations. He or she is also responsible for modeling proper safety habits and practices.

QUESTIONS FOR REVIEW AND DISCUSSION

1. Why are laboratories important in agricultural education instruction?
2. What are the factors for determining when and how to use agricultural education laboratories?
3. What are some differences in using laboratories with middle school and high school students?
4. How does the teacher influence laboratory activities?
5. What are different systems for storing supplies, tools, and small pieces of equipment?
6. What are the advantages and disadvantages of heterogeneous grouping of students according to ability levels?
7. Why are laboratory cleanliness and organization important?
8. Why is student behavior of special concern in laboratory situations?
9. What protective equipment and clothing might be necessary in an agriscience laboratory?
10. How are laboratory activities developed for effective instruction?

ACTIVITIES

1. Investigate the fire and health codes and requirements for agricultural education laboratories in your state. Begin by interviewing the safety coordinator or person in charge of safety in a local school district.
2. Choose an agricultural education laboratory facility, such as one for agriscience or agricultural mechanics. Design a learning centers rotation schedule. In addition, put together a list of supplies, tools, and equipment needed for the learning centers. Include quantity, cost, and potential suppliers.
3. Using the *Journal of Agricultural Education, The Agricultural Education Magazine,* and the proceedings of the National Agricultural Education Research Conference, write a report on research about laboratory instruction within agricultural education. Current and past issues of these journals can be found in the university library. Selected past issues can be found online at the American Association for Agricultural Education and the National Association of Agricultural Educators Web sites.

CHAPTER BIBLIOGRAPHY

Frick, M., and S. Stump. (1991). *Handbook for Program Planning in Indiana Agricultural Science and Business Programs.* Unpublished manuscript, Purdue University.

Indiana Department of Education. (1999). *Teacher/Local Team Self-study of Standards and Quality Indicators for Agriscience and Business Program Improvement.* Indianapolis: Author.

Lee, J. S. (2000). *Program Planning Guide for AgriScience and Technology Education* (2nd ed.). Upper Saddle River, NJ: Pearson Prentice Hall Interstate.

National Association of Agricultural Educators (NAAE). (2012). *NAAE Web Site.* Accessed on November 12, 2012.

National FFA Organization. (2012). *Local Program Success* (online). Accessed on November 12, 2012.

Phipps, L. J., and E. W. Osborne. (1988). *Handbook on Agricultural Education in Public Schools* (5th ed.). Danville, IL: The Interstate Printers & Publishers, Inc.

Phipps, L. J., E. W. Osborne, J. E. Dyer, and A. Ball. (2008). *Handbook on Agricultural Education in Public Schools* (6th ed.). Clifton Park, NY: Thomson Delmar Learning.

part four

Supervised Agricultural Experience, FFA, and Community Resources

22

Supervised Agricultural Experience

Jose, a freshman agriculture student, is learning about agricultural careers on the FFA.org Web site on his home computer. He is trying to decide on a career pathway in the field of agriculture.

Cheehlu, a sophomore agriculture student, is spending a Saturday following Ms. Matsumoto, the owner of Matsumoto Floral. Cheehlu would like to start her own floral shop when she finishes high school.

Asia, a junior agriculture student, is spending the evening working on completing the summary pages of his record book. Asia has a small flock of purebred Suffolk ewes and one ram. He plans to expand his sheep flock over the next few years and sell project lambs to 4-H and FFA members in his community.

Kelsie, a senior agriculture student, is devoting her weekend to writing a paper for her Agriscience Fair project. She is planning to attend college next year and will major in plant science. She plans to become a plant breeder and would eventually like to work for the USDA or a land-grant university.

What do these four students have in common? They are all participating in SAE activities that may eventually lead them into an agricultural career.

Experience is important to all of us. Employers want employees who have had experience. Learners like the notion of firsthand experience. Agricultural educators place emphasis on experience for students.

To experience something, an individual must be actively involved in sensing it or making it happen. Many students learn best by actually doing a particular activity or job. Learning by doing is often called

objectives

This chapter introduces supervised agricultural experience. It covers the importance and types of supervised agricultural experience and the teacher's role in supervising and evaluating supervised agricultural experience programs. The chapter also explains how supervised agricultural experience relates to the other components of the total agricultural education program. It has the following objectives:

1. Explain the meaning and importance of supervised agricultural experience
2. List and distinguish between the types of supervised agricultural experience
3. Describe how to develop supervised agricultural experience, including training plans and agreements
4. Discuss the role of supervision
5. Describe record keeping and the kinds of records kept
6. Explain how to evaluate supervised agricultural experience
7. Relate supervised agricultural experience to FFA achievement

terms

directed laboratory supervised agricultural experience
entrepreneurship/ownership supervised agricultural experience
experiential learning
exploratory supervised agricultural experience
FFA degree program
improvement project
job shadowing
National FFA Scholarship Program
paid placement supervised agricultural experience

proficiency award
research and experimentation supervised agricultural experience
Star Award
supervised agricultural experience
supervised agricultural experience program
supplementary projects (activities)
training agreement
training plan
training station
unpaid placement supervised agricultural experience

22–1. On-farm supervision and instruction of a student with equine supervised agricultural experience promotes proficiency development.

(Courtesy, Education Images)

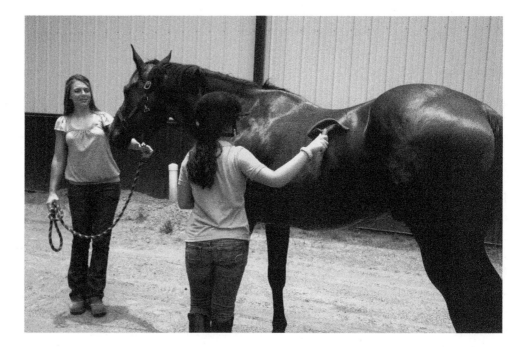

experiential learning. Learning through actual experience is usually fun, although students also learn from their mistakes.

Knowledge gained through experience is often easier to remember than that gained through memorization. For example, students can read about how to drive a car, but until they actually get behind the wheel and try to drive, they do not fully understand what is required to be a successful motor vehicle operator. The "behind the wheel" experience follows general instruction on how to operate a vehicle safely.

THE MEANING AND IMPORTANCE OF SUPERVISED AGRICULTURAL EXPERIENCE

In agricultural education, **supervised agricultural experience** (SAE) is the experiential learning component of the program, which includes the application of concepts and principles learned in the classroom to planned, real-life settings under the supervision of the agriculture teacher. SAE should improve agricultural awareness and/or the skills and abilities required for a student's career.

SAE is the part of agricultural education that provides students with the opportunity to actually apply and build on what they learn in the classroom. SAE activities range from growing plants and animals to using technology in conducting an in-depth study of a particular topic of interest to an individual student. More advanced SAEs include actual employment in a business or the creating and running of the students' own business operations. SAE is sometimes referred to as the work-based learning component of agricultural education.

The objective of SAE is to provide planned, goal-oriented, and practical activities that help students develop the skills needed to be successful in the workplace or in life. SAEs should be planned around the goals and career plans of the individual student.

22–2. Supervised agricultural experience provides the opportunity for students to develop hands-on skills.
(Courtesy, Education Images)

Teachers supervise the experiences to make them more meaningful and relate them to the classroom instruction. Teachers should involve and cooperate with parents and other adults to provide the most meaningful experiences for students. Older students may work in businesses and are supervised by their employers in cooperation with the teacher.

Students should be actively involved in deciding and planning the SAEs that are most interesting and important to them. Each student should list career goals before beginning to plan SAEs.

For detailed information, use the National FFA Organization Web site. The FFA Web site also contains information about SAE under the "Resources" section, which includes the *2000 SAE Best Practices Guide*. Resources available for purchase include (1) SAE Activities Wall Chart, (2) SAE Idea Cards, and (3) SAE Poster Series that can be ordered from the National FFA Organization. Another good source of information is the *Local Program Success Guide*, which can be accessed through its Web site.

Planned Programs

The **supervised agricultural experience program** (SAEP) is a series of individualized practical learning activities planned by the teacher, the student, the student's parents/guardians, and, if applicable, the employer. The program meets established minimum criteria, is supervised by a qualified agriculture teacher, and develops competencies related to the interests and career goals of the individual student. The things learned in the classroom are applied in real-life situations through SAEs. Carefully planned SAEs make in-class instruction more meaningful. SAE allows the agriculture teacher to expand the boundaries of the school classroom to include the whole community as an instructional facility. All the community's agriculturalists become potential resource persons. A number of incentives for supervised agricultural experience achievement are available to students. Many of these incentives provide FFA award recognition through contributions of National FFA Foundation sponsors.

SAE, like FFA, is designed to accomplish identified objectives of the agricultural education program and is conducted under the supervision of the agriculture teacher. It, too, is an integral part of agricultural education instruction and considered to be intracurricular. SAE adds to the instruction received during class time. It allows each student to enrich his or her education by focusing on the area or areas of greatest interest to the student.

Making decisions about the kind of SAE to focus on and then planning the activities of the SAE helps students learn about addressing real-life situations. Many students find it easier to remember classroom lessons when they apply the lessons though SAEs.

SAEs are important for increasing students' understanding of agriculture and for developing skills and abilities they need for careers. Through SAE programs, students learn to deal with problems and events that occur in everyday life. They often find they need to learn more about a particular topic and are directed to specific areas where they need additional study. Some students are motivated by the opportunity to earn money, some find SAE an enjoyable way to learn, and others gain personal satisfaction from their individual accomplishments.

Benefits of Supervised Agricultural Experience

SAE benefits students, teachers, employers, agricultural education programs, the local community, and the agricultural industry. Benefits for students are the most important. They include the following:

- Development of decision-making skills, including career and personal choices
- Improved self-confidence and human relations skills
- Application of knowledge learned in the classroom
- Knowledge of a variety of occupations and careers
- Development of time management and record-keeping skills
- Document of experience needed on job applications
- Discovery of areas of personal interest
- Practice of responsibility and development of independence
- Development of pride through personal accomplishment

How Supervised Agricultural Experience Came into Being

Since its inception in public secondary schools in 1917, agricultural education has included some form of experiential education as a teaching strategy. In the early days, when "vocational agriculture" students came from farms or ranches and were preparing to return there after completing their high school education, experiential learning usually consisted of production enterprises in livestock, poultry, crops, and so on, conducted on their home farms or ranches. The purpose of these "projects" was threefold: (1) to provide opportunity to develop through experience, supervised by the teacher, knowledge and skills required to conduct profitable production agriculture enterprises, (2) to demonstrate to the community modern agriculture practices, and (3) to help students (Future Farmers, as they were then called) to begin their actual establishment in farming.

Beginning with the 1950s, and especially after 1963, it became generally recognized that agriculture included more than farming. The U.S. Office of Education proposed classification of agriculture by occupational clusters. These were production agriculture (farming and ranching), agricultural supplies and services, agricultural mechanics, agricultural products and processing, ornamental horticulture, agricultural resources, and forestry.

22–3. Teachers often provide individual instruction and guidance with supervised agricultural experience.
(Courtesy, Education Images)

With this broadened concept of agriculture, agricultural education's mission was similarly expanded. It now had the task of preparing individuals for occupations in all seven occupational clusters, and the original threefold purpose of "agricultural education projects" was no longer valid.

Today's agricultural education programs offer a wide variety of SAEs from which students plan their individual SAE programs. The important concepts for modern SAE programs include (1) supervision by a qualified agriculture teacher, (2) quality practical experience that provides hands-on learning opportunities for the student, and (3) one or more separately identified activities planned and conducted around the student's personal interests and career goals.

Characteristics of Supervised Agricultural Experience

A modern SAE program has the following characteristics:

1. The activity is identified with or seeks to identify a specific agricultural enterprise, occupation, career, or problem and involves the student in hands-on experiences directly associated with an enterprise, occupation, career, or problem.
2. The student's involvement in this experience includes time outside of the school's usual agricultural education class hours.
3. Under some circumstances the student's experience may be located and/or conducted on the school premises.
4. The student plans the SAE with the assistance of an agriculture teacher and conducts it under the supervision of that teacher.
5. The agriculture teacher allocates a part of his or her time to the supervision of students' SAEs.
6. The student keeps records pertaining to his or her experience as prescribed by the teacher, and these records are periodically reviewed by the teacher.
7. The student may be engaged individually in SAE or be engaged cooperatively with other students.
8. The student's plan for the SAE includes goals and provisions for growth in scope and complexity.

Purposes of Supervised Agricultural Experience

Today, the agriculture teacher's main function is to serve as a manager, coordinator, or consultant of learning for guiding students as they seek careers in agriculture. Modern SAE programs can certainly lead to establishment in farming, but that is no longer the only goal. Purposes of today's SAE programs include the following:

- Providing opportunities for hands-on experience in skills and practices that lead to successful personal growth and employment in an agricultural career
- Providing opportunities to gain documented experience that can provide references for future education and/or agricultural employment
- Providing opportunities for students to identify, develop, and demonstrate personal characteristics required for successful personal growth and employment (e.g, initiative, responsibility, dependability, and self-reliance)
- Providing opportunities for students to observe and participate in the "world of work"
- Capturing, retaining, and focusing student interest in agriculture
- Providing opportunities for students to discover and deal with financial realities

For beginning agricultural education students, the selection of SAEs need not have a direct career goal relationship. Many of today's agricultural education students are seeking to establish their occupational goals. SAE for them can be exploratory in nature. SAE is beneficial in preparing a student for work even if it is not directly related to the job or jobs the student eventually attains.

TYPES OF SUPERVISED AGRICULTURAL EXPERIENCE

There are several types of SAEs that a student may incorporate into a SAE program. The SAE program might comprise a single type of SAE, or it might be a mixture of two or more types. The names and/or number of recognized SAE types are not consistent across the country and are likely different in your state. Recognized types of SAEs include the following:

1. Exploratory
2. Paid placement
3. Unpaid placement
4. Entrepreneurship/ownership
5. Directed laboratory
6. Research and experimentation
7. Improvement projects
8. Supplementary projects (activities)

Exploratory supervised agricultural experience is the type that provides an opportunity to investigate a number of areas in agriculture. The purpose is to help students understand and appreciate the field of agriculture and gain information to help them make decisions about their future education and careers. Beginning agricultural education students benefit the most from this type of SAE. One approach in providing exploratory SAE is job shadowing. *Job shadowing* is observing the work of an experienced person. The student spends time with an adult who is working in a specific occupation and observes the adult performing the work. The student may

22–4. Supervised agricultural experience requires careful attention to records. (Here is a teacher directing a student in keeping a daily log of activities with animals.) (Courtesy, Education Images)

spend one or more days with this individual and may also shadow several different workers. A major purpose of exploratory SAE is to expose students to the wide variety of agricultural career opportunities.

Paid placement supervised agricultural experience is the type in which the student is employed for compensation. A planned program of experiences is used. Placement may be on a farm or ranch, at an agricultural business, in a school laboratory, or at another community facility. The purpose is to provide practical experience and develop skills needed to enter and advance in a particular occupation. The location where a student is placed is called a *training station*. A student with this type of SAE may, for example, work as a floral designer, fish hatchery assistant, or farm supplies store clerk.

Unpaid placement supervised agricultural experience is the type in which the student works in a job for experience only and is not compensated in any other manner for hours of labor. Placement may be in the same or similar training stations as those used for paid placement. Younger students are often in unpaid placements and then later advance to paid placements as they gain experience in particular occupations. An example of this type of SAE is working without pay for a landscaper, farmer, or local agribusiness or in the school lab.

Entrepreneurship/ownership supervised agricultural experience is the type in which students develop skills needed to own and manage enterprises. The students own materials and other resources for their enterprises. These enterprises may be individually owned, or they may be partnerships, cooperatives, or enterprises involving other forms of group ownership. They may be conducted on school property or off. The key feature is that the students engaged in this type of

22–5. A lamb is part of this student's ownership SAE as well as an FFA showing activity.

(Courtesy, Education Images)

SAE have financial investment or risk in their enterprises. Examples of entrepreneurship/ownership SAE include owning a lawn care service, raising fish, growing and selling flowers, and operating a custom harvesting business.

Directed laboratory supervised agricultural experience is the type in which a group of experiences in a practical activity are planned by the agriculture teacher. Directed laboratory SAE is especially for students who are unable to engage in other forms of SAE. The experiences are usually on school property but may occur in other locations. They are often funded through some source other than the students' own funds. They may be conducted partially during class time but also involve out-of-class student labor. Students may or may not share in profits returned from these activities. An example is growing plants in a school greenhouse or fish in a school aquaculture facility. Some schools have land and livestock laboratories for group projects with field crops and livestock.

Research and experimentation supervised agricultural experience is the type in which the student carries out an investigation into a problem using scientific approaches. The experiences usually involve identifying a particular problem, searching for information, and then conducting a scientific experiment or using other research procedures to arrive at conclusions. The student then makes recommendations about how to solve the particular problem. Students conducting research and experimentation SAE programs often exhibit at FFA agriscience fairs and/or general science fairs. Research and experimentation SAEs may involve cooperating with a science teacher or a scientist in the local community. An example of this type of SAE is studying water pollution, feed nutrients, or tissue culture. Teachers interested in this type of SAE may also be interested in learning more about Science, Technology, Engineering, and Mathematics (STEM) education and can connect through the PBS Teachers STEM Education Resource Center.

An *improvement project* is a supplementary experience or group of experiences carried out in conjunction with one of the other types of SAE programs. Improvement projects usually involve home or community work. They may or may not involve competencies related to agricultural occupations and are usually unpaid. They may be group or individual projects that contribute to the agricultural knowledge and/or skill of the student. An example is constructing a pen for a livestock enterprise or a tank for an aquaculture enterprise. Landscaping the home and renovating the home lawn may also be considered improvement projects.

A *supplementary project (activity)* is a specific skill gained through experiential learning outside normal class time that adds to the agricultural knowledge and competency of the student. The skill is not related to the major SAE and it is usually accomplished in less than a day. An example is pruning a fruit tree. Helping another student dock lamb tails and changing the spark plug in a small engine may also be considered supplementary projects (activities).

PLANNING AND CONDUCTING SUPERVISED AGRICULTURAL EXPERIENCE PROGRAMS

SAE is the most challenging component of the agricultural education program model to implement, and researchers have suggested that new approaches be explored for delivering this part of the program. Large classroom enrollments, student diversity, teacher confusion about middle school SAEs, and availability of resources are often cited as barriers to student participation in SAE. Most agriculture teachers value SAE as an important part of the program but have difficulty with implementation. To overcome these barriers teachers need to be more creative and innovative in providing SAE opportunities. At the local level teachers can make their own determination about what constitutes an appropriate SAE for the students enrolled in their program. They will need to advise students seeking advanced FFA degrees and SAE award recognition about the requirements for the advanced degrees and FFA awards based on SAEPs. Today many teachers use group cooperative SAE projects involving plants, small animals, or other experiential learning materials. Exploratory SAEs and research-type SAEs are also becoming more common in today's agriculture programs.

The agriculture teacher's primary responsibility for SAE is to assure that it is an essential, effective component of the overall agricultural education program. The teacher should ensure that all agricultural education students are aware of its values, purposes, characteristics, and opportunities and that they participate in it.

The first step in assisting students is to teach an introductory unit on SAE. This unit should cover the kinds and types of SAE that will be most beneficial to the students enrolled in the class. The teacher should cover how SAE works and what students gain from it. The relationship of SAE and FFA should also be discussed. A list of possible SAEs available at the school and in the community should be developed and provided to each student. Many teachers have older students visit the class to discuss their SAE programs.

Because SAE is unique to agricultural education as a program requirement, students often don't understand it well enough to assume the initiative. They very likely will not know how or where to get started. This places other responsibilities on the agriculture teacher. First, the teacher has a responsibility for the development of SAEP opportunities. The teacher should find agricultural training stations (jobs) in the community that are available to agricultural education students. Farmers, ranchers, and agribusiness managers should be encouraged to provide work opportunities (not necessarily paid) for students referred to them by the agriculture teacher. Provisions of child labor laws should always be followed.

Besides arranging for job opportunities, the teacher should establish a reservoir of ideas and opportunities for individual and group agricultural projects for students to draw upon when they are unable to identify prospective activities by themselves. The teacher should also actively assist in helping students find, purchase, and transport project materials, equipment, and livestock.

A training agreement and a training plan should be completed for each SAE program. A *training agreement* is a written statement of the exact expectations, understandings, and arrangements of all parties involved in the SAE program. It is signed by the student, student's parent(s) or guardian(s), training employer, and teacher (on behalf of the school). A sample training agreement is shown in Figure 22–7.

A *training plan* is a written statement that documents the specific training activities in which the student is expected to participate. The experiences/competencies in the training plan are those to be carried out during SAE. They are often based on the occupational competencies needed for the student to enter his or her tentative occupational objective. The training plan provides space for recording when the experiences/competencies have been accomplished. It may be signed by the student, student's parent(s) or guardian(s), and employer. Copies of both the training agreement and the training plan should be made and provided to all parties involved. A sample training plan form is shown in Figure 22–8.

THE ROLE OF SUPERVISION

Teacher supervision of SAE programs should be a high priority in high-quality agricultural education programs. According to the *Local Program Success Guide*, teacher involvement is key to bridging the gap between the classroom and the workplace and has a direct correlation to SAE program quality and student

---

TRAINING AGREEMENT

Date _____

This agreement is to provide supervised experience for ___(name of student)___ , _(age)_ , *as* a part of the agricultural education program in _(name of school)_ . The experience will be provided by ___(name of employer)___ and will commence on _(date)_ . The experience will terminate on or about _(date)_ . The details included below apply to this agreement.

The usual working hours will be:

When in school: _____ When not in school: _____

Compensaton will be as follows: _____

The supervisor/responsible person will be _____

Liability insurance coverage is _____

The employer agrees:

* To provide the student with opportunities to learn and develop competencies for successful employment
* To instruct the student in the desirable work procedures
* To assess the student's performance and provide feedback to the student and the teacher
* To follow applicable provisions of the child labor laws of the Fair Labor Standards Act
* To notify parent(s)/guardian(s) and the school immediately in case of an accident or serious problem
* To permit the teacher to make supervisory visits to the work site

The student agrees:

* To perform job duties dependably, honestly, and productively
* To keep the employer informed well ahead of time of changes in work schedule, including necessary and justifiable absences from work
* To follow instructions carefully, practice safety, and care for equipment and materials
* To be courteous to, and considerate of, employees and customers
* To keep accurate records of work experiences and make reports as required by the school
* To groom and dress appropriately and refrain from the use of illegal and legal substances that may impair job performance
* To avoid visits by friends to the work site and avoid telephone or e-mail communications that are excessive or impair work

The parent(s)/guardian(s) agrees:

* To promote the value of supervised experience as a learning tool
* To support the student in successfully performing job duties
* To be sure of satisfaction with regard to the work conditions

The instructor, on behalf of the school, agrees:

* To visit the student at the training site at frequent intervals but with consideration of the schedule and interests of the employer
* To provide feedback to the student to assure appropriate job skill development
* To provide school-based instruction that supports learning at the training site

All parties agree:

* To follow an initial ten-day trial period to allow time for the student to prove himself or herself
* To terminate this agreement for cause at any time by either party following due notice if problems arise that might lead to a situation that does not support an appropriate experience
* To allow the school to serve as a mediator in the event problems arise that cannot otherwise be resolved

Signatures:

Student _____   Employer _____

Address _____   Address _____

Telephone No. _____   Telephone No. _____

Approved:

Parent(s)/Guardian(s) _____

Address _____   Telephone No. _____

Approved for the school:

Instructor _____   Telephone No. _____

22–7. An example of a training agreement.
(Note: This example is designed for use with placement SAE.) (Adapted from Phipps & Osborne, 1988)

TRANING PLAN

Name of Student _____ Teacher _____

Occupational/Educational Objective _____

Beginging Date _____ Ending Date _____

Training Station/Employer _____

Paid _____ Unpaid _____ School Name _____

Experience/Competency	Date Accomplished	School-Related Instruction	Date Completed

Signatures:

Student _____ Employer _____

Address _____ Address _____

Telephone No._____ Telephone No._____

Approved:

Parent(s)/Guardian(s) _____

Address _____ Telephone No. _____

Approved for the school:

Instructor _____ Telephone No. _____

22–8. An example of a training plan.

(Note: This plan shows that instruction in class is related to the activities in the training plan. Multiple copies of the form may be needed.)

success. However, with today's increasing student enrollments, lack of teacher release time, and decreasing funds for travel expenses, supervision of SAE has been weakened. Some states and schools still provide financial incentives to encourage teacher supervisory visits. For example, many schools provide the agriculture program with a vehicle (often a pickup truck) for use by the agriculture teacher(s). Other schools reimburse teachers for their mileage to make supervisory visits using personal vehicles.

Teachers should keep records of supervisory visits for income tax purposes, to provide documentation for maintaining financial support for visits, and to document the evaluation of student performance in SAEs. The *LPS Guide* contains forms to assist teachers in documenting SAE visits. Some states and local departments have their own forms. Regardless, always know and follow school policy.

After teaching the class about the types of SAEs available, the teacher should make individual visits to each student's home or placement station to help the student develop his or her own SAE plan. During these visits, the teacher should be prepared to offer SAE suggestions based on the student's particular interests, goals, and available opportunities. The teacher should provide information about how to find materials, facilities, and other items needed for the SAE program.

It is important to involve parents/guardians in planning the SAE program because a younger student will likely need their assistance in conducting the SAEs. Involving them early in the planning process allows them to assist in determining the best SAEs for the student. If parents/guardians are personally interested in their student's SAE, they are more likely to assist with ideas, transportation, financial support, and other items that will ensure a successful SAE program. A home visit to all new students entering the agricultural education program allows teachers to meet the parents, review available facilities, and discuss SAE requirements and opportunities.

After the student has established the SAE program, the teacher should schedule organized and purposeful visits to observe the student's activity and assure that the student's experiences are of high quality. The frequency of supervisory visits by the teacher will vary among the students according to the complexity of their SAE programs. However, the teacher should make periodic visits throughout the duration of the activity to help ensure that each student has a successful SAE.

For a student employed in an agricultural job as part of his or her SAE program, the teacher should consider the employer as a co-supervisor. The teacher and the employer should work together to ensure that the experience helps the student prepare for a career.

A student may conduct his or her SAE program at home. In this case, the teacher has an opportunity to include a parental visit with the task of observing the student's SAE activity. The teacher should take advantage of this opportunity to become better acquainted with the parents/guardians. For a student who conducts his or her SAE program away from home, the teacher should incorporate into the visitation schedule at least one parental/home visitation each year. The purposes of home visitations are as follows:

1. To demonstrate to parents/guardians that the teacher is interested in the development of their child
2. To form an alliance with parents/guardians for the career and personal guidance of their child
3. To become acquainted with home conditions that may have a bearing on the student's performance
4. To inform the parents/guardians of program purposes, of expectations and activities, and of their child's performance

When students carry out SAE activities in school facilities, the teacher is responsible for maintaining a safe environment in the facilities and for assuring that the students conduct themselves safely. Teachers often use classroom time on student sharing and discussion of experiences, since these experiences are an extension of the classroom instructional program.

In addition to scheduled supervisory visits, the agriculture teacher is often "on call" for students who have immediate need for assistance with their SAE programs. This is sometimes necessary because animals get sick, equipment breaks, and employer–employee relations become strained at unexpected and inconvenient times. Students frequently panic in these crises and desperately need the assistance of the teacher.

RECORD KEEPING AND KINDS OF RECORDS KEPT

The teacher is responsible for assuring that all agricultural education students incorporate record keeping as an important segment of their SAEPs. Such records document the experiences of students and serve as the basis for applying for various advancements and awards in FFA.

The teacher must teach students how to keep appropriate records related to their experiences and ensure that the records are kept current. Most states have adopted specific record systems for use by students. This is helpful when evaluating student records for both grading purposes and competitive awards at the local and state levels. Teachers should become familiar with their state's record-keeping system and develop lessons and activities to ensure their students keep accurate and up-to-date records. One of the most difficult tasks a teacher can undertake is trying to help students apply for advanced degrees when records are not accurate and up to date.

The records kept may vary from one state or school to another. Some states or schools require computer-based systems; others require specially prepared record books. Most state record systems include templates for training agreements and training plans.

Other items usually included in record systems are as follows:

- Student name and chapter
- Year in agriculture
- Period covered
- Teacher's name
- List of enterprises
- Budgets
- Accounts receivable
- Accounts payable
- Income
- Expenses

22–9. Supervised agricultural experience records may be kept using a computer-based record keeping system.

(Courtesy, Education Images)

- Inventories
- Financial statements
- Income summary

Besides the records kept by each student, the teacher should maintain SAEP records in the department files that include the following:

1. Individual student SAE agreements and training plans
2. Individual student records of kind, size, growth, and performance
3. Information on visitations, including dates, contacts, mileage, and major observations
4. School-wide summarization of student SAEs by kind and scope

The purposes for keeping records in the department files are (1) to provide documentation of the teacher's time and expenses incurred in making SAEP visits; (2) to assist the teacher in making preparations for visits by reviewing agreements, training plans, and past visitation recommendations; (3) to allow the teacher to summarize SAEP activities for reporting to the administration, local community, and state leaders; and (4) to provide documentation for grading student SAEP performance and identifying student potential for award recognition and advanced degrees.

EVALUATING SUPERVISED AGRICULTURAL EXPERIENCE PROGRAMS

Since the agricultural education activities conducted through SAE are intracurricular, the evaluation of student performance in agricultural education should include consideration of the student's level of involvement and performance in those activities.

Grading Supervised Agricultural Experience

A grading system for SAE should be based on the premise of every student enrolled in the agricultural education class being able to attain the highest grade possible. In multiple-teacher departments, the grading system should be agreed upon by the agricultural education teachers and uniformly applied. It should be possible for students to be informed at any time concerning their particular grade status. Visible records, such as grading charts, point award systems, or rubrics, may be used for this purpose. The grading system should be explained to every student enrolled in agricultural education so that it is thoroughly understood. The system should be a matter of record and be incorporated into the department's management plan. Because of the interrelationship of SAE and FFA activity to the instructional program, these areas may constitute a percentage of the total grade. (Chapter 19 discusses evaluation procedures in more detail.) A sample SAE on-site evaluation form can be found in the *Local Program Success Guide* located in "Resources" section of the National FFA Web site.

Out-of-class-time participation can be reasonably viewed as agricultural education homework. As such, full credit for the agricultural education course(s) in which the student is enrolled plus the grade earned in the related SAE activity should be dependent upon satisfactory, measured participation.

Student participation in classroom/laboratory, FFA, and SAE is essential in order for a student to have access to the full curriculum of the agricultural education program. This makes documentation of student progress in SAE and FFA just as important as documentation of student progress in the classroom.

Enhancing SAE Quality

In their eighty-five years of including SAE as an instructional strategy, agricultural educators have developed many proven practices, some of which have already been mentioned in this chapter. Additional proven practices are listed here for teachers desiring to increase the quality of their SAE programs.

- Prepare and distribute to students a handbook that describes the school's requirements for SAE, lists the kinds of projects that can be included in the SAE, explains how SAEs are evaluated, and gives examples of good-quality SAEs that show progress from year to year.
- Be aware that the term "SAE program" may not be understood by some students. A simpler term, such as the old standby "project," can be used, although it has limited meaning in the strictest sense.
- Require a written training plan for every student's SAEP. That plan should be reviewed annually by the student, the teacher, the student's parents/guardians, and, if applicable, the employer.
- Utilize National FFA agricultural proficiency and achievement award systems.
- Incorporate accomplishments into FFA chapter point award systems.
- Emphasize honor of FFA State and American Degrees; recognize chapter members who attain these degrees.
- Encourage participation in "project competition" programs—local level and above.
- Solicit local organizations to provide support, such as livestock "chains."
- Develop local sources for project funding—e.g., banks and credit institutions, booster club loan funds, etc.
- Provide school facilities for SAE programs.
- Encourage cooperative projects for students with limited resources.
- Maintain regular written and oral communications with parents/guardians.
- Provide project tours for parents/guardians, prospective students, and other interested persons.
- Adjust home visitation hours to coincide with times when parents/guardians are home.
- Involve parents/guardians in school laboratory workdays and improvement projects.
- Maintain a visible record of teacher supervision visits as a means of keeping SAE programs in the minds of students and visitors to the agriculture department.
- Plan a visitation schedule to assure equitable supervision of all students' SAE programs.
- Take beginning students on tours of successful projects.
- Utilize summer months to contact all first-year students and their parents/guardians to discuss SAE plans.
- Take steps to help assure the success of each student's first project.
- Use third- and fourth-year students as advisors to beginning students.
- Utilize the assistance and experience of other teachers whose students have successful SAE programs.
- Provide the school board and administration with special presentations.
- Invite board members and administrators to serve as local judges for project competition.

RELATIONSHIP OF SUPERVISED AGRICULTURAL EXPERIENCE AND FFA

Recognizing students for their SAE achievements is important. A major part of the recognition program is acknowledgment by the teacher, parents/guardians, employers, and other students. FFA provides additional opportunities to recognize students conducting quality SAE programs. For many students, FFA is their best opportunity for receiving recognition.

FFA Degrees and Proficiency Awards

The **FFA degree program** is designed to encourage students to establish and work toward career goals in the agricultural industry. Degrees are also designed to promote and recognize student achievement in leadership, personal growth, scholarship, and career development. A student's achievement in these areas should be documented in the SAE program records. One of the criteria for advancement in the FFA degrees at the high school level and beyond is based on planning and developing a quality SAE program to gain the skills needed to be successful in an agricultural career. The five FFA degrees are listed and described in Chapter 23.

A **proficiency award** is an award provided through FFA to a member who excels in the development of specialized skills in a specific agricultural area. Proficiency awards provide opportunities for students to be recognized at the chapter, state, and national levels in areas related to SAE programs. There are two major categories for proficiency awards. One is "placement," and students who are employed in agricultural businesses (including farms) are eligible to apply for the placement awards. The other category is "entrepreneurship," and members who own production enterprises or agribusinesses may apply for these awards.

Examples of proficiency award areas are agricultural communications, agricultural processing, agricultural services, agriscience research and experimentation, beef production, diversified crop production, fruit production, food science and technology, goat production, home and/or community development, landscape management, and wildlife management.

22–10. A student in an FFA swine Career Development Event (CDE) uses knowledge and skill acquired through supervised agricultural experience and classroom instruction. (Courtesy, National FFA Organization)

FFA Star Awards

A *Star Award* is considered the top award provided by the National FFA Organization to an individual student in FFA. The award is based on a student's overall achievements in FFA, including outstanding accomplishments in SAE.

Star Awards recognize outstanding FFA members at each degree level, beginning at the seventh or eighth grade with the Star Discovery Award. The recipient is the local member who has been most active and shown leadership.

The first Star Award at the high school level is the Star Greenhand Award. This award goes to a chapter's most active Greenhand member. The award is based on the member's SAE program and demonstrated leadership abilities.

The Chapter Star Farmer Award goes to a member with outstanding SAE in production agriculture who has demonstrated the most involvement in all chapter activities. The Chapter Star in Agribusiness Award is the same as the Chapter Star Farmer Award except the recipient's SAE must be in agribusiness rather than production agriculture.

The Chapter Star in Agricultural Placement Award goes to the top member with outstanding SAE involving placement in production agriculture, agribusiness, or directed laboratory that is not agriscience-based. The placement may be for either paid or unpaid experience.

The Chapter Star in Agriscience Award goes to the top chapter member who has outstanding SAE that involves natural resources and/or who has a research/experimentation-based or science-based laboratory experience.

At the state level, Star Awards are provided for State Star Farmer, State Star in Agribusiness, State Star in Agricultural Placement, and State Star in Agriscience. The state awards are also based on outstanding SAE and on participation in various FFA activities at the chapter level and above.

The National FFA Organization recognizes the American Star Farmer, American Star in Agribusiness, American Star in Agricultural Placement, and American Star in Agriscience. The national awards are based on a member's SAE and participation in activities from the chapter level through the national level.

The National FFA Foundation provides funds for plaques and medals for chapter Star Award winners. The foundation also provides funds for medals, plaques, and cash awards for all the Star Award winners at the state and national levels.

Scholarship Awards

The *National FFA Scholarship Program* provides opportunities for financial assistance for students continuing their education at the postsecondary level. Each scholarship has specific criteria defined by the sponsor when the scholarship is established. Some scholarships are based on student career goals. Outstanding SAE accomplishments may be included on the scholarship application form. Additional information about the National FFA Scholarship Program can be found at the FFA Web site.

Other Awards

Another FFA award program closely related to student SAE is the Agri-Entrepreneurship Awards. These awards are designed to encourage entrepreneurial thinking among agricultural education students. They are based upon business plans developed by students that relate a market opportunity overlooked by others. These cash awards are available at the chapter, state, and national levels.

The FFA Agriscience Fair recognizes middle school and high school students who are studying the application of science and technology in agricultural enterprises. Participation begins at the chapter level and continues to state and national levels. The FFA Agriscience Fair provides recognition opportunities for students conducting research/experimentation-type SAE.

Grants are available through the National FFA and provide small amounts of start-up cash to students with limited resources. A grant is awarded based on sponsor guidelines and must be used to establish or reinforce a student's SAE. Application information and other details about SAE grants can be found on the FFA Web site at **www.ffa.org**.

Recognition Is a Motivational Tool

While awards and incentives should not be the primary reason for conducting SAE, they do provide valuable opportunities for teachers to gain publicity for students and the agriculture program.

Recognition is an important motivational tool, and for many students, their SAE provides avenues for them to achieve recognition through FFA. It may be from exhibiting winning livestock or other agricultural products at a local, county, or state fair, or it may be from winning a local, state, or national proficiency or other FFA award. Good teachers excel at motivating students, and FFA provides many motivational tools for teachers to use.

REVIEWING SUMMARY

SAE is one of the three major interrelated components of the agricultural education program. It involves students, teachers, parents/guardians, and employers and extends the agricultural education program outside the classroom walls into the local community. When properly conducted, SAE is based on student interests, needs, and career goals, is supervised by a qualified agriculture teacher, is based on the curriculum, is cooperatively planned, is documented, and includes recognition for student achievement.

Several types of SAE programs are available, and students may choose the type best suited to their personal goals and career aspirations. Teachers are responsible for teaching students, parents/guardians, employers, and others about SAEs and their value. They are also responsible for providing overall supervision of student experiences, documenting and evaluating the experiences, and ensuring that students are recognized for their achievements.

QUESTIONS FOR REVIEW AND DISCUSSION

1. What is SAE?
2. What are the benefits of SAE programs?
3. What are the types of SAE programs?
4. Who are the key people involved in planning and conducting SAEs?
5. What are the purposes of modern SAE programs?
6. Why are SAE training agreements important?
7. What SAE records should be maintained in the department files?
8. What are some important considerations in evaluating SAE programs?
9. What are some ways to recognize students for SAE achievements?
10. How are FFA and SAE related?

ACTIVITIES

1. Use the Internet to investigate SAE programs. A site to start your search is the National FFA Organization Web site.
2. Prepare a term paper on SAE in your state. Topics could include types of SAE programs available and student participation by type, comparison of electronic and paper record systems, and the history of SAE programs in your state.

CHAPTER BIBLIOGRAPHY

Barrick, R. K., M. Hughes, and M. Baker. (1991). Perceptions regarding supervised experience programs: Past research and future direction. *Journal of Agricultural Education, 32*(4), 31–36. doi: 10.5032/jae.1991.04031.

Binkley, H. R., and R. W. Tulloch. (1981). *Teaching Vocational Agriculture/Agribusiness.* Danville, IL: The Interstate Printers & Publishers. Inc.

Camp, W. C., A. Clarke, and M. Fallon. (2000). Revisiting supervised agricultural experience. *Journal of Agricultural Education, 41*(3), 13–22. doi: 10.5032/jae.2000.03013.

Dyer, J. E., and E. W. Osborne. (1995). Participation in supervised agricultural experience programs: A synthesis of research. *Journal of Agricultural Education, 36*(1), 6–14. doi: 10.5032/jae.1995.01006.

Lee, J. S. (2000). *Program Planning Guide for AgriScience and Technology Education* (2nd ed.). Upper Saddle River, NJ: Pearson Prentice Hall Interstate.

National Council for Agricultural Education. (1992). *Experiencing Agriculture: A Handbook on Supervised Agricultural Experience.* Alexandria, VA: Author.

National FFA Organization. (2002). *Local Program Success Guide.*

National FFA Organization. (2013). *SAE Resources.* Retrieved from https://www.ffa.org/ffaresources/Pages/default.aspx.

Newcomb, L. H., J. D. McCracken, J. R. Warmbrod, and M. S. Whittington. (2004). *Methods of Teaching Agriculture* (3rd ed.). Upper Saddle River, NJ: Prentice Hall.

PBS Teachers. (n.d.). *PBS Teachers STEM Education Resource Center Web Site.*

Phipps, L. J., and E. W. Osborne. (1988). *Handbook on Agricultural Education in Public Schools* (5th ed.). Danville, IL: The Interstate Printers and Publishers, Inc.

Phipps, L. J., E. W. Osborne, J. E. Dyer, and A. L. Ball. (2008). *Handbook on Agricultural Education in Public Schools* (6th ed.). Clifton Park, NY: Thompson Delmar Learning.

Rayfield, J., and B. Croom. (2010). Program needs of middle school agricultural education teachers: A Delphi study. *Journal of Agricultural Education, 51*(4), 131–41. doi: 105032/jae.2010.04131.

Retallick, M. S. (2010). Implementation of supervised agricultural experience programs: The agriculture teachers' perspective. *Journal of Agricultural Education, 51*(4), 59–70. doi: 10.5032/jae.2010.04059.

Roberts, T. G., and J. F. Harlin. (2007). The project method in agricultural education: Then and now. *Journal of Agricultural Education, 48*(3), 46–56. doi: 10.5032/jae.2007.03046.

23

FFA

"To practice brotherhood, honor agricultural opportunities and responsibilities, and develop those qualities of leadership which an FFA member should possess" (National FFA Organization, 2012c). This excerpt from the official FFA Opening Ceremony describes what FFA members do when they get together to hold an official FFA chapter meeting. More important, it encapsulates the meaning of FFA to thousands of American youth enrolled in agricultural education. From the organization's official beginning in 1928 to today, millions of young people have benefited socially, emotionally, and professionally from their participation in the FFA experience.

FFA's mission is to make a positive difference in the lives of students by developing their potential for premier leadership, personal growth, and career success (National FFA Organization, 2012c). This mission is manifested in the leadership programs, Career Development Events (CDEs), scholarship programs, and community development programs offered to FFA members at the local, state, and national levels.

objectives

This chapter provides background information on FFA and its role in agricultural education. It has the following objectives:

1. Discuss the history and purpose of FFA
2. Explain how FFA is structured as an organization
3. Discuss FFA basics that are important in a local chapter
4. Explain the FFA degree system
5. Explain how to develop an FFA program of activities
6. Identify best practices in managing an FFA chapter

terms

American FFA Degree	Greenhand FFA Degree
chapter development	Henry C. Groseclose
Chapter FFA Degree	New Farmers of America
community development	program of activities
Discovery FFA Degree	Public Law 105-225
FFA Creed	State FFA Degree
FFA Emblem	student development
FFA Motto	

23–1. A group of officers of a local FFA chapter are planning an activity.
(Courtesy, Education Images)

WHY JOIN FFA?

At some point in an agriculture teacher's career, a student will ask them, "Why should I join the FFA?" The FFA is, after all, an organization based on voluntary membership. What would cause a student to want to be a part of an organization that often takes extra time and effort beyond the school day? The agriculture teacher should be ready to answer this question in terms that the student can understand, but the actual reason why FFA is important to the agricultural education program is rooted in human motivation theory. One of the most prominent theorists on human motivation was Abraham Maslow, and his hierarchy of needs theory of human motivation explains the value of the FFA experience for students.

The human being is an integrated organism. It is impossible to separate the various components of a person's self. When an individual experiences hunger, it is their whole self that is hungry and not just selected physiological components (Maslow, 1970). Until the basic physiological needs are met, the human is motivated to satisfy these needs. Once basic physiological needs are met, a different set of needs become evident. Maslow referred to these as the safety needs and characterized them as stability, security, and protection from harmful external conditions. If both the physiological and safety needs are met, then love and affection needs emerge. Maslow (1970) categorized these needs as the need for contact and intimacy and a sense of belonging.

The next level of need is identified by Maslow as the esteem needs. These needs are characterized by a person's desire for status, fame, dominance, and importance. The satisfaction of self-esteem needs will lead the individual to feel self-confident, worthy, and useful in their environment. Resting upon esteem needs are cognitive needs characterized by a person's search for knowledge and understanding. If these needs are met, the individual progresses to aesthetic needs that are identified as a person's desire for order and beauty. At the top of the hierarchy is self-actualization. Even if all of the other needs are met, the individual will develop a sense of restlessness and discontentment unless he or she is accomplishing goals true to oneself (Maslow, 1970).

By the time a young person reaches high school age, he or she has matured to the point in Maslow's hierarchy where there is a need for a sense of belonging and contact with others. A student may have progressed beyond these needs and reached a point whereby they have a strong need for status, recognition, and importance. FFA membership may allow students to fulfill this need for comradeship. In order to promote a sense of belonging and intimacy, the FFA has developed traditions that tend to bring young people together for a common purpose.

One particular method by which the FFA recognizes the commonalities among youth enrolled in agricultural education is through the perpetuation of various traditions of the FFA organization. One such tradition is the use of the FFA jacket. The FFA jacket is a blue corduroy garment with the student's name on the front, FFA emblems on the front and back, and the name of the student's FFA chapter and state association. At official FFA events where students meet other FFA members, the FFA jacket serves as a symbol of organizational pride and facilitates the meeting of new friends and acquaintances (National FFA Organization, 2012c). In addition, official FFA ceremonies are designed to focus members' attention on the common interests they share with other members.

For those students who progressed to the esteem needs level in Maslow's hierarchy, FFA programs and services might help to fulfill those needs. Competitive activities such as the FFA Proficiency Awards Program and Career Development Events reward students for their achievements. This may help fulfill a student's need for recognition and importance, and support their desire to learn.

HISTORY AND PURPOSE OF FFA

Boys and Girls Clubs for Rural Youth   The movement to develop organizations for agricultural youth began at the turn of the twentieth century. There is some question as to when boys and girls agricultural clubs were established in the United States. A. B. Graham organized boys and girls clubs in January 1902 in the Springfield Township school community in Clark County, Ohio. Club meetings were held once per month in an assembly room of the county building. These were corn clubs. Later the clubs were broadened to include vegetable projects (McCormick & McCormick, 1984).

Another possibility is that W. B. Otwell created the first boys and girls clubs in agriculture in Macoupin County, Illinois. Otwell created a corn yield contest for local boys as a way to encourage attendance at farmer's institutes. Farmers were normally reluctant to attend these training institutes, but the corn yield competitions sparked their interest. The first year's contest involved 500 boys. There is also evidence that the first boys club may have been organized in Holmes County, Mississippi, by W. H. Smith, the local school superintendent. Agricultural clubs for girls sprang up in South Carolina in 1910 (True, 1969).

Agricultural clubs were incorporated into public schools as a means to provide social interaction among youth and to encourage their interest in academic subjects. These early school clubs were highly organized, with monthly meetings focused on agricultural subjects. J. B. Berry (1924) reported that elementary children were participating in junior project clubs. In his handbook on agricultural education, Berry (1924) discussed the important role of the schoolteacher in advising and the activities of the clubs. Berry said, "The wise teacher utilizes pupil activities to as great extent as possible, thereby developing leadership qualities in pupils" (Berry, 1924, p. 196).

The Beginning

The movement to create the FFA organization started long before the passage of the Smith-Hughes Act in 1917 and the incorporation of FFA in 1928. Many states already had some type of system for providing training in the agricultural and mechanical arts to students below the college level. Clubs and organizations designed to encourage and support farm youth grew out of the agricultural education movement, but there was no national effort to coordinate the activities of the individual clubs.

A Student Organization Emerges   With the passage of the Smith-Hughes Act, the national coordination of agricultural education made it convenient for a student organization to emerge with the goal of encouraging farm boys and their families to adopt the best practices of agricultural production. Between 1917 and 1928, the agricultural education profession began the work to create a national organization.

One of the most notable efforts to begin an organization was in Virginia. Walter S. Newman, state supervisor for agricultural education in Virginia, was a pioneer in the development of FFA. He believed that the rural farm boys were not getting the same opportunities for personal growth and advancement as their city-bred counterparts. The isolation inherent to farm life made it difficult for farm youth to experience the cultural and social life so readily available to urban youth. This isolation bred discontent and stunted the development of a positive self-image in farm youth. Newman believed that some type of organization was necessary to provide opportunities for personal advancement and the growth of self-confidence in rural farm youth. In 1925, he joined with colleagues Henry C. Groseclose, Edmund Magill, and H. W. Sanders in developing the Future Farmers of Virginia.

Before the creation of FFA, regional and national livestock and dairy judging activities provided opportunities for competition among agriculture students. In 1926, federal agricultural education officials reached an agreement with the American Royal Livestock Show to host a livestock judging contest in Kansas City, Missouri. Later that same year, the National Congress of Vocational Agriculture Students was also established in Kansas City, Missouri. This three-day convention included tours of local agricultural businesses, banquets, and meetings designed to encourage students to learn more about the agricultural industry.

Future Farmers of America Formed   By 1928, the movement for a national agricultural youth organization had grown so strong that the Federal Board for Vocational Education was asked to assist in the development of the youth organization, its governance structure, and its bylaws. The National Congress of Vocational Agriculture Students and the American Royal Rodeo and Livestock Show also included a special convention held for the purposes of establishing the Future Farmers of America.

On November 20, 1928, thirty-three delegates from eighteen states officially adopted the constitution and bylaws of the new organization and elected Leslie Applegate of New Jersey as the national president. C. H. Lane was chosen to serve as the National FFA Advisor, and **_Henry C. Groseclose_** was elected to serve as the executive secretary/treasurer. The old Baltimore Hotel where the organization was founded is no longer standing, but a monument erected by the citizens of Kansas City, Missouri, to honor the establishment of FFA in that city exists on the site of the hotel. Until 1999, the National FFA Convention was held in Kansas City, Missouri.

Expansion of FFA   As the National FFA Organization grew, it became necessary to develop more effective business practices to keep pace with the needs of a growing number of members. Delegates to the 1938 National FFA Convention set aside funding to establish an FFA camp. In the process of doing this, they also created a board of trustees to oversee the construction and development of this camp and subsequent properties. In 1939, these funds were used to purchase 25.5 acres of land in Alexandria, Virginia, for the national FFA camp. This property, once owned by George Washington, was to serve as the headquarters of FFA for fifty-nine years. The National FFA Foundation, Inc., was established in 1944 to support the activities,

23–2. FFA represents opportunity for students to travel, learn, and develop important personal skills.
(Courtesy, National FFA Organization)

programs, and services of the National FFA Organization. In 1948, the National FFA Supply Service opened its doors on the Alexandria property, providing official FFA jackets and other paraphernalia for FFA members and chapters.

In 1952, the *National Future Farmer Magazine* began publication and all FFA members received it as part of their membership benefits. In 1953, the U.S. Postal Service honored the FFA with a special commemorative postage stamp. In 1959, a new headquarters building was constructed on the site of the old national FFA camp.

By the 1960s, FFA was maturing as an organization. Many new programs and services were being offered to its members, but the social climate of America had progressed to the point where it was impossible for FFA to continue to limit its membership to farm boys. In 1965, FFA and New Farmers of America (NFA) merged to become one organization. In 1969, girls were officially admitted to FFA membership, even though they had been unofficial members for years. One method by which FFA advisors were able to secure FFA membership for girls before 1969 was to list only their first initials and last names on the official FFA roster. The National FFA Organization was none the wiser that a certain "G. Bradley" of a local FFA chapter in North Carolina was actually a female agriculture student, Genie Bradley. Numerous such cases exist in the history of the FFA organization.

In 1971, the National FFA Alumni Association was established to encourage friends of FFA to continue their support of FFA through their time and resources.

The 1980s and 1990s were a period of rapid change for the National FFA Organization. In 1988, delegates at the National FFA Convention, in an effort to reflect the diverse nature of FFA members, changed the name of the Future Farmers of America to the National FFA Organization. In 1989, the *National Future Farmer* became *FFA New Horizons,* and in 1996, the National FFA Board of Directors voted to move the National FFA Convention to Louisville, Kentucky, in 1999 and the

23–3. The emblem of the New Farmers of America.
(Courtesy, Education Images)

National FFA Center to Indianapolis, Indiana, in 1998. In 1998, Public Law 81-740 was replaced by provisions in Public Law 105-225 to meet more accurately the future needs of the National FFA Organization. The convention location alternates between the cities of Indianapolis, Indiana, and Louisville, Kentucky.

THE NEW FARMERS OF AMERICA

The ***New Farmers of America*** (NFA) was the companion organization to FFA that served black Americans from 1926 to 1965. NFA was established during the winter of 1926 in Virginia at the urging of Dr. Walter S. Newman, state supervisor of agricultural education, and Dr. H. O. Sargent of the Federal Board for Vocational Education. Sargent's primary responsibility for the federal board was to coordinate agricultural education programs for black students. Both Newman and Sargent recognized the need for an organization for black students, and in May 1927, the New Farmers of Virginia held its first state convention in Petersburg, Virginia. In 1935, after forming a national organization out of the various state associations of New Farmers, the New Farmers of America held its first national convention at the Tuskegee Institute in Alabama.

The New Farmers of America was in most regards very similar to the Future Farmers of America. Both organizations had similar contest and award programs, degree programs, emblems, leadership development programs, business operations, and student leadership structures. The adult leadership for both NFA and FFA was provided through the Federal Board for Vocational Education.

For many black Americans, the New Farmers of America added dignity to the pursuit of agricultural occupations. While the United States struggled through years of segregation and the American education establishment operated under the concept of "separate but equal" schools, NFA offered young black students the incentive to pursue agricultural careers while building their self-confidence and technical skills (Wakefield & Talbert, 2003). Both NFA and FFA served their respective clientele for many years.

The passage of the Civil Rights Act of 1965 also marked the end of the coexistence of NFA and FFA. Both organizations merged into a single organization just as schools across America were desegregating. The merger was not easy. Members of NFA were proud of their heritage and felt that they had the most to lose in the merger. Many NFA members were afraid they would not receive adequate recognition and representation in the National FFA Organization (Wakefield & Talbert, 2000). State and federal leaders worked together to make the transition work, and at the National FFA Convention in 1965, the two organizations became one. Some would say that NFA members were justified in their concerns. Today, fewer black students are involved in agricultural education and FFA nationwide than there were prior to the merger of the two organizations (Wakefield & Talbert, 2000).

HOW FFA IS STRUCTURED

The foundation of FFA activity is the individual FFA member. Because FFA is an organization founded upon the premise of preparing young people for careers in agriculture, certain minimum qualifications must be met in order to be an FFA member. An FFA member must be enrolled in at least one agricultural education course per academic year in a school that has an FFA chapter in good standing and have a supervised experience program that prepares the member for a career in agriculture.

Intracurricular Organization

FFA is an intracurricular organization. An intracurricular organization is a component that is a part of the curriculum—not an organization outside the curriculum or an extracurricular activity.

In the early years of FFA, the SAE requirement was met without much ado. Students who enrolled in agricultural education courses also received instruction in FFA and were required to develop and maintain quality SAE programs. Thus, this minimum requirement was automatically met in most cases. Students in agricultural education completed a series of courses in the high school agricultural education curriculum and then pursued careers upon graduation from high school. In recent years, however, more and more students are pursuing higher education upon graduation from high school. Agriculture has evolved into a highly technical and specialized industry, and many agricultural education students need college educations for successful entry into agricultural occupations. To be competitive in the college admissions process, students must often complete additional college preparatory work while in high school. This additional course work could reduce the number of elective courses students can complete in high school, therefore making it difficult for students to enroll in agriculture courses and participate in FFA.

When a student is denied admission into an agriculture class because of conflicts with college preparatory classes or other valid reasons, the student has the option of completing a planned individualized course of study in an agriculturally related field. To be an FFA member, a student who cannot enroll in a regular agriculture class must complete a planned course of study. This could be an independent study course that includes instruction in an agricultural subject. It should also include the development of an SAE program for the student. There is a great deal of flexibility in developing this plan of work, but the best plans include the following:

- Work goals and objectives
- Clear assignments with due dates
- Schedule of visits by the agriculture teacher for individualized instruction
- List of responsibilities for all parties involved (i.e., what the student, parents, and teacher agree to do with regard to this plan)
- Grading scale
- Calendar of topics or competencies to be mastered by the student
- List of resources needed to complete lessons

Because a fair degree of flexibility exists within the planned course of study, the possibility also exists that the course of study will not be managed effectively. Often official course credit is not granted to a student who completes the planned course of study, so the experience may not be a quality one for the student. Teachers may also see this as a way to involve students in FFA competitive events without requiring them to take an agricultural education course. To protect the integrity of the teacher, the student, and the FFA organization, the planned course of study should be thoroughly prepared in written form and be agreed to by the student, the student's parents, and the agriculture teacher. Furthermore, the student and the teacher should keep detailed records of the student's progress through the planned course of study. It is important to note that some state FFA associations may have specific rules regarding the administration of the planned course of study.

In addition to enrollment in an agricultural education class and the SAE requirement, students must also participate in the activities of the local FFA chapter and conduct themselves in a manner congruent with the aims and purposes of FFA. FFA members are also usually required to pay annual dues. Annual membership dues fall into three categories: national, state, and local. Students usually pay unified dues to all three levels of the organization at one time.

Agricultural education students begin their journey through FFA by joining the local FFA chapter at their school. The local FFA chapter is where students develop interpersonal and leadership skills for agricultural careers. FFA members participate in local chapter activities, such as Career Development Events, proficiency awards, and community service projects. FFA members also participate in chapter recreational activities and attend chapter business meetings. The local FFA chapter is advised by the agricultural education teachers at the school and led by a team of chapter officers elected by the chapter membership.

Charters and State Advisement

The state FFA association charters local FFA chapters, and the National FFA Organization charters state associations. The state FFA association has the responsibility for administering all the state-level FFA activities and providing assistance to local FFA chapters. There are state FFA associations in all fifty states. Puerto Rico and the Virgin Islands also have FFA. A state FFA advisor and a state FFA executive secretary (may be the same person) administer each state association, and technical assistance is provided by the agency responsible for administering the agricultural education program in the state.

The state FFA advisor has traditionally been the chief administrator of agricultural education in the state. This person generally serves as the chief executive officer of the FFA association. The state FFA executive secretary serves as the chief operating officer and handles most of the daily administrative matters of the association.

A team of state FFA officers provides leadership in the youth activities of the state FFA association. The major duties of state FFA officers are member leadership development, recruitment, and retention. Boards of directors provide the leadership for some state FFA associations. A state board of directors usually includes FFA advisors, who represent the agricultural education profession, and state FFA officers, who represent the members. This board provides the overall leadership and policy for the state association according to the guidelines set forth in the state FFA constitution. Some states also have active FFA foundations that raise funds to support state and regional FFA activities.

A state FFA association plays an important role in determining the quality of the FFA experience for students in the state. State FFA leaders determine the standards of performance for Career Development Events, provide funding for the training and development of chapter FFA advisors and student leaders, and serve as a clearinghouse for national FFA membership dues. While state FFA associations cannot have lesser standards than those set forth in the National FFA Constitution, they can be more rigid and exacting.

The National FFA Organization requires that FFA members earn the State FFA Degree before they can receive the American FFA Degree. State associations can set very rigid standards for the state degree, thus effectively influencing the quality of supervised experience programs by which students qualify for the state degree and controlling the number of FFA members eligible for the award.

Another important role of state associations is to certify teams and individuals for participation in national FFA competitive events. Before any student can participate in a national FFA Career Development Event, the state FFA advisor or state FFA executive secretary must approve the student. The National FFA Organization also gives the state association the authority to grant and revoke the charters of local FFA chapters.

National FFA

The National FFA Organization provides the overall leadership and direction for FFA. As FFA grew in the years before World War II, questions began to arise as to how the organization should be administered. Planning and administering education was primarily the responsibility of the states, but the federal government has traditionally exerted some influence. The Smith-Hughes Act made the federal government a partner in the administration of agricultural education programs in the

23–4. The headquarters of the National FFA Organization is located in Indianapolis, Indiana.
(Courtesy, Education Images)

states, but the act did not specifically define FFA's role in agricultural education. The Smith-Hughes Act was signed into law eleven years before the National FFA Organization was founded, and local and state boards of education were faced with the problem of trying to determine how FFA should be administered. FFA members were participating in field trips, judging contests involving livestock, and other activities that created liability issues for boards of education.

It had been maintained since the inception of FFA that the organization was an integral part of instruction in vocational agriculture. However, state and federal employees were administering FFA even though it was a private organization. Before FFA, local agriculture clubs were not coordinated through a national organization. Once these clubs became FFA chapters, however, there was considerable concern as to the degree of responsibility and liability for FFA activities by schools and school officials. As pressure mounted from school boards and school officials, state and national leaders recommended that the Federal Board for Vocational Education assume administrative leadership for FFA and that the board appoint a special committee of state leaders in agricultural education to advise the national FFA officers and convention delegates (Tenney, 1977).

The Federal Board for Vocational Education adopted these suggestions. The comptroller general of the United States supported the action of the federal board by determining that FFA was an integral component of vocational agriculture and that it was a necessary, practical, and legitimate use of state and federal funds to allow state and federal officials to administer the FFA organization.

W. Tenney, in his history of FFA, wrote, "Education historically had been a primary responsibility of the states, but education which increases the productive capability of the nation makes it necessary for the federal government to define its role and determine how it will participate" (Tenney, 1977, p. 78). In 1946, the George-Barden Act (Public Law 79-586) further expressed the integral nature of FFA and agricultural education by authorizing the expenditure of funds for attending meetings of the FFA organization.

From its beginning in 1928, FFA had been a private organization incorporated in the state of Virginia. As the organization expanded across the United States, it became necessary to strengthen the relationship between FFA and the federal government. In 1950, Congress passed Public Law 81-740, which granted FFA a federal charter. This charter provided for incorporation of the National FFA Organization. Public Law 105-225 updated the charter in 1998.

FFA Incorporation   The FFA Board of Directors consists of the secretary of education, four staff members in the Department of Education, and four state supervisors of agricultural education. The secretary of education is chair of the board but appoints the chief federal officer for agricultural education to serve in that capacity. Since there are not four U.S. Department of Education staff members to appoint to the board, these four slots are filled by agriculture teachers, teacher educators, and other appropriate appointees. This board works with the national FFA officers as the administrative board of FFA.

Specifically, the incorporation

- Created the Board of National Officers. These six student officers advise the FFA Board of Directors on all matters but have no voting privileges.
- Set policies governing the powers of the Board of Directors.
- Protected the name "FFA" and the emblem from copyright infringement.
- Established the headquarters of the FFA Corporation in Washington, DC. Even though the National FFA Center is located in Indianapolis, Indiana, this federal law requires that the National FFA Headquarters be located in the District of Columbia.

In 1998, **Public Law 105-225** replaced Public Law 81-740, which had established a federal charter for FFA. These laws, and the events leading up to and surrounding their passage, created an inseparable bond between FFA and agricultural education.

The federal charter of FFA (PL 105-225) clearly defines the purpose of FFA in relation to agricultural education, but it falls short in firmly securing the integral nature of FFA and agricultural education in two specific areas. First, the federal charter does not explicitly state that every agricultural education program must have an FFA chapter. The law reads, "The purposes of the corporation are to create, foster, and assist subsidiary chapters composed of students and former students of vocational agriculture in public schools qualifying for Federal reimbursement under the Smith-Hughes Vocational Education Act (20 United States Code 11-15, 16-28) and associations of those chapters in the States, territories, and possessions of the United States" (National FFA Organization, 2012c).

Second, the Smith-Hughes Act of 1917 has been repealed, and no school now qualifies for federal reimbursement under that act. Agricultural education programs are allowed to have FFA chapters and participate in FFA activities, but federal law does not mandate this. Of the approximately 1 million students enrolled in agricultural education, only 523,309 were members of FFA (National FFA Organization, 2012a). Despite the intentions of the federal charter, some state departments of education and school boards do not interpret Public Law 105-225 as a compelling case for the inclusion of FFA in the agricultural education program (National FFA Organization, 2012c).

It is important to note that enrollment in agricultural education is estimated to be approaching 1 million students. Differences can be attributed to some individual students taking more than one agriculture class concurrently, some schools not having FFA chapters and not reporting, and some students being enrolled in private and home schools.

How does FFA serve as an integral part of agricultural education? The task of using FFA resources to provide high-quality instruction in the agricultural sciences falls to the agriculture teacher. Many Career Development Events and activities offered by FFA are cooperative learning strategies. The wise agriculture teacher incorporates these strategies in planning lessons and units of instruction.

FFA BASICS

One of the first things new students do upon entering the agricultural education class is learn the basics of the FFA organization. The assimilation process requires that students learn the history, ceremonies, and member responsibilities if they are to gain maximum benefits from membership.

Emblem

One of the best-known icons of the organization is the official emblem. The **FFA emblem** is a graphic symbol that represents or identifies the FFA organization. It has undergone very few changes over the years. The first emblem was derived from a design created for the Future Farmers of Virginia in 1927, but the origin of the design likely comes from a Danish agriculture organization in the 1920s. This emblem featured an owl perched on a spade with the rising sun in the background.

The FFA emblem today is composed of the image of an owl sitting on a moldboard plow in a field with the rising sun in the background. The owl represents knowledge and wisdom in the agricultural sciences, the plow represents labor, and the rising sun represents a new day in agriculture. This image, which rests on a background resembling the cross section of an ear of corn, is framed by the words "Agricultural Education" and displays the letters "FFA." Corn is a field crop common to all states,

and the cross section of corn on the emblem symbolizes the common bond created by agriculture for all students enrolled in agricultural education. The words "Agricultural Education" and the letters "FFA" are included together to represent the integral nature of FFA to agricultural education. An eagle at the top of the emblem symbolizes the national scope of the organization. In 1950, Public Law 81-740 ensured the sole rights of FFA to use the emblem, and registered trademarks continue to protect its use.

Colors

National blue and corn gold were adopted in 1929 as the official colors of the National FFA Organization. National blue is derived from the blue field on the U.S. flag and represents the national unity. Corn gold comes from the color of grain fields at harvest time and is symbolic of national pride in the highly productive agricultural industry. These colors appear in the official FFA jacket, official publications, and other uses deemed appropriate by the organization.

Dress

The Fredericktown FFA Chapter of Fredericktown, Ohio, was one of the first to produce and wear an FFA jacket. The official FFA jacket of today has a large official emblem centered on the back and is framed by the name of the state FFA association above the emblem and the name of the FFA chapter or organizational unit below the emblem. The front of the FFA jacket has a smaller FFA emblem on the left side and the name of the jacket's owner on the right. The FFA jacket was adopted as a component of official dress in 1933 at the National FFA Convention.

Official dress for young men is the FFA jacket zipped to the top, an FFA tie, a white shirt, black pants and socks, and black shoes. Official FFA dress for young women is the FFA jacket zipped to the top, an official FFA scarf, a white blouse, a black skirt or pants, and black shoes.

23–6. A chapter president in official dress presides over a meeting.
(Courtesy, Education Images)

In recent years, rapidly surfacing trends in clothing have caused concern among FFA leaders. A negative image has been created in the minds of FFA supporters by inappropriately clad FFA members receiving awards at conventions and conferences. Consequently, some chapters and state associations have undertaken to develop guidelines for appropriate dress. The North Carolina FFA Association, for example, has established minimum requirements for dress at FFA Career Development Events to be long pants (or a skirt of appropriate length for young women), a shirt with a collar, or an acceptable t-shirt with the FFA or school logo. The responsibility for being appropriately dressed should belong to the student and the student's agriculture teacher. Not wearing the appropriate clothing for FFA events may result in disqualification or point-reduction penalties.

FFA advisors do not have an official dress code to follow when attending FFA events. It is strongly recommended that teachers set a positive example for students by dressing professionally at all times. It does nothing for the professional image of a teacher to be underdressed in comparison with his or her students.

Creed, Motto, and Salute

The creed, motto, and salute are important components of routine FFA activities.

Creed The *FFA Creed* is a statement of beliefs held by every FFA member with regard to his or her place in the industry of agriculture. In the first two years of the organization's existence, FFA had at least two creeds for members to use in local chapter activities. The *Official FFA Manual* included one version of the creed that was dissimilar to the one used most frequently by members in chapter activities. A college professor from Wisconsin, E. M. Tiffany, wrote the version of the FFA Creed adopted by the National FFA Convention delegates in 1930. The FFA Creed has experienced little change over the years and is still recited by Greenhand candidates as they work toward attainment of the degree.

Reciting the FFA Creed is sometimes thought to be an excellent introduction to public speaking for young people. More important, reciting the FFA Creed gives some students their first introduction to character and values education. Agriculture teachers should take the time not only to have every student recite the FFA Creed but

23–7. The FFA Creed.
(Courtesy, National FFA Organization)

I believe in the future of agriculture, with a faith born not of words, but of deeds—achievements won by the present and past generations of agriculturists; in the promise of better days through better ways, even as the better things we now enjoy have come to us from the struggles of former years.

I believe that to live and work on a good farm, or to be engaged in other agricultural pursuits, is pleasant as well as challenging; for I know the joys and discomforts of agricultural life and hold an inborn fondness for those associations which, even in hours of discouragement, I cannot deny.

I believe in leadership from ourselves and respect from others. I believe in my own ability to work efficiently and think clearly, with such knowledge and skill as I can secure, and in the ability of progressive agriculturists to serve our own and the public interest in producing and marketing the product of out toil.

I believe in less dependence on begging and more power in bargaining; in the life abundant and enough honest wealth to help make it so—for others, as well as myself; in less need for charity and more of it when needed; in being happy myself and playing square with those whose happiness depends upon me.

I believe that American agriculture can and will hold true to the best traditions of our national life and that I can exert an influence in my home and community, which will stand solid for my part in that inspiring task.

also to have a class discussion or lesson that dissects the creed and explains the importance of values and character.

Motto The ***FFA Motto*** is a short expression of the character of FFA and represents a quick and easy way to remember the mission of FFA. It reflects the pragmatic philosophy of work prevalent when C. H. Lane, the first National FFA Advisor, recommended it to the National FFA Convention delegates.

Salute The FFA salute is the Pledge of Allegiance to the flag of the United States: "I pledge allegiance to the Flag of the United States of America and to the Republic for which it stands, one Nation under God, indivisible, with liberty and justice for all." The FFA salute is usually part of the ceremony used to close FFA meetings.

Code of Ethics

The FFA Code of Ethics helps guide FFA members as they go about activities and assures ethical conduct and appropriate representation to other FFA members and the general public. It helps members move toward adulthood. Adopted in 1952, the code sets the standard by which all FFA members should conduct themselves. The code was revised in 1995 to reflect an appreciation for diversity in the FFA organization.

Learning to Do,
Doing to Learn,
Earning to Live,
Living to Serve.

23–8. The FFA Motto.
(Courtesy, National FFA Organization)

FFA PROGRAMS

FFA programs fall into three basic categories: career development, leadership and personal development, and FFA chapter development. Under the career development umbrella are Career Development Events, proficiency awards, and the FFA degree program.

Career Development Events are competitive activities designed to reinforce the skills students learn from instruction in the agricultural sciences and through

We will conduct ourselves at all times in order to be a credit to our organization, chapter, school and community by:

1. Dressing neatly and appropriately for the occasion.

2. Showing respect for the rights of others and being courteous at all times.

3. Being honest and not taking unfair advantage of others.

4. Respecting the property of others.

5. Refraining from loud, boisterous talk, swearing and other unbecoming conduct.

6. Demonstrating sportsmanship in the show ring, judging contests and meetings.

7. Being modest in winning and generous in defeat.

8. Attending meetings promptly and respecting the opinion of others in discussion.

9. Taking pride in our organization, activities, supervised experience program, exhibits, and the occupation of agriculture.

10. Sharing with others experiences and knowledge gained by attending national and state meetings.

11. Striving to establish and enhance my skills through agricultural education in order to enter a successful career.

12. Appreciating and promoting diversity in our organization.

23–9. FFA Code of Ethics.
(Courtesy, National FFA Organization)

independent study in supervised experience. All the Career Development Events are competitive activities, and most are team activities. For instance, the Forestry Career Development Event requires a team of four FFA members to complete a series of activities relevant to the forestry industry. Team members are evaluated on their ability to identify important timber species, demonstrate forest inventory practices, complete a written test, and reveal through an interview their knowledge of current forestry issues. Participants with qualifying scores generally proceed through a local event to a regional or state event before reaching the national level. The National FFA Foundation, Inc., provides the funding and support for Career Development Events at the national level and, in some cases, funding at the state and local levels. In addition to the national Career Development Events, state associations have the option to create Career Development Events specific to their respective needs. These special state-created CDEs do not advance to national competition. Information on CDEs is provided each year by the National FFA Organization.

Students who have outstanding supervised experiences have the opportunity to compete in the FFA proficiency awards program. This program involves transferring data from SAE records to a proficiency award application and submitting that application for judging. Based upon the nature of their SAE, students choose to participate in the placement or entrepreneurship category of a proficiency award. The placement category is designed for those students who primarily earn wages through their supervised experience, while the entrepreneurship category is for students who own enterprises.

The applications in a particular area are judged, and the top-ranked ones proceed to the next level of judging. The process for judging proficiency applications varies from state to state.

The National FFA Organization judges those applications certified as state winners and recognizes national finalists and winners at the National FFA Convention each fall.

FFA DEGREES

The FFA degree program is designed to encourage students to establish and work toward career goals in the agricultural industry. The degrees of membership provide students with a measure of their progress toward agricultural careers and reward them for their efforts. There are five degrees of membership in FFA.

Discovery Degree

The **Discovery FFA Degree** is designed for students enrolled in a middle grades (7 and 8) agricultural education program. It is the first degree a student can earn in FFA. The Discovery FFA Degree is optional and not required for advancement to the next level. Besides submitting a written application for the degree, the student must meet the following requirements in order to wear the bronze and blue Discovery Degree pin (National FFA Organization, 2012c):

- Be enrolled in at least one agricultural education class for at least a portion of the school year while in grade 7 or 8.
- Pay the required membership dues.
- Participate in at least one extracurricular local FFA chapter activity.
- Have knowledge of careers available through the agriculture industry.
- Have some basic knowledge of the FFA chapter program of activities.

The Star Discovery Award is presented to a student who has excelled in leadership and participation in FFA at the middle grades level.

Greenhand Degree

Until the creation of the Discovery Degree, the **Greenhand FFA Degree** served for many years as the first degree a member could attain. The Greenhand Degree name is derived from an early American colloquialism that describes a novice on the farm. The term was applied to the newest FFA members as well because they are novices in the FFA organization. Greenhand members are usually first- and second-year students enrolled in a high school agricultural education program.

The Greenhand Degree is awarded by the local FFA chapter. Besides submitting a written application for the degree, the student must meet the following requirements in order to wear the bronze Greenhand Degree pin (National FFA Organization, 2012c):

- Be enrolled in agricultural education and have a planned supervised agricultural experience.
- Learn and explain the basics of FFA: the creed, motto, salute, mission statement, emblem, colors, and code of ethics, as well as the proper method of wearing the FFA jacket.
- Demonstrate knowledge of key events in FFA's history, knowledge of the chapter constitution and bylaws, and knowledge of the chapter program of activities.
- Have access to the *Official FFA Manual* and the *FFA Student Handbook*.

A local chapter or state association may set more stringent requirements for the degree but may not set less rigorous guidelines than those of the National FFA Organization. The Star Greenhand is awarded to Greenhands who demonstrate leadership, participation, and significant progress in SEPs during the membership year. Students who receive the Greenhand Degree are encouraged to attain the Chapter FFA Degree.

23–10. Greenhands in the Smyrna, Delaware, FFA chapter developed a bulletin board with autographed handprints.
(Courtesy, Education Images)

Chapter FFA Degree

The **Chapter FFA Degree** is the highest degree awarded by a local FFA chapter. It is earned by those students who have well-established SEPs and have distinguished themselves in FFA activities during the academic year. Recipients of the Chapter FFA Degree must have demonstrated leadership abilities and significant academic progress in all course work. In addition to submitting a written application for the degree, a candidate must meet the following qualifications (National FFA Organization, 2012c):

- Have earned and received the Greenhand FFA Degree.
- Have satisfactorily completed the equivalent of at least 180 hours of systematic school instruction in agricultural education at or above the ninth-grade level and be enrolled in an agricultural education course.
- Operate an approved supervised experience program and develop plans for continued growth and improvement in the supervised experience program.
- Participate in the planning and conducting of at least three official FFA chapter functions.
- Earn and productively invest at least $150 or work at least 45 hours in excess of scheduled class time, or a combination thereof.
- Lead a group discussion for 15 minutes and demonstrate five procedures of parliamentary law.
- Show progress toward individual achievement in both academic work and the FFA award programs.
- Complete at least ten hours of documented community service, above and beyond any hours completed as part of a supervised agricultural experience program.

For those students who have demonstrated exceptional leadership ability, outstanding progress in their SEPs, and growth in personal development through participation in FFA activities at the chapter level and above, the Star Chapter Awards await them. There are four areas in which a student can earn additional recognition through the Star Awards program.

The Star Awards program includes the following:

- **Chapter Star Farmer**—The Chapter Star Farmer is awarded to a student who excels in production agriculture.

- **Chapter Star in Agribusiness**—The Chapter Star in Agribusiness is awarded to a student who excels in agribusiness.
- **Chapter Star in Agricultural Placement**—The Chapter Star in Agricultural Placement is awarded to a student who excels in a wage-earning SEP.
- **Chapter Star in Agriscience**—The Chapter Star in Agriscience is awarded to a student who excels in an agricultural science SEP.

State FFA Degree

State FFA associations award the **State FFA Degree** to those students who have excelled in their FFA experience. Each state association sets the requirements for the State FFA Degree, provided those requirements are not inconsistent with the National FFA Constitution. Some state associations award the degree to students based upon their achievement of a standard set of requirements. In these states, every student who meets the minimum requirements will earn the degree. Other states choose to award the degree based upon a quota system that allows only a certain percentage of members to receive the degree each year. In these states, students may meet the minimum requirements but might not earn the degree if they are not in the top percentage of members submitting the application in a particular year. A candidate for this degree must meet the following qualifications (National FFA Organization, 2012c):

- Have earned and received the Chapter FFA Degree.
- Have been an active FFA member for at least two years (twenty-four months) at the time of receiving the State FFA Degree.
- Have completed the equivalent of at least two years (360 hours) of systematic school instruction in agricultural education at or above the ninth-grade level.
- Earn and productively invest at least $1,000 or work at least 300 hours in excess of scheduled class time, or some combination of hours and dollars that adds up to the minimum earnings requirement, in a supervised experience program.
- Perform ten procedures of parliamentary law and present a six-minute speech on an agricultural or FFA topic.
- Serve as an officer, as a committee chair, or as an active participant on a chapter committee.
- Have satisfactory academic grades as certified by the local agricultural education instructor and a school administrator.
- Participate in the chapter program of activities and in at least five different official FFA activities above the chapter level.
- Complete at least twenty-five hours of documented community service, above and beyond any hours completed as part of a supervised agricultural experience program. These twenty-five hours must be earned through participation in at least two community service activities.

State FFA Degree candidates with the best SEPs and stellar records of leadership and participation in FFA are awarded the State Star Awards. Like the Chapter Star Awards, the State Star Awards recognize excellence in four areas: production agriculture, agribusiness, placement, and agriscience.

American FFA Degree

The **American FFA Degree** is the highest degree awarded to an active FFA member. For many students, earning this degree is the crowning achievement of their FFA experience. The requirements for the degree are rigorous and challenging. A candidate for this degree must show significant growth and development in supervised experience and have a distinguished record of leadership and personal development that continues beyond the years in high school. The earliest a student could be awarded this degree is one year after graduation from high school. To receive this degree, a candidate must meet the following requirements (National FFA Organization, 2012c):

- Have earned and received the State FFA Degree.
- Have held active membership for the three years (thirty-six months) immediately preceding the application for the degree.
- Have actively participated in FFA activities at the chapter level and above.
- Have completed a minimum of three years (540 hours) of systematic secondary school instruction in an agricultural education program or, if less than three years of instruction were available at the school last attended, have completed the program in agricultural education offered by that school.
- Have graduated from high school. There is a waiting period of one academic year after high school graduation before the degree can be granted. This allows the student to become established in an agricultural career or in post-secondary education. A student may work toward the degree during the waiting period.
- Have a current supervised experience program through which the member has exhibited excellence in its management and operation by earning and investing at least $7,500. A student may also work some combination of paid and unpaid hours to earn the degree, provided the hours worked equal $7,500 in earnings. The value of unpaid hours is established at $3.33 per hour. This dollar amount may increase in subsequent years, as the degree requirements are revised.
- Have a record of high academic achievement and of outstanding leadership and community involvement.
- Complete at least fifty hours of documented community service, above and beyond any hours completed as part of a supervised agricultural experience program. These fifty hours must be earned through participation in at least two community service activities.

The top American FFA Degree candidates can receive American Star Awards in four areas: production agriculture, agribusiness, placement, and agriscience. Upon receipt of the American FFA Degree, a student's active membership in FFA is over. By receiving the American FFA Degree, the student has, in effect, "graduated" from FFA by achieving the highest honor available to any FFA member. Membership in the FFA Alumni now awaits the American Degree holder who wishes to continue to be a part of the organization.

PROGRAM OF ACTIVITIES FOR AN FFA CHAPTER

A *program of activities* (POA) is the map by which a local FFA chapter charts its course for the school year. This program serves as the curriculum guide for FFA by providing student activities that develop leadership, citizenship, and career skills. The program of activities is designed to provide activities for the individual member, the FFA chapter, and the school community. A common fallacy is that the program of activities is only a written document that lists all the things a chapter will do during the school year. This limited view of the program of activities tends to cause teachers and students alike to see the process as tedious. By its very nature, a program of activities is a living thing. It encompasses every activity that an FFA chapter does during the year and includes activities that are fun, educational, and of benefit to the community (National FFA Organization, 2012b).

The National FFA Organization has established criteria for appropriate chapter activities. Activities must provide the following:

- Educational experiences both inside and outside the classroom
- Opportunities for experiential learning
- Opportunities to learn and practice leadership skills
- Accessibility to all students, regardless of their present skills, abilities, and experiences
- Opportunities for students to develop a healthy self-image
- Experiences that connect students to agriculture and natural resources

A high-quality program of activities provides an FFA chapter with a basis of operation from year to year. The POA helps the chapter develop a sensible budget and calendar along with a list of human and fiscal resources needed to carry out a year of valuable activities and services for members. The National FFA Organization (2012b) divides the program of activities into three major areas: student development, chapter development, and community development.

Student Development

The ***student development*** part of a program of activities should provide members with opportunities to learn new leadership skills and develop teamwork skills through participation in leadership training experiences. Through careful execution of a POA, the FFA chapter provides opportunities for students to attend local, state, and national leadership conferences, such as the following:

- 360 Leadership Conference, a chapter leader training conference.
- 212 Leadership Conference, a personal leadership development conference.
- The Washington Leadership Conference (WLC), which is sometimes said to be the flagship of leadership training and development in FFA. This is a week-long leadership training conference held annually in Washington, DC. Students attending this conference learn essential leadership skills and have the opportunity to see leadership in action by visiting the U.S. Congress and other institutions in the nation's capital.
- State Presidents' Conference, which is a weeklong conference designed to prepare state officers for the crucial issues they will face as National FFA Convention delegates. Leadership training is also provided during this activity.
- National Leadership Conference for State Officers (NLCSO), which is a week-long conference that provides leadership training for state FFA leaders.
- Building Leaders and Strong Teams of Officers (BLASTOFF), which is a one- or two-day leadership conference providing basic skills training for state FFA officers.

State FFA associations and local FFA chapters also design and deliver leadership training according to the goals outlined in their respective programs of activities. The student development part of a program of activities also provides experiences for students that encourage a healthy lifestyle, scholarship, and career skill development. Examples of activities that an FFA chapter may deliver for students include the following:

- Attending an FFA camp or leadership conference
- Holding a local agriscience fair to allow students with agriscience SEPs to exhibit their work
- Establishing a scholarship program that pays the fee for students to attend leadership conferences
- Holding an FFA awards banquet

Chapter Development

The ***chapter development*** part of a program of activities stabilizes the FFA chapter and allows it to serve the members effectively. A "healthy" FFA chapter is in a better position to serve its members than a "sick" one. Chapter development activities leverage the resources of individual members and the resources of the school to carry out the FFA program. Examples of activities a chapter can do include the following:

- Developing a chapter budget and a three-year fiscal plan
- Recruiting activities, such as Food for America and children's agricultural fairs
- Conducting chapter leadership workshops
- Cooperating with the local FFA Alumni chapter
- Publicizing member activities and accomplishments in the local media
- Conducting an annual fundraiser to pay for FFA activities

Community Development

The ***community development*** area of a program of activities provides the framework for members to perform service-learning tasks that benefit the community. With careful planning, FFA members can learn the value of being engaged and responsible community members through community service projects. This is an expanding area of activity.

BEST PRACTICES IN MANAGING AN FFA CHAPTER

Success with the FFA component of agricultural education requires attention to selected areas of its programs as related to the local curriculum. Researchers have attempted to identify the "best practices" in managing an FFA chapter. Several best practices are covered here.

Local Program Success

The National FFA Organization has developed Local Program Success, which is an initiative to improve the overall quality of agricultural education programs. Local Program Success is designed to help teachers share knowledge about the best practices in teaching and learning. The National FFA Organization publishes the *Local Program Success Guide* and a *Local Program Resource Guide*, which are available to every agriculture teacher.

The FFA Advisor

The single greatest resource an FFA chapter can have is a highly motivated and well-qualified agriculture teacher as the advisor. The teacher is the essential key to the organization because he or she provides all the initial training and information about FFA to members. Furthermore, the role of the FFA advisor is specific to the agriculture teacher. No one else—parent, alumnus, sponsor, or administrator—can serve in that role. Because FFA is a component of instruction, the FFA advisor is the gatekeeper to all the things FFA members can do in the organization. Therefore, the best interest of students is served by having an advisor with a good understanding of FFA and the leadership skills necessary to help students grow personally and professionally through their FFA experience.

FFA advisors should develop the habit of setting aside time for professional development by attending workshops, reading the most current literature about FFA and agricultural education, and participating in a variety of FFA activities. FFA advisors should seek leadership positions in their professional organizations and on FFA

23–11. Teaching leadership skills involves instruction in proper handshakes and greetings.

(Courtesy, Education Images)

boards and committees to broaden their knowledge. An FFA chapter will be only as good as the FFA advisor. So, what do FFA advisors do?

FFA advisors supervise the activities of the FFA chapter

The FFA advisor is also the agricultural education instructor. Since the primary role of the teacher is to supervise all instructional activities, it naturally falls to the teacher to supervise all FFA activities as well. This is especially true when one considers that FFA activities are, in reality, instructional activities. FFA advisors may enlist the help of parents, colleagues, and alumni members to assist in supervising activities, but the primary responsibility for safe FFA programming falls on the shoulders of the agriculture teacher as FFA advisor.

FFA advisors are the primary resource for current FFA programs and services

Many students walk into their first agricultural education classroom with no idea of the depth and breadth of FFA programs available to them. It is the FFA advisor's responsibility to make the students aware of leadership, personal development, and career-related programs offered by the FFA. The FFA advisor trains a new FFA officer team every new school year. Furthermore, the FFA advisor serves as the "institutional memory" of the FFA chapter by explaining to new students how the FFA chapter operated in years past. This collected knowledge is useful to students as they plan future FFA activities. The experienced FFA advisor will have a network of resource persons who can be called upon to assist in FFA programs. The state and national FFA administrative staffs often communicate pertinent program information to students through the FFA advisor. The FFA advisor also provides information to parents, colleagues, school administrators, and program supporters about FFA programs. It is critical that the agriculture teacher/FFA advisor keep up to date on FFA programs, so that the local FFA chapter remains a contemporary and relevant organization for its members (National FFA Organization, 1998).

The FFA advisor encourages youth development through the FFA

One of the most important roles for the FFA advisors involves youth development. The FFA advisor provides instructional resources that develop the leadership potential for students. Students often need encouragement to develop their leadership potential. The FFA advisor takes the first step by providing leadership training for all students in the agricultural education program. The FFA advisor also provides opportunities for students to attend regional, state, and national leadership programs.

The FFA advisor builds support for the agricultural education program
The local agricultural education program needs the support of the community in order to provide a relevant instructional program for students. The FFA advisor builds support in the community by informing citizens about FFA program activities. It is also important to market the agricultural education program within the school as well, so that colleagues understand how the agricultural education program supports the overall school mission. The FFA has traditionally held an excellent reputation for leadership, professional development, and service. This reputation is a powerful tool in explaining the importance of the agricultural education program.

The FFA advisor includes all students in FFA programs
The FFA advisor encourages all students to be involved in FFA chapter activities. This is accomplished by making sure that a variety and sufficient number of FFA activities are provided for all students. No student can participate in every FFA activity, but every student should have the opportunity to participate in most FFA activities of interest to them. The FFA advisor also includes students in FFA through Career Development Events and awards programs. These competitive activities provide valuable leadership and career-based training for students. By offering a variety of Career Development Events, the FFA advisor can involve a significant number of students in instructional activities beyond the regular school day.

The Agricultural Education Program

Another important component of a successful FFA chapter is a "customized" agricultural education program. The agriculture teacher is bombarded almost daily with new programs, services, rules, or regulations. It does not take long for a teacher to feel buried in the avalanche of information that appears on his or her doorstep. To combat this and create an effective and relevant agricultural education program, the teacher, with the support of an advisory group, should choose the FFA activities, programs, and services to incorporate into the instructional program. If there isn't a dairy operation within 100 miles of the school, no student is currently interested in a career in dairy science, and dairy production is not part of the curriculum, then it is perfectly acceptable to eliminate the dairy production CDEs from the list of activities for an FFA chapter. Agriculture teachers are human, and there are limited hours in which they can and should supervise FFA activities. The best agriculture teachers carefully choose the most important FFA services and offer them to the members. One shortfall in teacher preparation seems to be in how to develop a customized agricultural education program that includes the right amount of FFA programming.

Some teachers tend to delegate responsibilities for some FFA services to others who are more knowledgeable and have more time to supervise the members. An example of this is having a local beef producer train the livestock judging CDE team. Delegating tasks to others is acceptable because it leverages the FFA chapter's capacity to serve the broader interests of students. It is important to note that delegation does not take the place of supervision. Agriculture teachers should continue to provide supervision over all FFA activities to assure program quality and student safety. There is a finite amount that can be delegated in an FFA chapter, so the FFA advisor should pay careful attention that his or her workload doesn't become unmanageable.

A Carefully Organized FFA Chapter

An FFA chapter that operates effectively and efficiently is a joy to advise, but a chapter does not become this way on its own. The first step is to have the chapter adopt a constitution and a set of bylaws that explain the mission, purpose, and direction of FFA at the school. These documents should clearly define the roles of leaders and the expectations of members and provide general guidelines for operation. Every FFA member should have a personal copy of the chapter constitution or have ready access to it. There is a sample chapter constitution online at the National FFA Organization Web site. From this starting point, the FFA chapter can develop a high-quality program of activities.

Trained Student Leaders

If part of the mission of FFA is to develop premier leadership through agricultural education, then being a chapter officer is really about learning how to lead within the context of FFA. Training chapter leaders is that extra class that teachers teach every day. When an FFA advisor looks at a new group of chapter officers and committee chairs, he or she has a choice. The advisor can choose to make these students his or her personal assistants in leading the FFA chapter, or he or she can choose to teach the students leadership skills so that they can be the leaders of the FFA chapter. There are several reasons an FFA advisor may choose to hold too tightly to the reins of leadership in the FFA chapter:

"As the teacher, I am the instructional leader in the classroom, and that thinking transfers into my role as FFA advisor."

> It is true that the teacher is the leader in the classroom, but no teacher would do a student's homework. Students who do not practice leadership skills have no real opportunity to hone them into something useful. The teacher should view the leadership opportunities available to students through FFA as independent practice of learned concepts.

"I am held responsible for the safety and welfare of students, and the transfer of leadership might be a significant risk."

> The reason this is such a popular position among agriculture teachers is that it is true. There is going to be a measure of risk anytime students are responsible for carrying out a program or activity. It is the same measure of risk that applies to a student who goes out onto the field during a high school football game. Protecting students from calculated risk is not a healthy thing for them, even though is it less stressful for the teacher. The teacher can reduce the risk involved in student-led activities by providing the necessary training and support that enables students to be successful.

"Students come and go, but I'll be here next year cleaning up the mess they make of the FFA chapter."

> Students who are well prepared for leadership roles and who are actively advised and trained by a highly motivated and qualified FFA advisor will not generally make a mess of things in the FFA chapter. The key is for the FFA advisor to be engaged in what the students are doing.
>
> The best FFA chapters do not always win awards at the state and national conventions. The best FFA chapters provide a high-quality

Table 23–1
ADVICE FOR PROVIDING A HIGH-QUALITY FFA EXPERIENCE

1. Link FFA activities to high-quality agricultural education curriculum.
2. Recruit and retain a diverse membership.
3. Inform every student about the diverse opportunities in FFA.
4. Elect capable officers and train them well.
5. Ensure that all members share responsibilities and have access to leadership and other opportunities.
6. Formulate a workable constitution and bylaws.
7. Develop a challenging program of activities.
8. Secure adequate financing.
9. Build school and community support.
10. Conduct fun, well-planned, regularly scheduled chapter meetings.
11. Maintain proper equipment and records.

learning experience for every member. This learning experience supplements and complements the classroom and lab instruction and the SAE. The best FFA chapters have well-prepared FFA advisors who are constantly seeking methods to teach FFA members in new and interesting ways. If FFA can remember its heritage and remain focused on its purpose, it will serve agricultural youth well into the twenty-first century.

REVIEWING SUMMARY

FFA is an integral part of an agricultural education program in a secondary school. The agriculture teacher is the FFA advisor and is responsible for having an active FFA chapter. The overall mission of FFA is to make a positive difference in the lives of young people.

The history of FFA is parallel with the emergence and development of agricultural education. Many efforts to form a student organization followed passage of the Smith-Hughes Act in 1917. Walter S. Newman, H. W. Sanders, Edmund McGill, and Henry C. Groseclose, of Virginia, are credited with much of the early planning and leadership for a student organization, specifically the Future Farmers of Virginia.

The primary organizational structure of FFA is at three levels: local, state, and national. The local level is in the local school system as a component of the agricultural education program. The state level is primarily operated through state education agencies. The national level is through the U.S. Department of Education, with the National FFA Organization providing important program leadership and support. The federal charter for FFA is said to be a major enactment that provides viability for FFA.

All teachers and students should be well aware of FFA basics, such as the emblem, colors, official dress, creed, motto, salute, and code of ethics. Some of these reflect the history of FFA but the code of ethics was created more recently to meet needs of students and teachers participating in FFA events.

FFA has Career Development Events (CDEs), leadership and personal development events, and chapter development activities. Students with outstanding SAE programs advance with FFA proficiency awards.

A system of FFA degrees is used to motivate participation and recognize the accomplishments of members. The degrees of membership are Discovery FFA Degree, Greenhand FFA Degree, Chapter FFA Degree, State FFA Degree, and American FFA Degree.

Each local chapter should prepare a program of activities. This is the plan that a chapter establishes to guide its work for the year. Students use a committee structure to prepare the program of activities. Most programs of activities have three major areas: student development, chapter development, and community development. Initiatives of the national and state FFA levels are often part of a local program of activities.

Local chapters tend to show the most promise if teachers use the best practices identified by leaders in agricultural education. The Local Program Success initiative is much a part of these best practices.

QUESTIONS FOR REVIEW AND DISCUSSION

1. How did FFA come into existence?
2. What is the mission of FFA? Is it relevant to current student needs?
3. How is FFA organized at the national and state levels?
4. What is the role of the National FFA Organization?
5. What is the importance of the federal charter for FFA (Public Law 105-225)?
6. Many agricultural education students are not members of FFA. Why?
7. What are the major parts of the FFA emblem?
8. What are the FFA colors?
9. Why is official dress used at FFA events? How relevant is the FFA jacket?
10. Why should members learn and recite the FFA Creed?
11. What is the FFA Code of Ethics? Why is it important?
12. What are the three major areas of FFA programs? Briefly describe each.
13. What are the FFA degrees? Identify and briefly explain each, including how requirements increase as the degrees become more advanced.
14. What is a program of activities? What are the three major areas of a program of activities?
15. What are some of the "best practices" in managing an FFA chapter?

ACTIVITIES

1. Write a paragraph that explains what PL 81-740 and its successor, PL 105-225, are designed to accomplish. In particular, explain the connection between classroom/laboratory instruction and FFA.
2. List on one page the top twenty moments in FFA history. Include the year and a one-sentence description of each.
3. Create a diagram that describes the organizational structure of FFA from the national level to the local level.
4. Prepare a paper that describes the history and development of New Farmers of America. Include information about its merger with FFA.
5. Prepare a one-page summary describing Career Development Events to new students in your agriculture program. Include the purpose for their existence.
6. Explore the National FFA Organization's Web site. Investigate the contents of the site as related to student members and to teachers. Also, investigate the links to other agricultural education organizations.

CHAPTER BIBLIOGRAPHY

Berry, J. B. (1924). *Teaching Agriculture*. New York: World Book Company.

Lambert, M. D. (2003). *Wall FFA Constitution*. Unpublished manuscript, North Carolina State University, Raleigh, NC.

Maslow, A. (1970). *Motivation and Personality* (2nd ed.). New York: Harper & Row.

McCormick, V., and R. McCormick. (1984). *A. B. Graham: County Schoolmaster and Extension Pioneer*. Worthington, OH: Cottonwood Publications.

National FFA Organization. (1998). *Agriculture Teacher's Handbook*. Alexandria, VA: National FFA Organization.

National FFA Organization. (2012a). *FFA Statistics*. Retrieved April 13, 2012.

National FFA Organization. (2012b). *National FFA Chapter Planning and Recognition: A Student Handbook*. Indianapolis, IN: National FFA Organization.

National FFA Organization. (2012c). *Official FFA Manual*. Indianapolis, IN: National FFA Organization.

Tenney, A. W. (1977). *The FFA at 50, 1928–1978*. Alexandria, VA: National FFA Organization.

True, A. C. (1969). *A History of Agricultural Extension Work in the United States: 1785–1923*. New York: Arno Press.

Wakefield, D. B., and B. A. Talbert. (2000). *Exploring the Past of the New Farmers of America (NFA): The Merger with the FFA*. Paper presented at the National Agricultural Education Research Conference, San Diego, CA.

Wakefield, D. B., and B. A. Talbert. (2003). A historical narrative on the impact of the New Farmers of America (NFA) on selected past members. *Journal of Agricultural Education, 44*(1), 95–104.

24
Community Resources

"There is no more agriculture in this community where I teach," stated Octavius Miller to the local career and technical education administrator. "I am going to leave teaching and sell automobiles," he further indicated. No doubt, his administrator was surprised, because Octavius had been there fourteen years.

Do you think Octavius had assessed the agriculture in the area? In fact, he had not. Octavius had a narrow view. He was thinking only of dairy farms. He had failed to consider new orchards, small animal businesses, flower shops, animal shelters, and the many other emerging agricultural businesses and resources that were within a short distance of the school.

As you become a teacher, how will you determine the resources in a community? Further, how will you use these resources to improve your agricultural education program?

objectives

This chapter introduces community resources. The chapter covers the importance of community resources in classroom/laboratory instruction, supervised agricultural experience, and FFA and how to identify resources available in the community. It also discusses resources provided by schools for supervised agricultural experiences and presents ways of relating the community resources to an agricultural education program. It has the following objectives:

1. Discuss the importance of community resources in classroom/laboratory instruction, supervised agricultural experience, and FFA
2. Explain how to identify community resources
3. Describe resources provided by the school for conducting supervised agricultural experiences
4. Describe ways of relating community resources to an agricultural education program
5. Describe how to promote and market an agricultural education program

terms

agricultural community
business community
community
community resource
resource
school community

24–1. Visiting local agricultural businesses, such as the veterinary clinic shown here, helps determine community resources that might be available.

(Courtesy, Education Images)

THE IMPORTANCE OF COMMUNITY RESOURCES IN AGRICULTURAL EDUCATION

A *resource* is something that can be turned to for support or help. A *community resource* is a resource found in the area served by the school that may be capable of supporting agricultural education in some way. Because of the hands-on "learning by doing" methodology of agricultural education and the diversity of the agricultural industry, agriculture teachers need multiple resources to be successful.

Future teachers should understand that they are not expected to know everything and do everything on their own. Using resources is not an indication of teacher weakness. The identification and wise use of resources are an indication that the teacher is informed and connected to local community expertise. Using available resources makes the teacher's job easier and more enjoyable.

Agriculture is a highly scientific and complex industry. As professional educators, agriculture teachers need the assistance of laypersons to provide high-quality educational programs. The increasing need for rapid change in an age of advanced technology complicates the teacher's role. Employment opportunities in agriculture are constantly changing, and new technologies are continually being developed and incorporated into the nation's agricultural industry and educational institutions. Effective teachers reach out into the surrounding community for assistance in keeping abreast of all these changes.

Teacher Support

In today's world, teachers need the support and cooperation of others in order to make a major difference in the lives of their students. Teamwork, partnerships, and cooperation are the keys to modern success.

Successful agriculture teachers are good managers of the resources available to them and don't try to do everything themselves. There are a number of key groups that can help "make or break" a successful agriculture program. Therefore, agricultural education teachers should find and meet the key people in the community and strive to involve them in the agricultural education program.

Training Opportunities

Students must be trained for today's jobs as well as new opportunities that become available. The need is constantly increasing for individuals trained in specialized technical occupations in agricultural fields. The *agricultural community* includes farmers, agricultural businesses, and others connected to the agricultural industry. This group of individuals can help teachers stay abreast of changing employment trends and opportunities. Developing and maintaining programs that include internships, workstudy, and other types of on-the-job training requires close coordination with industry representatives.

Parents are important partners. Their involvement and support is essential in planning and conducting classroom and laboratory instruction, supervised agricultural experience programs, and for supporting FFA events and activities. The parents of first-year students are especially important, as they are often the source of the financial, transportation, and motivational support required to start supervised agricultural experience programs and to get their children involved in FFA functions.

Employers are also key partners for providing agriculture students with training stations and other opportunities for real-world experiences. Employers are often major financial contributors to the local FFA chapter for scholarships, achievement awards, and other FFA activities. They are also excellent resource people to speak to students in classroom and laboratory settings.

Local government officials and civic organization leaders can provide support for the agriculture program and FFA chapter through community improvement opportunities. They may also be sources of supervised agricultural experience opportunities. The mayor and other key community members have an interest in community development. They are often eager to involve young people in learning about local government and in making the community a better place to live.

Alumni are another source of expertise. They can help train judging teams, speak to classes about the importance of FFA and supervised agricultural experience, serve as chaperones for trips, provide employment opportunities, and collectively provide financial and human resources for the local program.

24–2. Community resources often include animal species and opportunities for learning not readily available at school.
(Courtesy, Education Images)

24–3. Sponsored test plots are an example of strong community support for an agricultural education program.
(Courtesy, Education Images)

Partnerships

Creating partnerships within the school is also important. When mainstreamed into agricultural education courses, students who are disadvantaged or who otherwise have special needs often require teachers to reach out for assistance in providing them with quality educational programming. The special education teachers in the school system are key resource people for assistance in this area. Creating partnerships with school counselors often leads to increased enrollments for the agricultural education program. Collaboration with science teachers and other faculty members can increase the quality of the instructional program and increase student achievement.

Another high priority of the agriculture teacher should be to keep the school administration informed about the agricultural education program and its value to the school and the community. Most administrators want to be connected with successful programs in their school and are more generous in supporting the programs they value.

By developing community partnerships, the agricultural education teacher increases school and community support for the program, eases his or her individual workload, and improves the overall quality and impact of the agricultural education program.

IDENTIFYING COMMUNITY RESOURCES

The **community** is a group of people living in the same locality and under the same government. The community consists of parents, employers, employees, government officials, and others living in the area served by the school. Teachers need to become familiar with the community and meet the key resource people living there.

School Community

When thinking about the communities that can be most helpful to agricultural education, a teacher should first consider the resources available through the school. The **school community** includes the students and all employees working in a particular school and the students and personnel of all the schools in the entire school system.

The beginning teacher should explore the school community to gain knowledge of the people and places that can provide assistance. Information is needed regarding a school's personnel, departments, and buildings. Identifying individuals important to the success of the agriculture department is essential. These people may be at the teacher's school or at other schools in the community. The teacher should become familiar with the educational system from the elementary through the postsecondary levels. Of importance to the agriculture teacher are the district and school administrators, business office personnel, guidance and counseling personnel, clinic personnel, librarians, secretarial staff, cafeteria personnel, transportation staff, custodial and maintenance staff, other faculty members, and school support staff.

The key personnel can be identified from personnel directories published by the school system. Organizational charts show the position structure and lines of authority and are helpful in understanding how the school system works. A school map shows the school's departments, offices, and room numbers. If a map is not available, the agriculture teacher and/or agriculture students can develop one. A school district map shows the locations of the schools that make up the entire system.

When seeking resources outside the school community, the teacher should look first to those most closely connected to the students. Parents can be one of the most important allies of the agriculture teacher. The agriculture teacher can become acquainted with them through home visits, parent–teacher meetings, and school activities. Students' and parents' names and contact information should be collected by the teacher and filed for future reference.

After becoming familiar with the school community and the parents of the students enrolled in the program, the teacher should survey the surrounding community to determine the resources available in close proximity to the school. Several strategies can be used to become familiar with the community around the school. The members of the agriculture advisory committee and/or the FFA chapter officers can assist the teacher in learning about the local community.

Other strategies for identifying community resources include conducting a windshield survey by driving around the community to determine the names and locations of agricultural businesses. The teacher can check the yellow pages of the local phone directory, review the chamber of commerce Web site, read the local newspapers, attend community meetings, visit the mayor's office, and learn where and when the "movers and shakers" in the community meet in both formal and informal settings. Another good strategy is to meet with the county agricultural agent and ask about the agricultural organizations and key farmers in the community. As the teacher networks in the community, the teacher's list of contacts will increase until he or she is familiar with the entire community.

Business Community

The **business community** is the group of individuals who manage businesses in the area, including those closely affiliated with agriculture. This group is especially important because these individuals are a prime source of training stations for student supervised agricultural experience programs. They are also potential employers of the program graduates.

The agriculture teacher should meet as many of the community's employers as possible, with a priority on those related to agriculture. If the community has a chamber of commerce, the teacher should become acquainted with the executive director of the chamber. Becoming involved with the chamber of commerce provides opportunities to meet many employers.

Government and civic organization leaders are another important resource group. Some of these individuals may already be identified, as some will be parents, school employees, and/or businesspeople.

24–4. Community resources available for educational purposes vary with the agriculture found in the local area.
(Courtesy, Education Images))

Once the available resources have been identified, they should be entered into a database that can be continually updated. This will ensure easy access to the names and correct contact information for the agricultural education program partners.

RESOURCES SCHOOLS PROVIDE FOR SUPERVISED AGRICULTURAL EXPERIENCE PROGRAMS

Supervised agricultural experience programs provide a link between the agricultural education program and the local community. The entire school should value this linkage, as it contributes to positive school and community relationships. Therefore, it is important for the school to provide resources for conducting supervised agricultural experience programs.

The school's major responsibility for the supervised agricultural experience segment of the agricultural education program is to provide the services of the teacher for supervising students in their supervised agricultural experiences. The teacher should receive compensation for on-site supervised agricultural experience visitations. The individualized supervised agricultural experience instruction that the teacher provides during the summer months is the major justification for extending the employment of the agriculture teacher beyond the regular school year. Ideally, school districts should provide one school period each day for teachers to supervise student experience programs, and teachers should also be employed throughout the summer to supervise students who conduct supervised agricultural experience programs during the summer months.

Because supervised agricultural experience programs are located throughout the community, the agriculture teacher is required to travel to provide on-site supervision. The school should provide the transportation, either by furnishing a district-owned vehicle and fuel or by paying mileage for supervisory travel when the teacher furnishes the vehicle and fuel. Often a school provides a pickup truck for the agriculture teacher, because part of the teacher's responsibility is to assist students in obtaining livestock, feed, seed, fertilizer, and equipment used in their supervised agricultural experience programs.

The school should also provide specialized equipment and facilities required for the successful operation of supervised agricultural experience programs that may

24–5. Retail stores, such as those that sell companion animals and supplies, offer excellent opportunities for learning and gaining experiences.
(Courtesy, Education Images)

not be available to students from other sources in the community. Examples include portable scales, greenhouses, land, and livestock facilities. Many schools supplement the funds they have available for these items with additional resources from booster clubs, service clubs, and private donations.

RELATING COMMUNITY RESOURCES IN AGRICULTURAL EDUCATION

Community resources provide support for agricultural education. Following are several examples of such support.

Support of Parents (or Guardians)

Support of parents (or guardians) is important to students, the FFA chapter, and the entire agricultural education program. To involve parents, the teacher should first obtain their names and addresses. He or she should then contact each student's parents and ask them to assist in educating their child. All involved in parenting a student should be contacted, whether they live together or not.

The teacher should personally invite parents to visit during school open houses and parent–teacher conferences. Good teachers also offer to meet parents at other times to accommodate those unable to attend specific school functions.

Parents should be asked to help with field trips and other events held in the community. They should also be invited to the annual FFA chapter banquet and other official events. Parent volunteers should be welcomed to the classroom and to FFA activities. Other strategies for involving parents include creating a parent booster club or inviting parents to become involved with the FFA Alumni affiliate.

Educating parents about the agricultural education program should be a high priority of the teacher. This is necessary for gaining parental support for the teacher's guidance and direction of supervised agricultural experiences and FFA activities.

Support of Businesses

Working with the business community is extremely important. This group provides training opportunities for students, while the agriculture program provides the business community with current and future workers, along with future leaders of the agricultural industry and the local community. The business community is a source of guest speakers on a variety of topics; advisory committee members; competitive event judges; experiential learning opportunities, including supervised agricultural experience programs; field trip sites; student employment; graduate placement; borrowed resources (equipment, materials, books, etc.); and financial support. The teacher and the students should develop a plan for informing employers about FFA and supervised agricultural experiences to build awareness and gain commitment for the agricultural education program.

Teachers can create and maintain relationships with the agricultural community by meeting business, industry, and agricultural organization leaders, staying in contact on a regular basis, and keeping them informed about the agriculture program. Teachers and students should invite them to program events and functions, ask for assistance and offer assistance, and recognize and thank them as often as possible.

Support of Former Students and FFA Alumni Members

Former FFA members and others in the community can also provide assistance and support for the agricultural education program. Their contributions can often be maximized through an active FFA Alumni affiliate or Young Farmer chapter. The FFA Alumni's role should be to support the local FFA chapter in meeting the needs of its members and to assist the agricultural education program in achieving its vision, mission, and goals.

Alumni members and FFA members should be encouraged to work together. This interaction provides valuable learning experiences for students. Examples of alumni support include raising funds for various events, awards, and activities; providing transportation, chaperones, judges, guest speakers, coaches, mentors, and other human resources for agricultural improvement and FFA chapter programs; and providing advocacy for agricultural education at school board and community meetings.

Local National Young Farmer Education Association (NYFEA) may also be a resource for local FFA chapters. Joint meetings of FFA, FFA Alumni, and local NYFEA provide opportunities for students to gain additional "real-world" information and contacts. NYFEA members can provide practice facilities and coaching for FFA teams, serve as resource persons for various events and activities, offer opportunities for farm and agribusiness tours, and provide leadership for service projects that promote agriculture.

Support of Educational and Research Institutions

All high schools have access to the resources of colleges and universities. These resources may be related to teaching, research, or service. Branch campuses or experiment stations often have well-trained scientists, economists, or others who can assist as valued resource persons.

Scientists can be invited to the school to give demonstrations and lead discussions. Proper preparation of students is needed. This includes instructing students in how to relate to resource persons.

School groups can take field trips to laboratories, test plots, and other facilities to learn from the scientists and technicians who are carrying out the research.

It is not always necessary to travel, as the Internet can be used. Students may establish dialogue with scientists on particular projects, such as one being done in agriscience. Web sites may be used to acquire information recommended by the cooperating scientists.

PROMOTING AND MARKETING AN AGRICULTURAL EDUCATION PROGRAM

Promoting or marketing the agricultural education program is important for increasing access to the available community resources. Strategies used by successful agriculture teachers for marketing a local program include the following:

- Distributing teacher and FFA officer business cards
- Holding an agricultural education program open house
- Meeting and working with the local press
- Preparing and mailing agriculture program newsletters
- Working with the FFA reporter to prepare timely news releases
- Promoting the program with signs, posters, calendars, Web sites, etc.

A key strategy for maintaining ongoing involvement of the community in the local agricultural education program is to recognize and thank the individuals, groups, and organizations that contribute time, money, and other resources to the program. This can be accomplished through news releases, thank-you letters from the teacher and the students, appreciation certificates and plaques, honorary membership in the FFA chapter, and other special recognition. Providing recognition and appreciation to those contributing to the program should be an ongoing priority of the agricultural education department, and care should be taken to ensure that all who contribute to the program receive appropriate recognition and thanks.

REVIEWING SUMMARY

Identifying and utilizing community resources connects the agricultural education program with the community. Community resources are important for providing students with state-of-the-art learning opportunities that prepare them for employment or for postsecondary education. Community partners may be school based or be from the community surrounding the school.

The agricultural education teacher and students should place a priority on identifying resources available in the community and on educating key partners about the agricultural education program. This will generate support and cooperation in providing a variety of educational opportunities and experiences.

Creating and maintaining long-term partnerships increases the value and impact of the agricultural education program in the local community.

QUESTIONS FOR REVIEW AND DISCUSSION

1. Why are community resources important to the agricultural education program?
2. What are some ways for teachers to identify community resources?
3. What resources should the school provide for supervised agricultural experience programs?
4. How can teachers connect community resources with the agricultural education program?

ACTIVITIES

1. Use the Internet to investigate the agricultural businesses in a local community.
2. Prepare a paper describing potential supervised agricultural experiences available in that community.
3. Prepare a term paper that describes how to develop links between an agricultural education department and the local community.
4. Develop a marketing plan for building rapport between an agricultural education department and the local community.
5. Use the local telephone directory to develop a list of potential resources in the school district. Cluster the resources to meet the needs of your report. Examples of clusters include horticultural businesses, farms

and ranches, equipment dealers and repair services, forestry, environmental and natural resources, animal care, agricultural supplies, agricultural services, and agricultural products processing and marketing. Arrange to visit a small sample of the businesses you have identified.

CHAPTER BIBLIOGRAPHY

Caffarella, R. S. (2002). *Planning Programs for Adult Learners* (2nd ed.). San Francisco: Jossey-Bass.

Kahler, A. A., B. Morgan, G. E. Holmes, and C. E. Bundy. (1985). *Methods in Adult Education* (4th ed.). Danville, IL: The Interstate Printers & Publishers, Inc.

Lee, J. S. (2000). *Program Planning Guide for AgriScience and Technology Education* (2nd ed.). Upper Saddle River, NJ: Pearson Prentice Hall Interstate.

National FFA Organization. (2002). *Local Program Success Guide.*

Newcomb, L. H., J. D. McCracken, J. R. Warmbrod, and M. S. Whittington. (2004). *Methods of Teaching Agriculture* (3rd ed.). Upper Saddle River, NJ: Prentice Hall.

Phipps, L. J., E. W. Osborne, J. E. Dyer, and A. L. Ball. (2008). *Handbook on Agricultural Education in Public Schools* (6th ed.). Clifton Park, NY: Thompson Delmar Learning.

A

academic software kinds of software that are useful to teachers and learners in the teaching and learning processes.

academic subjects courses or areas of study required of all students, including English/language arts, mathematics, science, and social studies.

accountability the state of being accountable or answerable; the holding of schools, programs, or teachers responsible for student learning and achievement; standardized tests in the core academic areas have become popular instruments for measuring accountability in education.

achievement evaluation the practice of measuring how much students know about a subject or how well they are able to use their knowledge; this measurement usually comes in the form of standardized or teacher-made tests.

activity manual an ancillary resource organized around a basal resource, such as a textbook, and containing activities that reinforce the achievement of objectives in the basal resource.

administration the organization that manages or directs an institution, such as a school or a college, composed of administrators or individuals who manage the procedures used to implement policies in the institution.

adult an individual who has reached the stage in life to be personally responsible for himself or herself and who has assumed a productive role in society.

adult agricultural education education for men and women who are engaged in the ordinary business of a society and who typically have jobs in an area of the agricultural industry.

adult basic education (ABE) instruction intended to provide adults with basic and specific skills needed to function in society; instruction for individuals who did not achieve a certain level of, or participate in, education as children.

adult education the process of providing learning opportunities to improve adults in many areas of their lives.

advisory committee a group of laypersons empowered to give advice on educational programs but not actually to make decisions.

affective influenced or affected by the emotions.

agricultural awareness consciousness or knowledge of the agricultural industry; awareness enhanced by agricultural education programs that teach students about the nation's plants, animals, and natural resources systems.

agricultural community the community consisting of farmers, agricultural businesses, and others connected to the agricultural industry; the group of individuals with knowledge of changing agricultural trends and opportunities.

agricultural education a program of instruction in and about agriculture and related subjects commonly offered in secondary schools, though some elementary and middle schools and some postsecondary institutes/community colleges also offer such instruction.

agricultural education program the total offering of agricultural instruction in a school ranging from one to several subject areas or classes and including supervised experience and student organizations.

agricultural literacy education about agriculture; above-average knowledge pertaining to the plants, animals, and natural resources systems.

agriculture the science, art, business, and technology of the plants, animals, and natural resources systems.

Agriculture in the Classroom (AITC) one of the best-known examples of an agricultural literacy initiative; provides instructional materials, human resources, and teacher training opportunities; exists in every state and has a federal presence within the U.S. Department of Agriculture.

agriscience laboratory a facility used in teaching science and math principles and concepts associated with agriculture; should adhere to standard recommendations of floor space, storage, and available utilities.

American Farm Bureau Foundation for Agriculture a national foundation focusing on agricultural literacy information.

American Federation of Teachers (AFT) a national education organization with more than 1 million members; unlike NEA, AFT is a recognized union and part of the AFL-CIO.

American FFA Degree the highest degree awarded to an active FFA member; the requirements for this degree are rigorous and challenging; the earliest a student could be awarded this degree is one year after graduation from high school.

ancillary instructional resource secondary or supplementary instructional material that accompanies primary instructional material, most often a textbook.

andragogy the study of processes and practices in adult education; it is based on the needs and readiness of the adult learners and thus is more learner-focused, collaborative, and problem-centered.

aquaculture laboratory a facility used for education in fish farming and related areas; inside laboratories should be equipped with tanks or vats, while outside laboratories may use ponds or raceways.

area career and technical education center a specialized type of high school designed to prepare students for careers; this type of school tends to offer a great variety of career and technical education courses that are typically taught by teachers who have documented work experience in the areas in which they teach.

Aristotle a pupil of Plato's, introduced to the world around 335 BCE; sometimes called the father of scientists; his work yielded both the scientific method and the problem-solving method of instruction; he believed in realism rather than idealism.

articulation a joining together; the way in which educational activities are organized and connected with other educational activities; for example, concepts learned in a horticulture class should prepare students for applications in a landscaping class.

Association for Career and Technical Education (ACTE) the national professional organization for advancing the education of youth and adults for careers; consists of more than 30,000 members and has twelve divisions, one of which is agricultural education.

authentic assessment an evaluation procedure that places the student in a situation that closely mimics the workplace or other life setting and then assesses the student's performance.

autonomy self-government or self-direction; local autonomy has always been a part of the agricultural education program philosophy to ensure that agricultural education programs are structured around the needs and resources of their local communities.

B

basal instructional resource a material that provides the base or foundation for the organization of a subject—for example, a textbook.

behavioral learning theory a theory of learning that focuses on observable changes in outward behavior and on the impact of external stimuli to effect change; behavioral learning objectives are formulated based on this theory.

bell-shaped curve a graphic depiction in which student scores are distributed along a line in the shape of a bell; shows how a large number of scores would be distributed; used to compare individuals with the group and with other individuals; also called a normal distribution curve.

block scheduling a class scheduling system in which students take fewer classes but in which each class period occupies a larger block of time, usually between 85 and 100 minutes; in this system, students usually take four classes each day and earn eight Carnegie units per school year.

Bloom's taxonomy a theory put forth by Benjamin S. Bloom in 1956 that classified educational objectives into three domains of learning: cognitive, affective, and psychomotor; the three domains are further broken down into more specifically definitive categories that may help educators pinpoint student learning behaviors.

board of education a group, provided for by the laws of the state, that has the legal authority to take action and use resources in the operation of a school district; members may be elected or appointed.

brain-based learning theory a theory developed in the 1990s by Renate Nummela Caine and Geoffrey Caine that united thought from multiple disciplines, including biology and psychology; led to the further development of the twelve principles of brain/mind learning.

business community the group of individuals who manage businesses in an area, including those closely affiliated with agriculture; valuable in helping develop the students and employ the graduates of the agricultural program.

C

career cluster a broad grouping of occupations that have similar characteristics; all possible occupations are placed into one of sixteen career clusters, such as the cluster of Agriculture, Food, and Natural Resources.

career exploration the use of investigative activities to help students identify potential career interests and gain the information relevant to those interests; offers students the opportunity to learn about agricultural careers and the skills needed in those careers.

career guidance the process designed to help a person choose a career and proceed through the development stages necessary for that career; involves providing students with career information and helping them make wise career decisions based on their strengths and interests.

career ladder the steps in an individual's general progression in a career from entry to retirement; additional education, expanded career responsibilities, experience, and other conditions are beneficial to an individual's movement upward.

Carnegie unit the amount of credit a student is awarded for a course that meets one instructional hour per day for 180 school days; important for determining high school graduation and college entrance requirements; a course that meets for fifty minutes per day for the entire 180-day school year would be worth one Carnegie unit.

chapter development the part of a program of activities that empowers individual members and the school to leverage their own resources in carrying out the FFA program.

Chapter FFA Degree the highest degree awarded by a local FFA chapter; earned by distinguished students who meet a number of requirements.

charter school a publicly supported school given freedom to operate outside of state and local bureaucratic control and thus able to experiment and try educational innovations; such a school is still required to meet local and state performance standards for its students.

Chautauqua movement an educational movement in the late nineteenth and early twentieth centuries that combined secular and religious instruction; named for Chautauqua, New York, where the movement was born; died out in the 1920s.

classroom a school facility designed for group instruction; ideally equipped with a number of items, including desks, chairs, and at least one chalkboard or whiteboard.

cognitive relating to "stored" information; having a basis in factual knowledge.

cognitive development the development of an individual's thought processes; includes adaptation to the environment and assimilation of information.

cognitive learning theory a theory of learning that focuses on the internal mental processes, how they change, and how they affect external behavior changes.

Comenius, Johann Amos (1592–1670); philosopher who brought forth the idea that teaching should promote a student's natural tendency to learn and that learning should proceed from easy subjects to more difficult ones; one of the first to promote education for career preparation.

community a group of people living in the same area and under the same government; different types of communities include business communities and school communities.

community development the part of a program of activities that provides the framework for members to perform service-learning tasks that benefit the community.

community resource a resource found in the area served by the school that may be capable of supporting agricultural education in some way.

community survey a process to collect comprehensive and specific local information from various areas of the agricultural industry; the collected data is used to evaluate the local agricultural education program and help make a wide array of decisions concerning the needs of the program and its students.

comparative agricultural education the study of the theory and practice of agricultural education in different cultures, geographical or political entities, or over time.

comprehensive high school a high school that encompasses the full range of subjects and activities in academic, career and technical, citizenship, and personal development areas; also offers citizenship development activities, such as clubs and organizations, and personal development activities, such as sports and driver's education.

computer-based module instructional materials that involve the use of units or modules guided by a computer program.

connecting activity an activity that establishes relationships between school and life; FFA, for example, provides connecting activities for students of agricultural education.

constructed-response question a question that requires the student to develop his or her own answer; fill-in-the-blank, short-answer, and essay questions are constructed-response questions; sometimes known as open-ended-answer question.

constructivism a process of learning whereby the student constructs knowledge from his or her past experiences; a set of learning theories that emphasize how students actively make sense of the information they receive; teachers who utilize constructivism encourage their students to make sense of the outside world by actively engaging in their environment.

contextual learning instruction given within a perspective to which students can easily relate; effective agricultural education instruction often contextualizes content using plants, animals, and natural resources systems.

continuing education education that brings a participant up to date in a particular area of knowledge or skills; education received after an individual has completed a general level of formal education.

continuing renewal credits credits that fulfill or help fulfill requirements for the renewal of teaching licenses; awarded for attendance and participation at workshops, conferences, and seminars and for participation in other inservice activities.

corrosive material a material with the potential to cause chemical reactions strong enough to damage metal and injure people; there are three kinds of corrosive materials: acids, bases, and miscellaneous materials such as iodine.

counseling advice and guidance for persons with problems by appropriately trained individuals, and help in finding satisfactory answers; an important service and component in today's school systems.

course of study a written guide that aids in teaching a particular subject or course; usually contains objectives, an outline of content, a list of learning activities, means of assessment, and suggested instructional materials.

craft committee an advisory group that is specific to a particular area, such as horticulture or forestry; often a subcommittee organized within a program advisory committee.

credentialing the process of determining and certifying that a person can perform to required standards and meet competencies necessary to their profession.

creed a system of general beliefs, principles, or opinions that guide individuals in a profession or organization.

criterion-referenced test a test that compares the score of an individual student with established criteria or standards; designed to determine how students compare with established criteria rather than how they compare with others.

cultural pluralism the social philosophy that the culture of a country, such as the United States, is a product of the cultures of that country's various immigrant groups; a teacher using a pluralistic approach would be an advocate for diversity in a school.

curriculum the list of all courses offered in a school; also a group of related courses, such as the agricultural education curriculum.

curriculum guide a written description of courses or areas/units of instruction, along with suggested lesson objectives; may also include scope-and-sequence outlines, suggested teaching–learning activities, content outlines, and lesson resources.

D

demeanor the outward manner of an individual; the way in which an individual behaves; individual attributes that can be changed by the individual if necessary.

demonstration a teaching process that involves showing students how to do something before they do it for themselves or that involves showing them what would happen as the result of a particular action.

desktop computer a personal computer with limited portability, essentially for use in the same location all of the time.

development the orderly, lasting change that occurs in humans as a result of maturation, learning, and life experiences; advancement from a simpler to a more complex stage.

De Witt, Simeon (1756–1834); in 1819, proposed colleges for teaching agricultural subjects and conducting experimental research; introduced the concept of agricultural colleges.

Dewey, John (1859–1952); a pragmatist; believed in "learning by doing"; believed in inquiry as an orderly and scientific process that consisted of hypothesizing a problem, producing possible solutions to the problem, and attempting to solve the problem using those solutions; the father of progressivism.

directed laboratory supervised agricultural experience the type of supervised experience in which experiences in a practical activity are planned by the agriculture teacher; geared toward students who are unable to engage in other forms of supervised experience; often funded through some source other than the students' own funds.

disability a disadvantage or deficiency; the inability to do something specific, such as hear or see.

discipline teaching structure, routine, and behaviors to students to facilitate classroom learning.

Discovery FFA Degree the first degree a student can earn in FFA; designed for students enrolled in a middle grades agricultural education program; this degree is optional and not required for advancement to the next level.

discrimination the act of treating one group differently from another group; often the result of prejudice; most forms of discrimination are illegal in the school system, as stated in such laws as Title IX, the Civil Rights Act of 1964, and the No Child Left Behind Act of 2001.

discussion a teaching process less formal than a lecture that depends on student involvement to disperse information; typically, both teacher and student ask and answer questions, offer opinions, and debate.

display panel an electronic presentation device connected to a computer, camera, or other equipment for the presentation of visual information.

diversity the variety of differences within a category or classification; most often refers to differences of gender, ethnicity, and socioeconomic status, though other forms of diversity, including geography, religious beliefs, and language, need to be considered.

E

e-book reader a mobile electronic device designed for reading of digital e-books and periodicals.

e-learning electronically based teaching and learning that usually involves some application of computer technology.

elective course/program a course or program that is not mandated by state policies or standards and that students are not required to take to graduate; courses or programs in agricultural education are electives, and whether they are offered is up to the local board of education.

electronic-based material instructional material accessed by the student or teacher through the use of a computer; typically includes CDs or DVDs.

elementary school typically a school with grades 1 through 6.

engagement act of a student making a psychological investment in their own learning

enrollment barrier an element in the school or community that prevents enrollment in agricultural education; some barriers include schedule conflicts, stricter enrollment requirements, and negative perceptions of the program.

entrepreneurship/ownership supervised agricultural experience the type of supervised experience in which students develop skills needed to own and manage enterprises; students engaged in this type of supervised experience have financial investment or risk in their enterprises.

equipment implements used to aid in the completion of a task; includes such instructional resources as welding machines, radial arm saws, trimmers, soil mixers, and components of irrigation systems.

Erikson, Erik (1902–1994); proposed the psychosocial theory of development, which is based on the idea that people have the same basic needs and that a purpose of society is to provide for those needs; he theorized that all people go through eight stages of psychosocial development, with the outcome of each stage having an effect on later stages.

evaluation in education, a process to analyze educational effectiveness and student achievement by using measurement tools, such as tests, observations, and performances.

experiential learning learning by doing; knowledge gained through experience.

experiment a trial that uses a definite and planned procedure; the process of conducting a test for the purpose of demonstrating a truth or examining a hypothesis.

exploratory supervised agricultural experience an educational experience that provides the student with an opportunity to investigate a number of areas in agriculture; designed to help students gain information for making decisions about their future education and careers; job shadowing is an example of an exploratory supervised experience.

F

facility a building, area, or item that is not movable and that is built or established to serve a particular purpose.

feeder school a middle school or junior high school whose students will attend a specific high school for which the feeder school provides preparation.

FFA Alumni an adult group within the National FFA Organization that supports agricultural education and is open to anyone who has an interest in supporting FFA; it currently has approximately 42,000 members; its mission is very similar to that of FFA.

FFA Creed a statement of beliefs held by every FFA member with regard to his or her place in the industry of agriculture; in 1930, E. M. Tiffany, an educator from Wisconsin, wrote the version of the FFA Creed adopted by the Third National FFA Convention.

FFA degree program a program of the National FFA Organization to encourage students to establish and work toward career goals in the agricultural industry; promotes advancement in FFA.

FFA Emblem a graphic symbol that represents the FFA organization; the first emblem was derived from a design created for the Future Farmers of Virginia in 1927; FFA has sole rights to use the heavily symbolic emblem.

FFA Motto a short expression that reflects the character and mission of FFA; it also reflects the pragmatic philosophy of work prevalent when C. H. Lane, the first National FFA advisor, recommended it to the National FFA Convention delegates.

field trip a learning experience that involves traveling away from the school site, for example to an agricultural experiment station to observe professional research projects.

fixed-response question a common type of test question that requires the student to select the correct answer from given choices; includes multiple-choice, true/false, and matching questions; sometimes known as forced-answer question.

flammable material   a material that easily catches on fire and burns; gasoline and kerosene are two examples.

Food for America   a program sponsored by FFA that provides elementary and middle school students with instruction on a variety of agricultural literacy topics, including food, fiber, and natural resources.

formal education   education or training provided in an orderly, logical, planned, and systematic manner and often associated with school attendance.

formative evaluation   evaluation occurring before or during instruction; used to determine the starting point for instruction and to provide feedback for students and teachers during the instruction.

free elective   an optional course that does not fulfill any educational requirements.

G

grade   a mark or descriptive statement that rates the extent to which a student has achieved a standard; typically assigned by either letter (A–F) or number (0–100); the format and structure of grading are typically specified in school system policy manuals.

Greenhand FFA Degree   until the introduction of the Discovery Degree, this was the first degree a member could attain; Greenhand members are usually first- and second-year students enrolled in a high school agricultural education program.

Groseclose, Henry C.   (1892–1950); the first National FFA Executive Secretary/Treasurer, elected on November 20, 1928; sometimes said to be the father of FFA because of his work in authoring important FFA documents.

grouping   placing students together to accomplish a common task or goal.

group teaching method   a teaching approach that instructs students together as a group; advantages include overcoming such teaching concerns as time and resource requirements; group instruction teaches all students the same content.

guidance   a process consisting of different techniques designed to help students make wise decisions regarding choices or changes; counseling is a popular technique for providing guidance to students.

H

Hamlin, Herbert M.   (1894–1968); professor of agricultural education at the University of Illinois; felt that decisions about local agricultural education programs should be made only with citizen input; author of *Public School Education in Agriculture: A Guide to Policy and Policy-Making* (1962).

handicap   a physical or mental disability that creates a disadvantage in certain situations; for example, blindness is a handicap if a situation requires sight.

Hatch Act   an act passed in 1887 that established agricultural experiment stations and educated the public about the implications of the research conducted at these experiment stations.

hierarchy of human needs   developed by Abraham Maslow in 1970, this hierarchy prioritized human needs into seven layers, starting with the most necessary: physiological needs, safety needs, belonging needs, esteem needs, and self-actualization.

high school   a school that includes grades 9 or 10 through 12; most high school agriculture courses are designed to develop specific skills in agriscience, agribusiness, technology, leadership, and human relations.

home school   education carried out by parents and others outside of traditional school buildings and typically in the home until completion of high school.

horizontal articulation   articulation within the same grade level among the various courses being taken by the student; allows continuity from subject to subject, such as from science to language arts.

Hughes, Dudley   (1848–1927); sponsor of the Smith-Hughes bill in the U.S. House of Representatives.

I

idealism   the action or practice of envisioning things in an ideal form; the theory that something is in its purest and ideal form as an idea.

improvement project   an experience or group of experiences, usually involving home or community work, carried out in conjunction with one of the supervised experience programs.

independent study   a learning activity that, for any of a variety of reasons, is not part of an organized class learning experience; in almost all independent study activities, the student takes on more responsibility for his or her instruction.

individual teaching method   an approach that recognizes that each student learns at a different pace and possibly at a different level; a primary advantage of individual instruction is the potential for greater student interest and deeper learning.

infuse   to introduce or instill one thing into another so as to affect it; in relation to agricultural literacy, to include agricultural lessons, examples, and topics in courses for other subjects.

instructional material material that contains the information being taught, for example, a textbook, transparency, electronic presentation, or magazine.

instructional materials adoption process used by school districts and state education agencies in selecting materials and allocating resources.

instructional resource material that a teacher uses to provide instruction and that students use to learn; examples are lab instruments, supplies, and safety equipment.

instructional resource guide a collection of material, often containing such resources as detailed lesson plans and sample tests, that helps a teacher plan and deliver instruction.

instrument a device used for a particular purpose, usually to facilitate work, for example, a caliper used to make precise measurements.

integration the process of combining academic curriculum with career and technical education curriculum so that learning is more relevant and meaningful to students; designed to eliminate distinction between academic and career and technical education.

interactive whiteboard a presentation medium that uses a computer that is connected to a projector to display materials on a white surface that acts as a touch screen; components may be connected wirelessly or via USB or serial cables.

intracurricular literally, "within the curriculum"; an integral part of the program or curriculum, as opposed to an extracurricular program or club; FFA is an intracurricular component of agricultural education.

inventory a complete and itemized list of all equipment, tools, and supplies; should be taken at least yearly; helpful in monitoring the condition of items and their frequency of use.

J

James, William (1842–1910); asserted that pragmatism is about searching for truth, a departure from Charles Peirce's belief that pragmatism is about finding meaning in a concept; believed that the test of an idea is in its truth or workability.

job shadowing spending time with and observing an experienced person or persons performing work in a specific occupation.

junior high school a school that includes grades 7 and 8 and sometimes grade 9; specific agriculture instruction may be offered at this level.

L

laboratory an area equipped for student experimentation, research, or study; different educational courses may require different kinds of laboratories.

laboratory activity an application of concepts and principles that includes experiments, hypothesis testing, demonstrations, learning exercises, application exercises, and skills practice.

laboratory instruction instruction occurring in laboratories that provides freedom of movement, psychomotor skill development, the use of special tools and equipment, and student self-directed learning; educational emphasis placed on application and hands-on learning through the conducting of experiments and the simulation of real-world experiences.

laboratory superintendent a teacher's assistant (usually a student) who helps with various laboratory duties, such as making sure students do their tasks and making sure that areas are clean, with tools and supplies properly put away.

Lane, Charles Homer (1877–1944); the first National FFA advisor, chosen November 20, 1928.

laptop a personal computer that is easily transported for mobile use and has the capabilities of a typical desktop computer; may have a rechargeable battery.

layperson an individual who is not an agricultural educator or other professional or certificated person employed by the school.

LEAP Licensure in Education for Agricultural Professionals; a certification program in agricultural education that is delivered online; administered by the Agricultural and Extension Education Department at North Carolina State University; accredited by the National Council for Accreditation of Teacher Education (NCATE).

learning the experience of gaining knowledge or skill that results in a permanent change of behavior.

learning center an area designed for one activity or lab that allows small groups of students to complete the activity or lab at the same time; typically self-contained, with all the equipment, tools, supplies, and instructions available in one place.

learning disability any of a variety of disorders, such as dyslexia, that interfere with a student's ability to learn; often leads to the failure or incapacity of the student to function at the age-appropriate level in language, reading, spelling, mathematics, and/or other areas of learning.

learning environment the various conditions that promote or impair teaching and learning; the teacher's responsibility is to provide a quality learning environment for the students.

learning module   a self-contained activity that a student can complete independently; most often used for cognitive and psychomotor development; advantages of learning modules are that students can work at their own pace and instruction can be modified to meet the needs of individual students.

learning objective   a carefully written statement of the intended outcomes of a learning activity; always written in terms of what the student will be able to do at the conclusion of the instruction.

learning stop   a location in a lab where students stop and read information, do a task, or observe some phenomenon.

learning style   a description of how a person prefers to learn and within what environment he or she prefers to learn; learning styles vary from student to student and can include a variety of factors, such as whether a student is intrinsically or extrinsically motivated and whether the student prefers to learn by auditory, visual, or kinesthetic means.

least restrictive environment   the idea that students with disabilities be allowed the closest-to-normal educational setting possible; required under the Individuals with Disabilities Education Act of 1975, which aimed to end in schools the discrimination against students with disabilities.

lecture   an oral presentation by the teacher or other individual that may include a wide range of techniques; an instructional method of delivering information to an audience or class.

lesson plan   a road map used by teachers to ensure effective, efficient, and empowering instruction.

lesson plan library   a collection of teaching plans focused on a broad area, such as landscaping or wildlife management, that aids the teacher in delivering instruction; lesson plans include summaries of content, suggested instructional strategies, sample tests, lab sheets, and transparencies or electronic-presentation images.

limiting factor   a resource whose unavailability or time requirements may make conducting a lab impossible or require its modification.

literacy   the condition of being able to read and write; the quality of having more than average knowledge in a particular field.

Local Program Success   (LPS) a national initiative designed to enhance the quality and success of local agricultural education programs through the employment of four strategies: (1) program planning, (2) partnerships, (3) marketing, and (4) professional growth.

lyceum   an institution introduced in 1826, when Josiah Holbrook founded the American lyceum in Millbury, Massachusetts; a society where the arts, sciences, agriculture, and public issues were presented and discussed; the lyceum movement faded after the Civil War.

M

magnet school   a specialized high school designed to attract students for a specific purpose, such as intense studying of the arts, science, mathematics, or even agriculture; usually, one must apply to attend a magnet school, and competition for enrollment is high.

mastery learning   an educational approach in which students are given frequent opportunities to practice and multiple opportunities to improve their scores; often associated with criterion-referenced tests.

middle school   a school that commonly includes grades 6 through 8; agriculture instruction at this level is broad, focusing on helping students understand the plants, animals, and natural resources systems and helping them make informed choices about their future.

misbehavior   a person's inappropriate, yet purposeful and motivated, attempt to cope with situations in the environment; behavior that violates social norms within the classroom.

mission statement   a written statement detailing the purpose of an organization and how the organization will go about achieving that purpose.

Morrill, Justin   (1810–1898); Congressman from Vermont and proponent of the American farmer and the working-class American; worked to pass a landmark land-grant bill that would be known as the Morrill Act of 1862.

Morrill Act of 1862   act signed into law on July 2, 1862, by President Abraham Lincoln; led to the creation of colleges for the teaching of the agricultural, mechanical, and military arts; named after Congressman Justin Morrill, who worked the legislation through Congress.

Morrill Act of 1890   act that set aside funds for teacher education in agricultural and mechanical arts; restricted funds to only those colleges where no distinction was made on the basis of race or color in student admissions; appropriated funds from the sale of public lands.

motivation   the energy and direction given to behavior; an inducement or incentive, whether intrinsic or extrinsic.

multicultural education   education that provides all students with equal opportunities to learn, regardless of their race, class, or culture; a reform idea that promotes equality and the use of various perspectives in education; a major goal of multicultural education is the improvement of academic achievement for all students.

N

National Association of Agricultural Educators (NAAE) the national professional organization for agriculture teachers; provides advocacy for agricultural education and for all agricultural educators; consists of more than 7,600 members in fifty state associations.

National Board for Professional Teaching Standards (NBPTS) an independent, nonprofit organization that provides advanced certification to teachers who meet its standards; the standards are based on five core propositions.

National Council for Accreditation of Teacher Education (NCATE) an organization that accredits college and university teacher education programs; uses a performance-based system and verifies that certain standards have been met; successful completers receive teaching licenses from the state of North Carolina. They are then eligible to teach in almost all fifty states through a process called reciprocity.

National Education Association (NEA) the largest national organization for educators, with more than 2.7 million members; NEA is involved on local, state, and national levels doing such work as bargaining, government monitoring, and lobbying.

National FFA Scholarship Program an initiative of the National FFA Organization to recognize excellence by awarding scholarships for students to pursue education beyond high school.

National Middle School Association (NMSA) an organization founded in 1973 that is dedicated to improving the education of young adolescents; a leading advocate for middle grades education.

National Postsecondary Agricultural Student Organization (PAS) an organization associated with agriculture, agribusiness, and natural resources in approved postsecondary institutions offering associate's degrees or vocational diplomas or certificates; one of the ten student organizations approved by the U.S. Department of Education as integral parts of career and technical education.

National Young Farmer Education Association (NYFEA) the official adult student organization for agricultural education as recognized by the U.S. Department of Education; its stated mission is to "promote the personal and professional growth of all people involved in agriculture."

needs assessment a program planning process used to collect and analyze community data, including statistics from various sources on such things as local agricultural products, employment, and employers.

New Farmers of America (NFA) the companion organization to FFA that served black Americans from 1926 to 1965; offered young black students the incentive to pursue agricultural careers while building their self-confidence and technical skills; the Civil Rights Act of 1965 brought an end to NFA by requiring that it merge into a single organization with FFA.

Newman, Walter Stephenson (1895–1978); a state supervisor for agricultural education in Virginia and a pioneer in the development of FFA; believed that some type of organization was necessary to provide opportunities for personal advancement and the growth of self-confidence in farm youth; in 1925, developed the Future Farmers of Virginia with the help of colleagues Henry C. Groseclose, Edmund Magill, and H. W. Sanders; later urged the development of NFA.

netbook a category of small and inexpensive laptop computer that has found some favor for educational purposes.

non-formal education education or training offered outside of school curricula and in settings other than those that involve the implementation of structured standards and fulfillment of diploma or degree requirements.

norm-referenced test a test that compares the achievement of an individual student with an established group norm; constructed to yield a bell-shaped curve, in which the majority of students have scores in the average range, with a few students having scores in the high and low ranges.

O

occupational specialist teaching license an alternative teaching license for an individual with substantial occupational experience but little or no advanced educational training; in many states, this license allows the teacher to teach only those courses that match his or her occupational specialization.

operant conditioning behavior modification in which the desired behavior is encouraged through positive or negative reinforcement.

OSHA Occupational Safety and Health Act; a federal act that requires certain fire, health, and safety standards be met in the school environment.

oxidizer a material that can initiate and promote combustion; chlorine is an oxidizer.

P

Page, Carroll (1843–1925); Vermont senator in the early 1900s with a passion for vocational education; believed it was the duty of the federal government to provide for job-specific training for young people; instrumental in establishing the Smith-Hughes Act.

paid placement supervised agricultural experience   a supervised experience in which the student is employed for compensation while gaining practical experience and developing skills needed to enter and advance in a particular occupation.

paper-based material   any material printed on paper, either in color or in black and white, and almost always bound in some way.

Peirce, Charles Sanders   (1839–1914); the first well-known proponent of pragmatism; believed true meaning evolves by way of a three-stage process through which one moves from an unconscious understanding of a concept to awareness of the characteristics of a concept and, finally, to the application of what one knows about a concept to other situations involving that concept.

performance-based evaluation   the evaluation of how well students can do something under given conditions; students provide tangible evidence of their learning to teachers, who observe and judge the evidence using tools such as rubrics, checklists, rating scales, or product scales.

Perkins, Carl   (1912–1984); senator from Kentucky and sponsor of the Vocational Education Act of 1963; in part responsible for the Perkins Acts (Federal Vocational Education Acts of 1984, 1990, 1996, 1998, and 2006), which attempted to modernize vocational education further and to expand its emphasis as career and technical education.

Pestalozzi, Johann Heinrich   (1746–1827); developed the principle of informal education, such as hands-on learning in the sciences.

Phi Delta Kappa International   a professional organization for educators on all levels, including undergraduates preparing to become teachers; an international association with more than 600 local chapters, most of which are based on college campuses.

philosophical foundations   the basic beliefs that guide people's actions; the guide by which educational programs are designed and delivered; developed by individual thinkers or philosophers.

philosophy   a discipline that attempts to provide general understanding of reality and interpret the meaning of what is observed; a system of values held by an individual or group.

Piaget, Jean   (1896–1980); a Swiss psychologist who developed a theory that there are four stages of cognitive development: sensorimotor, preoperational, concrete operational, and formal operational.

Plato   (circa 428–347 BCE); an early Greek philosopher and star student of Socrates who continued in the dialectic tradition through his writings and teaching; author of *Republic*, which attempted to define the nature of the ideal state; established the first known college of higher education, called the Academy, where he taught idealism.

policy   a general guiding principle, usually based on a law; established by individuals or groups that have the authority to do so, for the purpose of guiding the management, procedures, and especially the decision making of government bodies, such as school boards.

portfolio   a collection of materials that represent a student's progress and accomplishments, usually contained in a notebook or a folder or on a CD-ROM or a Web site.

postsecondary education   typically defined as education offered at technical schools, community colleges, or junior colleges, though it sometimes includes college- and/or university-level education.

postsecondary program   a program that is sequenced after the high school level, such as a program at a community college or at a career and technical center; agricultural education in one of these programs is more specialized and provides students with advanced technical competencies to enter and progress in specific agricultural careers.

pragmatism   a philosophy centered on the concept of "knowing by doing"; a relatively young philosophy, it views knowledge as a practical activity and understands questions about meaning and purpose within this context.

prejudice   a preconceived preference or idea formed without complete information or examination of the facts.

Premack principle   the psychological principle that places the less preferred activity before the preferred activity and uses the preferred activity as positive reinforcement for the less preferred activity.

pretest   a test given before instruction is begun to determine previous student learning in the subject area; helpful in deciding which material needs to be covered in detail and which can be either lightly reviewed or omitted altogether.

principal   a person who has controlling authority or is in a position of presiding rank, especially the head of a local school site.

procedure   the way a policy is to be implemented or carried out; a set of established forms or methods for conducting the affairs of an organized body, such as a school board.

product scale   a scale on which the different levels of quality are represented by actual objects, with each object having an assigned numerical or letter grade, depending on its position on the scale; for example, in agricultural mechanics the teacher can develop a product scale containing welds of varying qualities, from exceptional to inferior, by which to grade the students' welds.

profession   an occupation that requires training and advanced study in a specialized field and adheres to a code of ethics to guide the individuals in the occupation; the body of qualified persons of one specific occupation or field.

professional   an individual who has acquired the necessary traits ascribed to the profession he or she practices; an individual who has an assured competence in a particular field or occupation.

professional code of ethics   a statement of ideals, principles, and standards related to the conduct of individuals within a profession.

proficiency award   an award through the National FFA Organization to recognize a student who excels in skill development.

program advisory committee   a committee that may serve to provide advice for the overall educational program; may consist of several subcommittees, sometimes called craft committees, that focus on specialty areas within agricultural education.

program components   constituent elements of a program; the three main agricultural education components are classroom and laboratory instruction, supervised experience, and involvement with FFA.

program development   the process of identifying and developing the necessary areas of formal instruction and other components in a local secondary school agricultural education program.

program evaluation   an assessment of progress in achieving stated goals and outcomes of the program plan; effective evaluation includes the collection of necessary data, the comparison of that data against the goals and outcomes set forth in the program plan, and recommendations, complete with action steps and implementation strategies, for improvement.

program of activities (POA)   the curriculum guide for FFA that works to develop leadership, citizenship, and career skills in students; provides activities for the individual member, the FFA chapter, and the school community.

program planning   a process that documents all the activities needed to design and implement local school education.

progressivism   educational theory marked by emphasis on the individual child, informality of classroom procedure, and encouragement of self-expression; based on John Dewey's ideas of inquiry and the creative use of relevant theory in solving problems.

project   an activity of educational value with one or more definite goals; advantages of projects include the opportunity to maximize student interest by allowing the student to design his or her own project and the ability to cater to the needs of the student.

promotional material   a product or approach, such as a brochure or a newspaper article, used to elicit desired responses from people; an effective way to relay information to a target group.

Prosser, Charles   (1871–1952); an elementary school teacher, principal, and school superintendent in New Albany, Indiana, who eventually became the secretary of the National Society for the Promotion of Industrial Education, a position from which he would contribute to the establishment of the Smith-Hughes Act; appointed director of the Federal Board for Vocational Education, which ensured that the provisions of the Smith-Hughes Act were being carried out according to law.

protective clothing   clothing that serves to reduce the risk of danger or personal injury; examples include gloves, headgear, chaps, and aprons.

proximity control   the shaping of student behavior by being present in the learning environment; often the simple presence of the teacher will cause the students' behavior to improve.

psychomotor   a domain of behavior marked by an individual's skill to manipulate something physically.

Public Law   105–225 passed in 1998 by the Congress of the United States providing technical amendments to Public Law 81-740. PL 81-740, passed in 1950, recognized FFA as an integral part of agricultural education and granted a Federal Chapter to the FFA.

punishment   a penalty imposed to discourage socially unacceptable behavior; different acts of misbehavior call for different modes of punishment.

R

reactive   a material that reacts with water or other materials; sodium is a reactive.

readiness level   the status of a student's emotional, mental, behavioral, and physical state; knowing a student's readiness level allows a teacher to decide whether that student is able to complete a particular task or receive particular instruction.

realism   the doctrine that the truth of an idea rests in its form; a philosophy that urges the thinker to seek the cause of a concept, on the belief that the true meaning of the concept lies in the cause.

reciprocity   the recognition by one state of the validity of a teaching license issued by another state; does not relieve a teacher from state-specific requirements.

recruitment the process of seeking and soliciting students to enroll in agricultural education courses; this process consists of six key variables: the agriculture program, the recruitment program, student characteristics, parents, school support, and community support.

reference material that refers to something more advanced than students would normally use; may be found in such formats as books, manuals, and DVDs.

required courses courses needed to meet core requirements for graduation or admission to an institution of higher education; these courses are determined through local school board policy.

research and experimentation supervised agricultural experience the type of supervised experience in which a student carries out an investigation into a problem using scientific approaches and then makes recommendations about how to solve the particular problem; results are often exhibited at FFA agriscience fairs.

resource something that can be turned to for support or help.

resource people experts in a particular area who can provide knowledge and insights that the teacher may not have; are also valuable in lending credibility to information or instruction that students doubt or about which they have heard conflicting opinions.

retention re-enrollment of a student in agricultural education classes; leads to increased enrollment in the program and to the student accumulating experience, knowledge, and skills relative to agricultural education.

role-playing the act of an individual assuming the real or imagined role of another individual; enables the individual to assume different perspectives and better comprehend different experiences.

S

safety a condition or state in which there is a reduced risk of danger or personal injury; safety must be a primary concern in agricultural education, as many items used for instruction in this program can be dangerous if misused.

salary schedule a list or table that shows the salary at each experience level and educational level; increments are added for each year of teaching experience, though some schedules may not go beyond fifteen or twenty years; the typical basis for teacher compensation.

scaffolding the help, prompts, and questions an adult gives a child to assist the child's cognitive development.

schema an abstract guide used to organize an experience or concept and assist in responding to it.

scheme an adaptable mental representation an individual forms of objects and events of the world.

school community the students and employees in a particular school; on a larger scale, the students and personnel of all the schools in the entire school system.

school district a geographical area under supervision of a given school board.

school management software a computer-based program used in schools to achieve a wide range of functions related to delivering education, recording data, and reporting information.

scope the extent, depth, or range of the curriculum; an important concept in preparing a course of study; for example, when determining the scope of a curriculum, one thing that will be decided is whether emphasis should be placed on a wide breadth of topics or whether fewer topics should be taught in greater depth.

scoring rubric a set of guidelines for judging student work; also called a rating scale.

secondary school typically a school with grades 7 through 12.

self-efficacy the belief that one can accomplish something and have a positive outcome.

sequence the order or progression of the material covered; sequence in agricultural education may be determined in any of several ways, such as from simple to complex or by production cycle; an important concept in preparing a course of study or a curriculum.

Seven Cardinal Principles educational objectives put forth by the Cardinal Principles Report of 1918 that dramatically shaped secondary education; the principles are command of the fundamental processes, worthy home membership, health, vocation, citizenship, worthy use of leisure time, and ethical character.

smartphone a mobile telephone with a computing platform that allows features such as Internet and e-mail capability far beyond telephone conversations.

Smith, Hoke (1855–1931); sponsor of the Smith-Hughes bill in the U.S. Senate.

Smith-Hughes Act act passed in 1917 that organized agricultural education so that it would be available to every student who wished to study it; among other provisions, it provided funding to the states for the purpose of training and employing teachers in agricultural education, industrial arts education, and home economics education.

Smith-Lever Act act passed in 1914 that created the Cooperative Extension Service, a partnership between the federal government and the land-grant colleges established for the purpose of extending knowledge about the best practices in agriculture to rural communities.

Socrates (circa 469–399 BCE); the best known of the early Greek philosophers; creator of the "Socratic method" of teaching, also known as the dialectic method, in which truth and understanding were arrived at through a series of questions designed to challenge answers.

stakeholder an individual or group that has a stake or strong interest in an enterprise or program, such as a local school; family members, taxpayers, and businesses all hold stakes in the quality, accountability, and overall educational impact of the local school.

standard a level of requirement; criterion; the expectation of what students should know and be able to do after completing a given area of instruction.

standard teaching license a license issued upon successful completion of an accredited teacher education program; additional requirements on both the national and state levels may include examinations, the submission of portfolios, and background checks; typically renewable every four to ten years for teachers of agriculture.

Star Award top award for an FFA member who excels in supervised experience, FFA participation, and educational achievement.

State FFA Degree awarded by state FFA associations to students who have excelled in their FFA experience; the requirements for the State FFA Degree are set by each state association, provided those requirements are not inconsistent with the National FFA Constitution; there are also national requirements that a candidate for this degree must meet.

stereotype a conception or opinion resulting from the assignment of oversimplified characteristics to an entire group.

storage facility a secure location, such as a cabinet or locker, where materials and equipment can be inventoried, organized, and protected; different materials and equipment may require different storage facilities.

student advisement the process of offering assistance to students regarding their career and educational decisions.

student-centered instruction a teaching method in which the processes of instruction are designed by the students and facilitated by the teacher; because the students create the processes themselves, they are more likely to be self-motivated to acquire the knowledge; the teacher, however, is ultimately responsible for determining what is best for the students.

student development the part of a program of activities that provides FFA members with opportunities to learn leadership and teamwork skills through participation in various training experiences.

student enrollment the total number of students within a school's agricultural education program; indicative of the impact the agricultural education program has within the school.

student-use material an instructional resource carefully written and otherwise prepared for use by students; examples include textbooks, activity manuals, computer-based modules, and record books.

summative evaluation evaluation occurring after instruction, often in the form of a grade; can provide feedback to determine the next level at which a student should be placed and what changes can be made the next time a topic is taught.

superintendent the one who has executive oversight and the authority to supervise and direct; the chief executive for the school system; provides for the preparation and administration of the school budget and the hiring of other school administrators, including school principals.

supervised agricultural experience the application of the concepts and principles learned in the agricultural education classroom in planned, real-life settings under the supervision of the agriculture teacher; should improve agricultural awareness and/or the skills and abilities required for a student's career. Also called *supervised experience.*

supervised agricultural experience program a series of individualized practical learning activities planned by the teacher, the student, the student's parents/guardians, and, if applicable, the employer; supervised by a qualified agriculture teacher; should develop competencies related to the interests and career goals of the individual student.

supervised study a method of instruction in which students are directly responsible for their own learning while under the direction of the teacher; can enhance student interest and develop problem-solving abilities.

supplementary projects (activities) specific projects or skills gained through experiential learning outside normal class time that add to the agricultural knowledge and competency of the student; these skills are not related to the major supervised agricultural experience and they are usually accomplished in less than a day.

supply a consumable material used to carry out instructional activities and/or work; supplies vary with the nature of the instruction and, depending on the field, may include potting soil, insecticide, metal, lumber, and oil, to name a few.

T

table of specifications a procedure used to chart test content, its relative emphasis, its educational level, and the types of questions used.

tablet computer a device with features of both a small computer and a smartphone, with an operation platform similar to that of a smartphone; apps can be used to achieve specific goals.

teach to impart knowledge or skill; to give instruction to.

teacher one who teaches, especially a person who is hired to teach.

teacher's manual a teacher-use material prepared to accompany a textbook and other student-use material; provides useful information, such as content summary and answers to questions in the student edition text, for the teacher to use in delivering instruction.

teacher-use material an instructional resource designed to aid the teacher in planning, delivering, and evaluating instruction; teachers' manuals, instructional resource guides, lesson plan libraries, and reference materials are examples of teacher-use materials.

teaching the art and science of directing the learning process; also a precept or doctrine.

teaching calendar a specialized teaching schedule that provides additional details on the time when units of instruction will be taught; the teaching calendar's increased specificity provides advantages over the basic teaching schedule, such as allowing a teacher to order supplies and materials, schedule field trips, and involve resource persons in a more timely and effective manner.

teaching license a certificate issued by a legally authorized state office documenting that an individual is qualified to teach and the conditions under which that person may teach; such conditions typically include subject area(s) and grade level(s).

teaching method the overall means, composed of a number of techniques and details, that a teacher uses to best facilitate student learning.

teaching schedule a plan showing the scope and sequence of content within an individual agricultural education course; typically includes such useful information as the broad units of instruction listed in the order in which they should be taught and the suggested number of class periods to spend on each unit.

technology-based development the use of curriculum databases, related matrixes, and technology-based access to assist in planning and developing the agricultural education program; improves the speed and ease of various processes, including standardization within a state and customization of curriculum to meet local needs.

tertiary system colleges and other institutions of learning after high school.

textbook a book systematically designed for use by students that deals with a specific subject, such as agriscience,

horticulture, or landscaping; often contains certain characteristics, such as chapter or unit organization, motivational approaches, and suggestions for hands-on activities.

Thales an ancient Greek philosopher, from around 580 BCE, who found that all matter could be reduced to the quintessential element water, a finding that would later be debunked.

theorem a statement or proposition that can be demonstrated as a truth; often part of a larger theory.

theory of multiple intelligences the proposal by Howard Gardner in 1983 that there are seven dimensions of intelligence: bodily-kinesthetic, interpersonal, intrapersonal, linguistic, logical-mathematical, musical, and spatial; puts forth the belief that everyone possesses intelligence in all seven dimensions, though most people excel in only a few.

tool a handheld device, such as a saw, that aids in performing manual or mechanical work.

toxic material a substance that is poisonous to humans, other animals, and plants; should be used carefully and stored in containers marked with a skull and crossbones.

traditional scheduling a class scheduling system that consists of six or seven one-instructional-hour class periods for five days per week for the entire school year; accordingly, students earn six or seven Carnegie units per school year.

training agreement a written statement, signed by the necessary parties, of the exact expectations, understandings, and arrangements of all parties involved in the supervised experience program.

training plan a written statement that documents the specific training activities in which the student is expected to participate; often based on the occupational competencies needed for the student to enter his or her tentative occupation.

training station the location on the job site where a student is placed as part of the supervised experience.

transition-to-teaching program a route to licensure for individuals who hold content degrees and have pursued careers in the content field; typically consists of intense education courses and an internship.

trimester scheduling an alternative class scheduling system that divides the school year into three twelve-week trimesters; each class period may be seventy to eighty minutes in length, with students taking five classes each day; classes taken for two trimesters equal one Carnegie unit each.

True, Alfred Charles (1853–1929); director of the Office of Experiment Stations for the federal government and an early proponent of agricultural education in public schools; his work led to increased funding for agricultural education and to its continued growth in schools below the college level.

Turner, Jonathan Baldwin   (1805–1899); a professor of classical literature at Illinois College, he brought the idea of agricultural colleges to American awareness; believed the federal government should pay for the establishment of universities to provide higher education in the arts and sciences.

U

unpaid placement supervised agricultural experience   the type of supervised experience in which the student works in a job for experience only and is not compensated in any other manner for hours of labor.

V

vertical articulation   articulation focused on preparing the student for higher grade levels; for example, students in grade 7 learn skills that will prepare them for grade 8, while students in grade 8 learn skills that will prepare them for grade 9.

Vocational Education Act of 1963   act that broadened agricultural education; revised the Smith-Hughes Act of 1917 by expanding agricultural education in the secondary schools to include a wide range of nonproduction agriculture, such as horticulture; opened the door for vocational training in other areas as well, such as business education.

voucher   a system that allows parents to choose the school their children attend, whether public or private, and receive public funds through a voucher arrangement.

Vygotsky, Lev   (1896–1934); a Russian psychologist who developed the sociocultural theory of cognitive development, a theory that emphasizes the role people play in a child's cognitive development; proposed that children develop through their interactions with adults; also proposed the "zone of proximal development."

W

work-based learning   also known as supervised experience; a part of education that allows students to practice in a workplace what they have learned in the classroom or laboratory; a key component of agricultural education.

Z

zone of proximal development   the level of proficiency in which students are unable to accomplish a task independently but can accomplish that same task with assistance.

Note: Page numbers followed by "f" indicate figures; and those followed by "t" indicate tables.